Asteroseismic Data Analysis

PRINCETON SERIES IN MODERN OBSERVATIONAL ASTRONOMY

David N. Spergel, SERIES EDITOR

Written by some of the world's leading astronomers, the Princeton Series in Modern Observational Astronomy addresses the needs and interests of current and future professional astronomers. International in scope, the series includes cutting-edge monographs and textbooks on topics generally falling under the categories of wavelength, observational techniques and instrumentation, and observational objects from a multiwavelength perspective.

Statistics, Data Mining, and Machine Learning in Astronomy: A Practical Python Guide for the Analysis of Survey Data, by Željko Ivezić, Andrew J. Connolly, Jacob T. VanderPlas, and Alexander Gray

Essential Radio Astronomy, by James J. Condon and Scott M. Ransom

Asteroseismic Data Analysis

FOUNDATIONS AND TECHNIQUES

Sarbani Basu and William J. Chaplin

PRINCETON UNIVERSITY PRESS • PRINCETON AND OXFORD

Copyright © 2017 by Princeton University Press
Requests for permission to reproduce material from this work should be sent to
Permissions, Princeton University Press

Published by Princeton University Press, 41 William Street,
Princeton, New Jersey 08540

In the United Kingdom: Princeton University Press, 6 Oxford Street,
Woodstock, Oxfordshire OX20 1TR

press.princeton.edu

ISBN: 978-0-691-16292-8
Library of Congress Control Number: 2017942607

British Library Cataloging-in-Publication Data is available

This book has been composed in Minion Pro with Univers Light Condensed for display
Printed on acid-free paper. ∞

Typeset by Nova Techset Private Limited, Bangalore, India
Printed in the United States of America

10   9   8   7   6   5   4   3   2   1

Contents

9 Interpreting Frequencies of Individual Modes: Other Diagnostics

10 Inverting Mode Frequencies

Preface and Acknowledgments

The year 1962 marked the beginning of the field of helioseismology, the study of the Sun using solar oscillation frequencies. The routine study of other stars using data on their oscillations had to wait for decades. The CoRoT and *Kepler* missions have provided asteroseismic observations of thousand of stars, and with that has come the need to find robust ways of analyzing and interpreting long, near continuous high-quality timeseries data. While there are many good textbooks that cover the physics of stellar oscillations, as well as the myriad types of stellar oscillators, there is nothing that describes the different techniques that are used to analyze the data. This book is meant to fill that gap, specifically for cool main sequence, subgiant, and red-giant stars that show solar-like oscillations. This book is aimed at those who want to enter the field or who simply want to know how asteroseismic analyses are carried out. We assume that readers already have some knowledge of stellar evolution and of Fourier transforms. In some respects this book can be thought of as an instruction manual for readers to learn about asteroseismology and, if they so desire, to start working in this discipline. We provide reading lists and provide problem sets to try out what is learned. Links to the data needed for the problem sets can be accessed via http://press.princeton.edu/titles/11170.html.

We have concentrated on how to analyze individual stars in detail. However, the growing nature of the field means that even as of the writing of this book many automated, machine-learning based tools are being developed to determine basic characteristics of stars. These new tools will be particularly useful for the analysis of the large amounts of data expected from future asteroseismology missions like TESS and PLATO. We do not cover these new tools here, partly because most of the underlying components of these systems are the ones used in analyzing single stars. However, it is quite possible that better single-star analysis techniques will also be developed in the future. Our book should give the readers a good grounding on the techniques that exist today and therefore enable them to pick up on newer methods as and when they are developed.

This book would not have been possible without the generous help of our colleagues. Our thanks go to Mikkel Lund for providing several figures; H. M. Antia and Kuldeep Verma for kindly redrawing their published figures in black-and-white for this book; and Jørgen Christensen-Dalsgaard, Sebastien Deheuvels, Ronald L. Gilliland, and Eric Michel for permission to use published figures. We also thank Diego Bossini, Jørgen Christensen-Dalsgaard, Thomas Kallinger, Anwesh Mazumdar, Andrea Miglio, Daniel Reese, and Ian Roxburgh for generously providing frequencies and structures of stellar models. Our particular thanks go to Ian Roxburgh for taking the time to explain the phase-matching method of analyzing and inverting frequencies.

And finally, we thank Earl Bellinger, Tiago Campante, Guy Davies, Yvonne Elsworth, Rachel Howe, Caitlin Jones, James Kuslewicz, Mikkel Lund, and Lucas Viani for reading early drafts of different chapters and giving very useful suggestions and the two anonymous reviewers for their comments on the first full draft.

Without the above help, this book would have been very incomplete.

Asteroseismic Data Analysis

1 Introduction

Asteroseismology is the study of stellar pulsations. Stellar pulsations give us a unique window into the interiors of stars and provide us with the means to probe the internal structure, physical processes, and state of evolution of a star. This makes it possible to build a reliable, tested, and calibrated theory of stellar evolution. Asteroseismic data can give us precise estimates of the mass, radius, and ages of stars; in fact these data are the only means by which we can, in principle, obtain model-independent estimates of masses of single field stars.

In addition to the study of stellar physics, asteroseismic data have a range of applications in other fields of astrophysics. One of the most successful applications has been in the study of exoplanetary systems, where accurate properties of the host star are needed to fully characterize the newly discovered exoplanets. Another application is in the study of the structure and evolution of the Galaxy—Galactic archaeology—using the properties of red giants. Red giants are intrinsically bright and can be seen over large distances. Estimates of the radii of these stars can tell us about their luminosity and hence distance; estimates of their masses can be used to estimate their ages. Thus these stars can be used to determine the ages of stellar populations in different parts of the Galaxy.

1.1 THE DIFFERENT TYPES OF PULSATORS

Stellar pulsations may be detected by observing the variations of a star's brightness as a function of time. Radial velocity observations are also used in certain cases, though most pulsating stars have been studied using brightness variations. There is a long history of observing pulsating stars. One of the first known oscillating stars, o Ceti, was discovered in 1596. P Cygni was discovered soon thereafter in 1600. The first Cepheids, δ Cephei and η Aquilæ, were discovered in 1784. And the other well-known pulsating star, RR Lyræ, was discovered in 1899. Since then, pulsating stars have been observed throughout the Hertzsprung-Russell (HR) diagram, and it appears that oscillations are a ubiquitous feature of stars.

The most well-known pulsating stars that have been observed over the years have two features in common: they have large amplitudes (and hence are easy to observe from the ground); and more importantly, most of the stars lie along a well-defined narrow strip on the $T_{\rm eff}$–luminosity plane, where $T_{\rm eff}$ is the effective temperature. The preponderance of pulsating stars in this region of the HR diagram has led it to be called the *instability strip*. These large-amplitude oscillators are usually self-excited (i.e., layers in the stars act as a heat engine). These layers can trap heat when the star is contracting and release it in the expansion phase, cooling the star and causing contraction again and thus allowing the cycle of pulsation to continue. For this mechanism to work, the pertinent layer has to be positioned at a suitable depth inside the star, which can happen for certain combinations of temperature and luminosity, with metallicity also playing a role. The most common cause of these oscillations is the position of the helium ionization zone—unlike other layers of stars, ionization zones become more opaque when heated. The position of iron ionization layers is important for pulsations in massive stars.

The focus of this book is, however, not the study of these well-known pulsators but stars with solar-like pulsations—the small-amplitude oscillations that are continually excited (in a stochastic manner) and damped by turbulence in the outer convection zones of the stars. Most stars with outer convection zones that have been investigated show evidence of these oscillations. Since they are not forced, the oscillations have very low amplitudes (e.g., a few parts per million in the case of stars like the Sun, up to a few parts per thousand for red giants). This is in stark contrast to some of the classical pulsators, as can be seen in Figure 1.1, where we show the light curves and oscillation spectra of a classical pulsator and a Sun-like star. Because of the weak oscillation amplitudes, asteroseismic studies of solar-type stars are a relatively new endeavor. Oscillations of this type were first detected on the Sun, and as a result stars showing them are usually termed *solar-like oscillators*. Some of the well-studied classes of oscillators are listed in Table 1.1.

Stellar oscillations are three dimensional in nature and thus to describe them, we need functions of radius, latitude, and longitude (i.e., r, θ, and ϕ). The angular dependence of the different modes of the small-amplitude, solar-like oscillations are usually described in terms of spherical harmonics, since these functions are a natural description of the normal modes of a sphere. The radial function is more complicated, as we shall see in Chapter 3. The oscillations are usually labeled by three numbers: the radial order n, the angular degree l, and the azimuthal order m. The quantities l and m characterize the spherical harmonic Y_l^m. Modes with $n = 0$ are the so-called fundamental or f modes, and are essentially surface gravity modes. The degree l denotes the number of nodal planes that intersect the surface of a star, and m is the number of nodes along the equator. The radial order n can be any whole number and is the number of nodes in the radial direction. Positive values of n are used to denote acoustic modes, that is, the so-called p modes (p for pressure, since the dominant restoring force for these modes is provided by the pressure gradient). Negative values of n are used to denote modes for which buoyancy provides the main restoring force. These are usually referred to as g modes (g for gravity). Modes with $l = 0$ are the radial modes in which the stars expand or contract as a whole (often referred to as *breathing* modes), $l = 1$ are the dipole modes, $l = 2$ the quadrupole modes, $l = 3$ the octopole modes, and so on. For a spherically symmetric star, all modes with the same degree l and order n have the same frequency. Asphericities, such as rotation and magnetic fields, lift this degeneracy and give rise to frequency splitting of the modes, making the frequencies m dependent. It should be noted that in the context of classical pulsators, the lowest-order radial (i.e., $l = 0$) mode is often referred to as the *fundamental* mode. These modes should not be confused with the f modes.

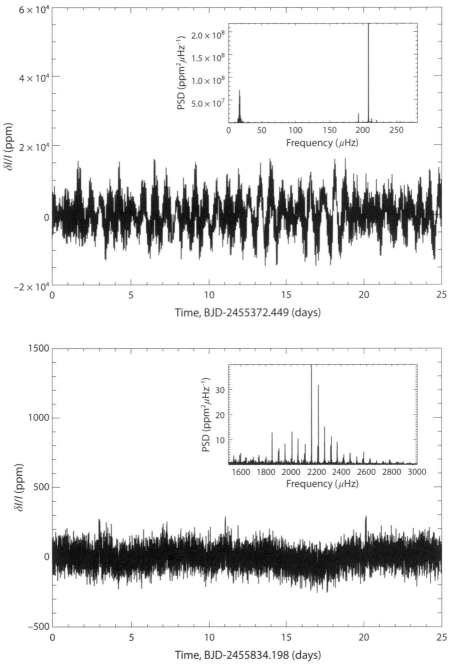

Figure 1.1. Top panel: A short segment of the *Kepler* lightcurve of a classical pulsator (KIC11145123) that shows both δ Scu and γ Dor oscillations (see Table 1.1 for definitions) and a zoom of the oscillation spectrum. Bottom panel: A similar length segment, and oscillation spectrum, of the main sequence star 16 Cyg A (HD 186408). Note the much larger amplitudes and longer periods of the detected pulsations in the classical pulsator compared to the Sun-like star. Unlike the case of the classical pulsator, for 16 Cyg A the variance of the lightcurve is not dominated by the oscillations; the biggest contribution is from shot noise, with similar amplitude contributions arising from oscillations and surface convection (i.e., granulation).

TABLE 1.1. TYPES OF PULSATORS

Name	Properties
On or near the main sequence	
Solar-like pulsators	Main sequence stars like the Sun that are cool enough to have an outer convection zone; Low-amplitude, multiperiodic p-mode pulsators with periods of the order of minutes to tens of minutes
γ Dor	Multiperiodic, periods of order 0.3–3 days; Lie at the intersection of the classical instability strip and the main sequence; Low-degree g-mode pulsators
δ Sct	Population I, multiperiodic, short periods of order 18 min to 8 hours; Lie at the intersection of the classical instability strip and the main sequence; Radial as well as nonradial p modes are observed
SX Phe	Same as δ Sct, but Population II stars
roAP	Rapidly rotating, highly magnetic, chemically peculiar A-type stars in the instability strip where δ Scuti stars are located; Multiperiodic, p-mode pulsations with periods of 5–20 min
SPB	Slowly pulsating B stars. Young Population I stars of about 8–18$M_\odot$; Similar to γ Dor pulsators, with slower periods due to their larger size; g-mode pulsators with periods of 0.5–5 days
β Cep	A short-period group of variables of spectral types B2–B3 IV–V; Can be multi-periodic; Periods of around 2–8 hours. Both p- and g-type pulsations are observed
Pulsating Be Stars	Oscillating Population I B-type stars that show Balmer-like emission; Generally have one dominant period but can be multi-periodic; Have high rotation rates and are often considered to be high rotation-velocity analogs of SPB and β Cep stars
Evolved stars	
Solar-like pulsators	Stars in the subgiant, red giant or red-clump phases; Low-amplitude, multi-period oscillators with periods of minutes to hours; Stars oscillate in p modes and mixed modes (i.e., modes that are p-like at the surface and g-like in the core)
RR Lyræ	Low-mass Population II stars in the core-He burning stage; Radial mode pulsators with periods of about 0.3–0.5 days; Generally mono-periodic; Classified into three groups according to the skewness of the light-curve; Found in the classical instability strip
Cepheids	High-mass pulsators in the core-He burning phase; Generally mono-periodic radial-mode pulsations with periods of about 1–50 days, but can sometimes have have two periods; Found in the classical instability strip
W Virginis	Population II analogs of Cepheids; Found on the instability strip evolving toward the asymptotic giant branch; Divided into groups by period, with groupings of 1–5 days, 10–20 days, or longer than 20 days
RV Tauri	F to K supergiants; Can be called long-period W Virginis stars, but with some differences; Periods of 30–150 days; Lightcurves show alternating deep and shallow minima; Radial pulsations

TABLE 1.1. TYPES OF PULSATORS (*CONTINUED*)

Name	Properties
Mira stars	Found near the tip of the giant branch, $1–7\times10^3 L_\odot$, with temperatures of 2,500 to 3,500 K. Generally long-period (hundreds of days to years) pulsations in the radial fundamental mode
Semiregular variables	Giants or supergiants of intermediate and late spectral types showing noticeable periodicity in their light changes accompanied by irregularities; Periods can be long, from 20 days to longer than 2,000 days; Usually subdivided into SRa, SRb, SRc, and SRd types according to periods and amplitudes; SRa stars are similar to Mira variables except that unlike the former, they oscillate in an overtone
Compact stars	
sdB stars	He burning stars that end up on the extreme horizontal branch because of mass loss; Masses less than $0.5 M_\odot$, $T_{\rm eff}$ in the range $23–32\times10^3$ K, and $\log g$ between 5 and 6; There are two classes of these multiperiodic pulsators: p-mode pulsators with periods of 1–5 minutes, and g-mode pulsators with periods of 0.5–3 hours
White dwarfs	
GW Vir	DO type white dwarfs that pulsate; These stars have high $T_{\rm eff}$, in the range of $70–170\times10^3$ K; Multiperiodic, with periods of 7–30 min; Pulsate in g modes
DB stars	Cooler than GW Vir type stars, $T_{\rm eff}$ between 11 and 30×10^3 K; Periods of 4–12 min; Pulsate in g modes
DA stars	Also known as ZZ Ceti stars; Very narrow $T_{\rm eff}$ range around 11,800 K; Periods from less than 100 sec to longer than 1,000 sec; Pulsate in g modes
Pre-main sequence stars	
PreMS δ Scu	Pre-MS stars whose tracks cross the instability strip and show pulsations like δ Scu stars
PreMS γ Dor	Pre-MS stars whose tracks cross the instability strip and show pulsations like γ Dor stars
PreMS SPB	High-mass pre-MS stars that show SPB-like pulsations

1.2 A BRIEF HISTORY OF THE STUDY OF SOLAR-TYPE OSCILLATIONS

The study of stochastically excited oscillators began with the Sun. Solar oscillations were first detected in the early 1960s. Later observations indicated that the detected oscillations were not merely surface phenomena. However, only in 1970 was a theoretical basis presented that interpreted the observations as standing waves trapped below the solar photosphere, a hypothesis fully confirmed by resolved observations of the Sun taken in the mid-1970s—observations that founded the field of helioseismology. Observational confirmation that the oscillations displayed by the Sun included truly global whole-Sun, core-penetrating pulsations followed in the late 1970s from observations made of the Sun as a star, (i.e., from observations in which the solar disc was not resolved).

The field of helioseismology developed rapidly beginning in the early 1980s. Dedicated ground-based networks were established to provide long, near-continuous observations of the Sun, allowing accurate and precise estimates of oscillation frequencies to be made and seismic studies of the solar activity cycle to be performed.

The first two dedicated networks made Sun-as-a-star observations. These observations do not resolve the solar disc, and hence their data are the most similar to the data we can get for other stars. The Birmingham Solar-Oscillations Network (BiSON) is still operating and continues to accumulate what is now a unique seismic record of the Sun stretching over more than three solar activity cycles. The International Research on the Interior of the Sun (IRIS) project collected similar data over several years. Next came the Global Oscillations Network Group (GONG), a network of six telescopes that make resolved-disc observations of the Sun, which therefore observes oscillations with short spatial scales on the solar surface. GONG started observing in 1995 and continues to operate.

Dedicated, long-term space-based observations followed soon after with the launch of the ESA/NASA Solar and Heliospheric Observatory (SoHO) in 1995. SoHO had three helioseismology-related instruments on board: the Michelson Doppler Imager (MDI), the Variability of solar Irradiance and Gravity Oscillations (VIRGO) package and the Global Oscillations at Low Frequencies (GOLF) spectrometer. Space-based observations are now being continued by the Heliospheric and Magnetic Imager (HMI) on board the Solar Dynamics Observatory (SDO).

Helioseismology has revolutionized our knowledge of the Sun. We know the structure and dynamics of the Sun in great detail thanks to these data. The data also show that the Sun changes with time; in particular, solar rotation in the outer layers changes with varying levels of solar activity. Describing what we have learned from helioseismology is beyond the scope of this book, but we have provided a reading list at the end of this chapter for those who may be interested in knowing more about this field.

The asteroseismic study of other solar-type stars took longer to develop because of the inherent difficulties of detecting extremely small variations of starlight from the ground, through the Earth's atmosphere. The first ground-based attempts were naturally focused on trying to detect pulsations in the very brightest main sequence and subgiant stars. In hindsight, the first detection of oscillations in a solar-type star was probably made in the early 1990s, when excess variability attributable to pulsations was observed in the subgiant Procyon (α CMi). At the time it was not possible to identify individual frequencies, and it would take observations made in the mid-1990s of another subgiant, β Hyi, to finally reveal unambiguous evidence for individual oscillations in another solar-type star. Clear detections of oscillations in a Sun-like main sequence star, α Cen A, followed soon after.

Ground-based observations made at large telescopes present several logistical challenges, and only recently has a dedicated network for the study of solar-like oscillations been conceived—the Stellar Observations Network Group (SONG)— which is now being developed. However, it is the advent of dedicated, long-term observations by space-based missions that has truly revolutionized the observational field of asteroseismology by providing data of hitherto unseen quality on unprecedented numbers of stars. The story of space-based asteroseismology started with the ill-fated Wide-Field Infrared Explorer (WIRE). The satellite failed, because coolants meant to keep the detector cool evaporated. However, it was soon realized that the star tracker could be used to monitor stellar variability and hence to look for stellar oscillations. This led to observations of α UMa and α Cen A. The Canadian mission Microvariability and Oscillations of Stars (MOST) was the first successfully launched mission dedicated to asteroseismic studies. Although it was not very successful in studying solar-type stars, it was immensely successful in studying giants, classical pulsators, and even starspots and exoplanets.

The major breakthroughs for studies of solar-like oscillators came with the CNRS/ESA CoRoT Mission and the NASA *Kepler* Mission. CoRoT observed many cool red giants and showed that these stars show nonradial pulsations; the mission observed some cool subgiants and main sequence stars for asteroseismology. *Kepler* increased significantly the numbers of subgiants and main sequence stars with detected oscillations, increased the sample for red giants, and also heralded an epoch of ultraprecise analyses made possible by observations lasting up to the 4-year duration of the nominal mission. Studies of pulsations in these types of stars are continuing with the repurposed *Kepler* mission, K2, which is providing data on targets in many different fields near the ecliptic plane.

Asteroseismology is going through a phase of rapid development in which we are still learning the best ways to analyze and interpret the data. With two more space missions being planned that will provide more exquisite asteroseismology data—the NASA Transiting Exoplanet Survey Satellite (TESS; launch 2018), and the ESA Planetary Transits and Oscillations of stars (PLATO; launch 2025) missions —this field is going to grow even more rapidly.

1.3 OVERVIEW OF THE DATA

While various aspects will be explored in detail in subsequent chapters, our aim in what remains of this chapter is to give a brief introduction to the basic appearance and properties of the pulsation spectra of solar-like oscillators.

As noted above, stars with near-surface convection zones show solar-like oscillations, modes that are stochastically excited and intrinsically damped by the near-surface convection. While this process limits the amplitudes of these intrinsically stable modes, many overtones are often excited to detectable levels. The modes in cool main sequence stars are predominantly acoustic in nature (i.e., they are p modes). The top panel of Figure 1.2 shows the frequency-power spectrum of the *Kepler* lightcurve of the G-type main-sequence star 16 Cyg A (HD 186408), the more massive component of the Sun-like binary system 16 Cygni. When oscillations are clearly detected, as is the case here, we see a rich pattern of peaks at the frequencies of the high-overtone (order n), low-degree (low-l) p modes of the star. Both radial and nonradial modes are present (bottom-left panel). However, because of geometric cancellation, only modes of low degree can be observed, and even then some of the rotationally split m components may be unobservable, depending on the angle of inclination of the rotation axis of the star. If we extend the plotted range of the spectrum to lower frequencies (bottom-right panel), we also see evidence of

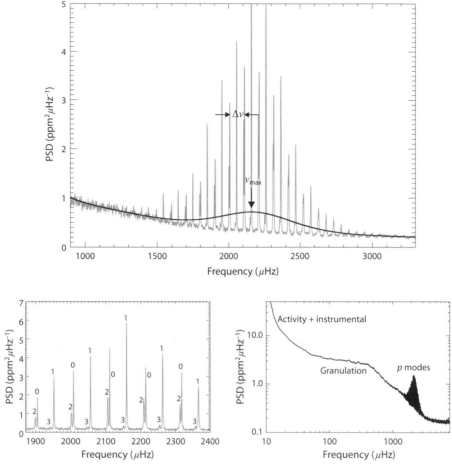

Figure 1.2. Frequency-power spectrum of the *Kepler* lightcurve of 16 Cyg A (HD 186408). Top panel: Lightly smoothed spectrum in gray, heavily smoothed in black, showing the Gaussian-like power envelope. Bottom-left panel: Zoom of the central part of the spectrum, with annotation showing the angular degree *l* of each mode. Bottom-right panel: Logarithmic plot of a wider frequency range, showing contributions from granulation, activity, and instrumental noise.

signatures of the surface patterns of convection (granulation) and the evolution of starspots (magnetic activity), in addition to contributions due to instrumental noise (e.g., drifts).

The power due to the oscillations is modulated in frequency by an envelope that typically has a Gaussian-like shape. The oscillation spectrum of 16 Cyg A is a classic example (top panel, Figure 1.2). We note in passing that there is growing evidence suggesting that oscillation power envelopes in the hottest (F-type) and coolest (K-type) solar-like oscillators tend to be flatter in shape. The frequency at which the observed power is strongest is called ν_{max}. This global asteroseismic parameter is a useful diagnostic of the physical properties of the near-surface layers of the star. As we will see in Chapter 3, it is related to the acoustic cutoff frequency of a star which,

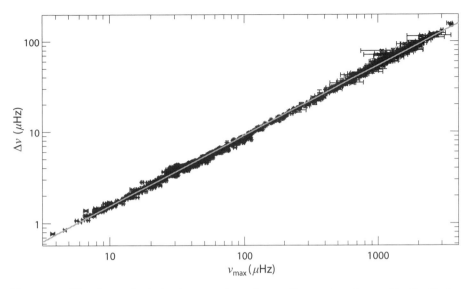

Figure 1.3. The observed relation between $\Delta\nu$ and ν_{max} of some stars observed by the *Kepler* mission. The gray line is a least-squares fit to the data. The fit implies that $\Delta\nu \propto \nu_{max}^{0.77}$.

with a few assumptions, implies a dependence on intrinsic properties of the form $\nu_{max} \propto g\,T_{eff}^{-1/2}$.

Another important global seismic parameter is the average large-frequency separation, $\Delta\nu$ (top panel, Figure 1.2). It is an average of the frequency differences

$$\Delta\nu_{nl} = \nu_{n+1,l} - \nu_{n,l} \qquad (1.1)$$

between consecutive overtones n of the same degree l. The observed average separation scales to very good approximation as $\bar{\rho}^{1/2}$, with $\bar{\rho} \propto M/R^3$ being the mean density of a star of mass M and surface radius R. This dependence follows if we assume that (1) the observed average separation is close to the separation in the asymptotic relation for p modes discussed in Chapter 3 and (2) we may treat stars as homologously scaled versions of one another. The average large separation and the frequency of maximum power are related, as can be seen in Figure 1.3. This is not completely surprising since both $\Delta\nu$ and ν_{max} depend on the global properties of a star. We shall examine the theoretically expected relation between the two parameters in Chapter 3.

The near-regular nature of the patterns of frequencies shown by main sequence stars may also be shown in the form of an échelle (ladder) diagram, like the one in Figure 1.4. It was made by dividing the spectrum of 16 Cyg A into frequency segments of length $\Delta\nu$. The segments were then stacked in ascending order, and the frequencies of the low-l modes were marked in each segment on the resulting diagram (see the different plotting symbols in the figure). Formally, we have plotted ν_{nl} against the wrapped (reduced) frequencies ($\nu_{nl} \bmod \Delta\nu$). Note that when the diagram is constructed in this way, modes in a given order n lie on lines that slope upward, as shown by the dotted lines. The orders can be made horizontal if the frequencies plotted on the vertical axis are instead those at the center of each order.

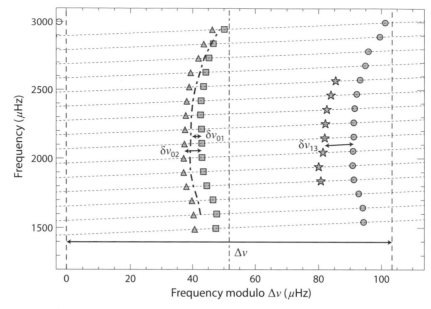

Figure 1.4. Échelle diagram of the oscillation frequencies of 16 Cyg A, with $l = 0$ modes plotted as squares, $l = 1$ modes as circles, $l = 2$ modes as triangles, and $l = 3$ modes as stars. Dotted lines follow the sloping orders. The heavy dot-dashed line follows the exact halfway frequencies between adjacent $l = 1$ modes. The vertical dashed lines mark zero and Δv on the abscissa, and the thin dot-dashed line marks $\Delta v/2$. Characteristic frequency separations are also marked on the plot (see text).

The diagram shows clear *ridges*, each one comprising overtones of a different degree. Departures from strict regularity of the frequency spacings are revealed by curvature in the ridges. As we shall see later, this carries important information on the underlying structure of the star. In addition to Δv, we have also marked three characteristic small frequency separations on the diagram. Shown are the separations between adjacent $l = 0$ and 2 modes [$\delta v_{02}(n)$], and between adjacent $l = 1$ and $l = 3$ modes [$\delta v_{13}(n)$]; $\delta v_{01}(n)$ measures deviations of the $l = 0$ modes from the exact halfway frequencies of the adjacent $l = 1$ modes ($\delta v_{10}(n)$ gives the separations with the mode degrees swapped). These small-frequency separations depend on the gradient of the sound speed in the deep interior of the star, and in main sequence stars provide a sensitive diagnostic of age (for a given assumed physics and chemical composition).

As main sequence stars evolve, the observed oscillations move to lower frequencies (i.e., v_{max} decreases), largely in response to the decreasing surface gravity. Figure 1.5 shows frequency spectra of five solar-like oscillators observed by *Kepler* (including 16 Cyg A, the second star down), which all have similar masses to that of the Sun. The stars are arranged from top to bottom in order of decreasing v_{max} (i.e., increasing evolutionary state). The top two stars are on the main sequence. The third and fourth stars are subgiants and therefore have finished burning hydrogen in their cores; the fifth star lies at the base of the red-giant branch.

Once stars leave the main sequence and enter the subgiant phase, they begin to show detectable signatures of gravity (g) modes, where the effects of buoyancy

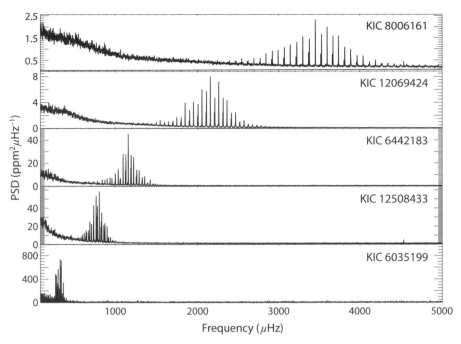

Figure 1.5. Frequency-power spectra of five solar-like oscillators observed by *Kepler*, which all have similar masses to that of the Sun. The stars are arranged from top to bottom in order of decreasing ν_{max}.

provide the restoring force. In a stable (nonconvective) layer, a small displacement of a parcel of fluid will cause it to oscillate with a frequency known as the Brunt-Väisälä frequency (or buoyancy frequency). As we shall see later, the Brunt-Väisälä frequency is the highest frequency that a g mode can have, and in the deep stellar interior it increases during the post–main sequence phase, eventually extending into the frequency range of the detectable, high-order p modes. When the frequency of a g mode comes close to that of a nonradial p mode of the same degree l, the modes interact, or couple, and undergo an avoided crossing (analogous to avoided crossings of atomic energy states). Interactions between the modes not only "bump" (shift) the observed frequencies but also change the intrinsic character of the modes so that they take on mixed p- and g-mode characteristics, having g-mode-like behavior in the deep interior and p-mode behavior in the envelope.

Figure 1.6 shows the échelle diagram of the oscillation frequencies of HD 183159 (KIC 6442183), the third star down in Figure 1.5. Note how we have copied, or replicated, the échelle into the ranges $\{-\Delta\nu, 0\}$ and $\{\Delta\nu, 2\Delta\nu\}$, where the frequencies are plotted with open symbols. For this star we have a single pure g-mode—which has a frequency of about 1,000 μHz—that has moved into the detectable p-mode range, and hence we see the effects of several $l = 1$ p modes coupling to it. The interactions manifest in the frequency pattern as an avoided crossing, with some of the $l = 1$ overtones being shifted significantly from the putative undisturbed $l = 1$ ridge. This includes two modes that have been shifted all the way across to the $l = 0$ ridge. There is one extra $l = 1$ mode at each avoided

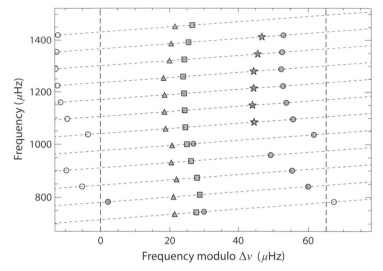

Figure 1.6. Échelle diagram of the oscillation frequencies of HD 183159 (KIC 6442183; third star down in Figure 1.5), with $l = 0$ modes plotted as squares, $l = 1$ modes as circles, $l = 2$ modes as triangles, and $l = 3$ modes as stars. Modes are plotted with open symbols where the échelle repeats.

crossing, implying a reduction in the frequency spacings between the relevant dipole modes. The frequency of the avoided crossing corresponds to the pure g-mode frequency that the star would show if it were comprised only of the central g-mode cavity; this frequency is a sensitive diagnostic of the core properties and the exact evolutionary state (again, for a given assumed physics and chemical composition).

Whereas in the above example we had a single g mode coupling to several p modes, as stars evolve into the red-giant phase, the increased density (in frequency) of g modes in the high-order p-mode regime leads to a much richer set of interactions and observable signatures. As we will see later, these signatures may be used to discriminate different advanced phases of evolution. As a foretaste, Figure 1.7 shows the frequency spectrum and échelle diagram of a star on the red-giant branch. *Kepler*-432 is a low-luminosity red giant that hosts a transiting Jupiter-sized planet, as well as a nontransiting gas giant revealed by ground-based Doppler velocity observations, which lies in a wider orbit. The annotation in the left panel of Figure 1.7 shows where the modes of different degrees of the star lie in its oscillation spectrum. In each order, we have several closely spaced g modes coupling to a single p mode, which gives rise to clusters of mixed $l = 1$ modes occupying the gray regions. The échelle diagram shows clearly that there are now several mixed modes in each order.

The quantities $\Delta\nu$, ν_{max}, and the individual frequencies form the basis of all asteroseismic analyses. While the two parameters $\Delta\nu$ and ν_{max} are usually enough to determine the mass and radius of the star, the individual frequencies provide more refined estimates and are also needed for modeling the interior. In this book we describe how we use these data to determine properties of a star.

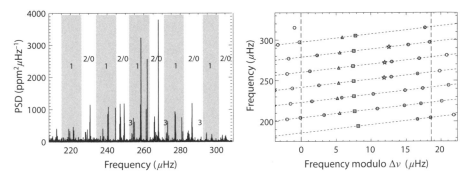

Figure 1.7. Left panel: Frequency-power spectrum of the planet-hosting star *Kepler*-432. The annotation marks where modes of different degree lie, including clusters of mixed $l = 1$ modes (gray regions). Right panel: Échelle diagram of the frequencies of *Kepler*-432, with $l = 0$ modes plotted as squares, $l = 1$ modes as circles, $l = 2$ modes as triangles and $l = 3$ modes as stars.

1.4 SCOPE OF THIS BOOK

The analysis of asteroseismic data obtained by *Kepler* has resulted in the development of specialized techniques to analyze and interpret the data. This book is a distillation of some of the techniques that are used. It is aimed at students and researchers who want to enter the field and at other astrophysicists who simply want to know how asteroseismic analyses are carried out. Instructors of senior undergraduate and graduate classes will also find this book useful. Although we do discuss the equations governing stellar structure and evolution, we expect that readers already have some knowledge of stellar theory and hence are familiar with the different stages of evolution of a star.

We start the main part of the book with a brief discussion of the equations of stellar structure in Chapter 2 and then derive the equations that govern stellar pulsations in Chapter 3. We also discuss some properties of the oscillations. Since there are texts that deal solely with the equations of stellar pulsations and their properties, we have limited ourselves to what we consider to be the bare minimum that a student needs to know to embark on this field. Chapters 2 and 3 on theory are followed by Chapters 4–9 on data analysis and interpretation. In Chapters 4–6 we discuss the basic content of observed time series of stellar brightness (or velocity) data on stars, why these data look like they do, and how they are analyzed to extract the asteroseismic parameters. In Chapters 7–9 we explain the different ways in which stellar properties are determined. These chapters are followed by Chapter 10 on inversions. Inversions have been successfully used by helioseismologists to construct a detailed picture of the internal structure and dynamics of the Sun and also some aspects of variations of its structure over time. Inversions of frequencies and frequency splittings of other stars are still in their infancy. However, because inversions give model-independent estimates of stellar structure, we include a fairly detailed description of inversion techniques.

Perhaps the most rapidly evolving part of this field is red-giant asteroseismology. Until comparatively recently it was assumed that red giants would not show nonradial oscillations. What CoRoT and then *Kepler* have shown is not only that nonradial oscillations are present, but also that they are detectable in such numbers that make the oscillation spectra notably richer and more complicated than those shown by main sequence stars. While we have not treated red giants in a separate chapter, wherever necessary we have discussed how red giants require different analysis techniques. Because of the rapid evolution of the field, the discussion on red giants will inevitably be incomplete.

1.5 FURTHER READING

This book deals exclusively with stochastically excited pulsators. The ways in which data from heat-engine pulsators are analyzed is often very different from the analysis techniques presented in this book. Readers interested in learning more about heat-engine pulsators are referred to the following books:

- Christensen-Dalsgaard, J., 2003, *Lecture Notes on Stellar Oscillations*, Aarhus University. Available online at http://astro.phys.au.dk/~jcd/oscilnotes/.
- Balona, L. A., 2010, *Challenges in Stellar Pulsation*, Bentham Publishers.
- Aerts, C., Christensen-Dalsgaard, J., and Kurtz, D. W., 2011, *Asteroseismology*, Springer.
- Catelan, M., and Smith, H. A., 2015, *Pulsating Stars*, Wiley-VCH.

For first-hand accounts of the development of helioseismology, readers can look at articles in:

- Jain, K., Tripathy, S. C., Hill, F., Leibacher, J. W., and Pevtsov, A. A., eds., 2013, *Fifty Years of Seismology of the Sun and Stars*, ASP Conference Series, **478**.

A nontechnical introduction to helioseismology can be found in:

- Chaplin, W. J., 2006, *Music of the Sun: The Story of Helioseismology*, Oneworld Publications.

Most available helioseismic data have been extracted from disc-resolved observations of the Sun, and as a result, the techniques used to extract helioseismic frequencies are very different from what we shall discuss in the following chapters. Readers interested in how such data are analyzed are referred to the following articles:

- Anderson, E. R. Duvall, T. L., and Jeffries, S. M., 1990, *Modeling of Solar Oscillation Power Spectra*, Solar Physics, **364**, 699.
- Hill, F. et al., 1996, *The Solar Acoustic Spectrum and Eigenmode Parameters*, Science, **272**, 1292.
- Larson, T., and Schou, J., 2011, *HMI Global Helioseismology Data Analysis Pipeline*, J. Phys. Conference Series, **271**, 012062.
- Reiter, J. et al., 2015, *A Method for the Estimation of p-Mode Parameters from Averaged Solar Oscillation Power Spectra*, Astrophys. J., **803**, 92.
- Korzennik, S. G., et al., 2013, *Accurate Characterization of High-Degree Modes Using MDI Observations*, Astrophys. J., **772**, 87.

Helioseismology has revealed details of solar structure and dynamics. Helioseismic analyses have also allowed us to use the Sun as a laboratory. The following reviews describe what helioseismology has taught us about the Sun:

- Christensen-Dalsgaard, J., 2002, *Helioseismology*, Rev. Mod. Phys., **74**, 1073.
- Gizon, L., and Birch, A. C., 2005, *Local Helioseismology*, Living Rev. Solar Phys., **2**, 6.

- Chaplin, W. J. and Basu, S., 2008, *Perspectives in Global Helioseismology and the Road Ahead*, Solar Phys., **251**, 53.
- Howe, R., 2009, *Solar Interior Rotation and Its Variation*, Living Rev. Solar Phys., **6**, 1.
- Basu, S., 2016, *Global Seismology of the Sun*, Living Rev. Solar Phys., **13**, 2.

To learn a little about the development of asteroseismology, the following articles are useful:

- Brown, T. M., and Gilliland, R. L., 1994, *Asteroseismology*, Ann. Rev. Astron. Astrophys., **32**, 37.
- Christensen-Dalsgaard, J., 2004, *An Overview of Helio- and Asteroseismology*, Helio- and Asteroseismology: Towards a Golden Future, Proc. SoHO 14/GONG 2004 Workshop, ESA SP 559, European Space Agency.
- Chaplin, W. J., and Miglio, A., 2013, *Asteroseismology of Solar-Type and Red-Giant Stars*, Ann. Rev. Astron. Astrophys., **51**, 353.

1.6 EXERCISES

1. Find the mean density of stars with $\Delta\nu$ values listed below. Assume that $\Delta\nu_\odot = 135\,\mu\mathrm{Hz}$:

(a) 123.41 μHz
(b) 120.04 μHz
(c) 80.81 μHz
(d) 55.72 μHz
(e) 29.44 μHz
(f) 12.09 μHz
(g) 9.630 μHz

Can you identify the dwarfs and giants in the sample?

2. Given the $\Delta\nu$ and $\nu_{\max}$ scaling relations, derive expressions for mass and radius. Derive the uncertainty in the estimated mass and radius in terms of uncertainties in $\Delta\nu$, $\nu_{\max}$, and T_{eff}.

3. Frequencies of evolutionary sequences of stellar models of masses $1.40\mathrm{M}_\odot$ and $1.5\mathrm{M}_\odot$ are given in archive `frequencies_ov2.tar.gz`[1]. The properties of the models may be found in file `properties_ov2.txt`. Draw the échelle diagram for each model, and explore how evolution changes the diagram.

4. Mode-frequencies of several stars observed by *Kepler* may be found in the archive `obs_freq.tar`. Draw the échelle diagram for each star and using the diagrams, infer which evolutionary stages the stars are in.

---

[1]Links to this and other online materials adjust to this book can be found at http://press.princeton.edu/titles/11170.html.

2 Theoretical Underpinnings: Modeling Stars

The process of modeling and interpreting stellar oscillation data usually begins with the construction of stellar models. This requires a knowledge of the equations that govern stellar structure. In this chapter we give a brief description of the process of constructing models.

We normally model stars assuming that they are spherically symmetric (i.e., the stellar properties at any given time depend only on the radial distance from the center). This essentially means that we can describe a three-dimensional star as a one-dimensional object. A second common assumption is that a star does not change its mass as it evolves. This assumption is generally true for the mass range of stars that show solar-like oscillations. While a star does lose a lot of mass in the early stages of star formation and also at very evolved stages (e.g., at the tip of the giant branch), at other stages mass loss is generally small. The Sun loses about 10^{-14} of its mass per year. Thus in its expected main sequence lifetime of about 10 Gyr, the Sun will lose only about 0.01% of its mass. In contrast, the radius of a star can change enormously over its lifetime. The equations of stellar structure are therefore written with mass as the independent variable. A star is divided into many mass shells, from the center at $m = 0$ to the surface at $m = M$, where M is the total mass of the star, and a model is obtained by solving the set of equations at each mass shell in a self-consistent manner.

2.1 THE EQUATIONS OF STELLAR STRUCTURE AND EVOLUTION

The equations of stellar structure are basically statements of conservation—of mass, momentum, and energy—along with the conditions required for thermal equilibrium. Additionally we need to describe energy generation and how energy is transported from the core to the surface. We also need to describe how elemental abundances change as a function of time.

The first equation defines conservation of mass: the mass dm of a shell at radius r and thickness dr is simply the volume times the density ρ of the shell. Since we assume spherical symmetry, the volume is just $4\pi r^2 dr$, or in other words,

$$dm = \rho 4\pi r^2 dr, \qquad (2.1)$$

which, since we want the equation in terms of m rather than r, can be recast as:

$$\frac{dr}{dm} = \frac{1}{4\pi r^2 \rho} \, . \tag{2.2}$$

The next equation governs the conservation of momentum, which in the case of stars is the equation of hydrostatic equilibrium describing how the outward force due to gas pressure P balances the inward force of gravity:

$$\frac{dP}{dr} = -g\rho, \tag{2.3}$$

g being the acceleration due to gravity, or in other words,

$$\frac{dP}{dm} = -\frac{Gm}{4\pi r^4} \, , \tag{2.4}$$

with m being the mass enclosed in a shell of radius r, and G being the gravitational constant.

Conservation of energy is next. Stars produce most of their energy in the core through nuclear reactions. At equilibrium, we assume that an amount of energy $L(m)$ flows through a shell at a mass m per unit time as a result of the energy generated. If ϵ is the energy released per unit mass per second by nuclear reactions, and ϵ_ν is the energy lost by the star because neutrinos can stream out of the star without depositing their energy, then:

$$\frac{dL(m)}{dm} = \epsilon - \epsilon_\nu. \tag{2.5}$$

The total luminosity of a star is then $L = L(M) = \int_0^M L(m) dm$, where M is the total mass of the star. This picture is, however, incomplete. Different layers in a star can expand or contract during evolution; for example, cores of stars in the sub-giant and red-giant stages contract rapidly while the outer layers of the stars expand. Thus Eq. 2.5 needs to be modified to include the energy released through gravitational contraction or the energy used up because of expansion. After working through the thermodynamics, these processes lead to

$$\frac{dL(m)}{dm} = \epsilon - \epsilon_\nu - C_P \frac{dT}{dt} + \frac{\delta}{\rho} \frac{dP}{dt} \, . \tag{2.6}$$

Here C_P is the specific heat at constant pressure, t is time, and δ is a quantity that depends on the equation of state and is defined as

$$\delta = -\left(\frac{\partial \ln \rho}{\partial \ln T} \right)_{P, X_i}, \tag{2.7}$$

where X_i denotes the mass fraction of element i. The last two terms of Eq. 2.6 are often grouped together and called ϵ_g (g for gravity), because they denote the gravitational release of energy.

The next equation describes the variation of temperature inside a star. Using the equation of hydrostatic equilibrium, it can be written trivially as

$$\frac{dT}{dm} = -\frac{GmT}{4\pi r^4 P}\nabla,$$

(2.8)

where ∇ is the dimensionless *temperature gradient* $d\ln T/d\ln P$. The difficulty with this equation lies in calculating ∇. The expression for ∇ depends on whether energy is transported by radiation or convection (conduction is negligible in most normal stars, but it can easily be included). We shall discuss the issue of ∇ in more detail a bit later in Section 2.2.

The next set of equations deals with elemental abundances as a function of position and time. There are three major ways that the chemical composition can change:

1. nuclear reactions,
2. changes in the boundaries of convection zones, and
3. diffusion and gravitational settling (usually simply referred to as diffusion) of helium and heavy elements.

The change of abundance because of nuclear reactions can be expressed as

$$\frac{\partial X_i}{\partial t} = \frac{m_i}{\rho}\left[\sum_j r_{ji} - \sum_k r_{ik}\right],$$

(2.9)

where m_i is the mass of the nucleus of isotope i, r_{ji} is the rate at which isotope i is formed from isotope j, and r_{ik} is the rate at which isotope i is lost because it turns into a different isotope k. Nuclear energy generation rates are inputs to models.

Abundances also change because the boundaries of convection zones in stars change as a star evolves. Convection zones are chemically homogeneous—eddies of moving matter carry their composition with them, and when they break up, the material gets mixed with the surroundings. This happens on timescales that are short compared to the timescale of a star's evolution. If a convection zone exists in the region between two spherical shells of masses m_1 and m_2, the average abundance of any species i in the convection zone is

$$\bar{X}_i = \frac{1}{m_2 - m_1}\int_{m_1}^{m_2} X_i\, dm\,.$$

(2.10)

Therefore the rate at which $\bar{X}_i$ changes will depend on nuclear reactions in the convection zone as well as the rates at which the boundaries m_1 and m_2 change. Thus we have

$$\begin{aligned}
\frac{\partial \bar{X}_i}{\partial t} &= \frac{\partial}{\partial t}\left(\frac{1}{m_2 - m_1}\int_{m_1}^{m_2} X_i\, dm\right) \\
&= \frac{1}{m_2 - m_1}\left[\int_{m_1}^{m_2}\frac{\partial X_i}{\partial t}\, dm + \frac{\partial m_2}{\partial t}(X_{i,2} - \bar{X}_i)\right. \\
&\quad \left. -\frac{\partial m_1}{\partial t}(X_{i,1} - \bar{X}_i)\right],
\end{aligned}$$

(2.11)

where $X_{i,1}$ and $X_{i,2}$ are the mass fractions of element i at m_1 and m_2 respectively. The first term within the square brackets in Eq. 2.11 is the change in abundances due to nuclear reactions, the second and third terms are the change due to changes in the position of convection zones.

The heavier elements, such as helium, sink down inside a star as a result of gravitational settling as well as diffusion due to composition and temperature gradients. All three processes are generally referred to as *diffusion* and may be described by a classic diffusion equation of the form

$$\frac{\partial X_i}{\partial t} = D\nabla^2 X_i \, , \tag{2.12}$$

where D is the diffusion coefficient, and ∇^2 is the Laplacian operator. The diffusion coefficient hides the complexity of the process. Other mixing processes, such as those induced by rotation, are often included the same way by modifying D. The coefficient D depends on the isotope under consideration; however, it is not uncommon for stellar evolution codes to treat helium separately and use a single value for all other heavier elements. D is an external input for stellar structure codes.

Although most modern stellar evolution codes can follow the evolution of many different isotopes, it is still customary to describe the composition of a star in terms of three variables: the mass fraction of hydrogen, denoted by X; the mass fraction of helium, denoted by Y; and the mass fraction of everything else, denoted by Z. The *metallicity* of a star is the heavy element fraction Z, though it is more common these days to denote the metallicity as Z/X to make comparisons with observations easier. Most observers determine the metallicity of a star from a few spectral lines and denote it on a logarithmic scale with respect to the solar metallicity. It is usually written as [M/H]:

$$[\text{M/H}] = \log(N_m/N_\text{H}) - \log(N_m/N_\text{H})_\odot, \tag{2.13}$$

where N_m and N_H are the number densities of metals and hydrogen, respectively. Very often, observers only provide [Fe/H]. In most cases, stellar modelers assume that the relative abundance of different elements in a star is the same as for the Sun and translate the observed metallicity as

$$[\text{Fe/H}] = \log(Z/X)_\text{star} - \log(Z/X)_\odot, \tag{2.14}$$

where $(Z/X)_\odot$ is the observed heavy element abundance of the Sun.

2.2 THE QUESTION OF ∇

We still need to define ∇ in Eq. 2.8. This depends on how energy is transported.

When energy is transported through radiation (i.e., in the radiative zones), ∇ is calculated assuming that energy transport can be modeled as a diffusive process.

In such a case the flux $\mathbf{F}$ can be written in a manner that is analogous to the conduction equation, that is,

$$\mathbf{F} = -k \, \nabla T, \tag{2.15}$$

where ∇T is the spatial temperature gradient, and k is the *conductivity*, but in this case k depends on the local conditions (e.g., temperature and pressure). A full derivation yields the result

$$\nabla = \nabla_{\mathrm{rad}} = \frac{3}{64\pi \sigma_B G} \frac{\kappa L(m) P}{m T^4}, \tag{2.16}$$

where σ_B is the Stefan-Boltzmann constant; and κ is the Rosseland mean opacity, which is an external input to stellar models.

Convection is more difficult to handle. This is, by its very nature, a three-dimensional phenomenon, yet our models are one-dimensional. Even if we constructed three-dimensional models, it is currently impossible to include convection realistically and to evolve the models at the same time. This is because convection takes place over timescales of minutes to hours, while stars evolve over millions to billions of years. As a result, drastic simplifications are used to model convection. Deep inside a star, the temperature gradient is well approximated by the adiabatic temperature gradient $\nabla_{\mathrm{ad}} \equiv (\partial \ln T / \partial \ln P)_s$ (s being the specific entropy), which is determined by the equation of state. This approximation cannot be used in the outer layers, where convection is not efficient and some of the energy is carried by radiation. In these layers one has to use an approximate formalism, since there is no "theory" of stellar convection as such.

One of the most common formulations used to calculate convective flux in stellar models is *mixing-length theory* (MLT), in which heat transport by convection is assumed to be analogous to heat transfer by particles: the transporting particles are the macroscopic eddies, and their mean free path is the *mixing length, l_c*. The main assumption in the usual mixing length formalism is that convective eddies move an average distance equal to l_c before giving up their energy and losing their identity. Different mixing length formalisms have slightly different assumptions about what the mixing length is. The mixing length is usually defined as

$$l_c = \alpha H_P, \tag{2.17}$$

where α, a constant, is the so-called *mixing-length parameter*, and H_P is the pressure scale height given by

$$\frac{1}{H_P} = -\frac{d \ln P}{dr}. \tag{2.18}$$

There are additional assumptions that involve the geometry of the fluid element and the average distance traveled by it. One way to approach this is the following. Assuming that ∇ is the actual gradient and ∇_{rad} is the radiative gradient in Eq. 2.16, we can then define three quantities:

$$U = \frac{3ac\,T^3}{C_P \rho^2 \kappa l_c^2} \sqrt{\frac{8H_P}{g\delta}}, \tag{2.19}$$

where a is the radiation constant, and c is the speed of light;

$$W = \nabla_{\text{rad}} - \nabla_{\text{ad}}, \tag{2.20}$$

where ∇ is the actual temperature gradient; and

$$\xi^2 = \nabla - \nabla_{\text{ad}} + U^2. \tag{2.21}$$

Like ∇_{ad}, the quantities C_P and δ (defined in Eq. 2.7) are determined by the equation of state. At any point in the star, one can calculate U. For a given value of U, we can get ξ by solving the cubic equation

$$(\xi - U)^3 + \frac{8}{9}U(\xi^2 - U^2 - W) = 0. \tag{2.22}$$

Since we know U and ∇_{ad}, solving for ξ gives us the value of ∇, which is what we need to substitute in Eq. 2.8. Other formulations of the mixing-length approximation differ in the expression for U which incorporates geometrical factors that describe the shapes of the convective eddies.

The problem with the mixing-length approximation is that there is no a priori way to determine what α is, and it is one of the free parameters in stellar models. The usual way to determine it is to use the α needed to model the Sun. How α is determined for the Sun is discussed in Section 2.7. While using the solar value of α is the norm, the assumption that all stars have the same convective parameters as the Sun is increasingly coming into question. There are variants of MLT that were formulated without free parameters; however, none of them can model the Sun properly without the introduction of a mixing-length-type parameter.

Of course, before we can use either radiative or convective temperature gradients, we have to determine whether a region is radiative or convective. Energy is transported either through convection or radiation, and the first step is to determine which applies. This is done usually using the so-called Ledoux criterion: one has convection when

$$\nabla_{\text{rad}} > \nabla_{\text{ad}} + \frac{\phi}{\delta}\nabla_\mu, \tag{2.23}$$

where

$$\nabla_\mu \equiv \left(\frac{d\ln\mu}{d\ln P}\right)_s, \quad \text{and} \quad \phi \equiv \left(\frac{d\ln\rho}{d\ln\mu}\right)_{P,T}, \tag{2.24}$$

with ϕ being determined from the equation of state. In a chemically homogeneous region, $\nabla_\mu = 0$, and hence the condition for convection reduces to the *Schwarzschild criterion*:

$$\nabla_{\text{rad}} > \nabla_{\text{ad}}. \tag{2.25}$$

2.3 OTHER PHYSICAL PROCESSES

The equations in Section 2.1 describe the most simple processes that are absolutely essential to model a star. Certainly, other processes take place that are not commonly included. These include rotation and the effects that rotation has on stars (e.g., rotation-induced mixing) and effects of magnetic fields. However, one process that most codes try to handle is convective overshoot.

The Schwarzschild criterion (Eq. 2.25) used to determine whether convection occurs is strictly speaking a condition on the acceleration of convective eddies. Eddies stop accelerating at $\nabla = \nabla_{ad}$, but since they carry momentum, they should, in principle, penetrate the radiative zone. This is usually called *overshoot*. Overshoot from convective cores is important, because it brings in extra hydrogen, thereby prolonging the life of a star. Overshoot below convective envelopes can expose cooler material from the outer parts of the star to higher temperatures and thereby change the abundances of such elements as lithium and beryllium that are destroyed easily at high temperatures. There is, however, no first-principles formulation of overshoot, which is not surprising, since there is no first-principles formulation of convection. Thus overshoot is treated in a rather ad hoc manner in the codes, using the assumption that the extent of overshoot can be written as $\alpha_{ov} H_P$, where α_{ov} is a free parameter. The overshooting region itself is handled in different ways in different codes. In many codes the temperature gradient of the overshooting region is kept as the radiative gradient, but the region is chemically mixed with the convection zone; such formulations are often referred to as *overmixing*. Other codes assume that while the overmixing takes place, the temperature gradient of that region also changes, and it is usually set to the adiabatic gradient.

2.4 BOUNDARY CONDITIONS

Equations 2.2, 2.4, 2.6, and 2.8, together with the equations relating to change in abundances, form the full set of equations that govern stellar structure and evolution. In most codes, Eqs. 2.2, 2.4, 2.6, and 2.8 are solved for a given X_i at a given time t. Time is then advanced, Eqs. 2.9, 2.11, and 2.12 are solved to give new X_i, and Eqs. 2.2, 2.4, 2.6, and 2.8 are then solved again. Thus we have two independent variables, mass m and time t, and we look for solutions in the intervals $0 \leq m \leq M$ (stellar structure) and $t \geq t_0$ (stellar evolution).

One should note that we have five equations (counting all composition-change equations for each species as one) in six unknowns (i.e., r, P, T, ρ, $L(m)$, and X_i). Thus these equations cannot be solved without an additional input. That input is provided by the equation of state that connects ρ, P, T, and X_i.

Four boundary conditions are required to solve the stellar structure equations in the interval $0 \leq m \leq M$. It would have been simple had we been able to apply all boundary conditions at either $m = 0$ or $m = M$. As it happens, the boundary conditions are split: the radius and luminosity boundary conditions are applied at $m = 0$, and the pressure and temperature conditions at $m = M$.

The behavior of the four quantities r, $L(m)$, T, and P near $m = 0$ dictates whether boundary conditions can be applied at the center. The mass equation can be integrated in a small sphere around $m = 0$ that has a constant central density of ρ_c to get

$$r = \left(\frac{3}{4\pi\rho_c}\right)^{1/3} m^{1/3}, \tag{2.26}$$

yielding the boundary condition

$$r = 0 \quad \text{at} \quad m = 0. \tag{2.27}$$

The luminosity equation can be expanded around $m = 0$ to give

$$L(m) = (\epsilon - \epsilon_\nu + \epsilon_g)_c \, m, \tag{2.28}$$

the subscript c denoting central conditions. It follows from Eq. 2.28 that

$$L(m) = 0 \quad \text{at} \quad m = 0. \tag{2.29}$$

The temperature and pressure equations, unfortunately, do not tell us directly what the quantities at $m = 0$ are, but only that we get an expansion around $T = T_c$ and $P = P_c$, T_c and P_c being the temperature and pressure at $m = 0$. Since we do not know a priori the values of T_c and P_c, we need to apply boundary conditions on T and P at the surface.

The simplest boundary conditions that one can apply at the surface are

$$P = 0, \quad T = 0 \quad \text{at} \quad m = M. \tag{2.30}$$

These conditions are obviously incorrect, since we know that stellar surfaces are hot—we would not be able to see them otherwise. And hot gases have pressure, and thus pressure is not equal to zero either. The surface is actually defined as the radius R at which the effective temperature T_{eff} is such that the total luminosity L of the star is

$$L = 4\pi R^2 \sigma_B T_{\text{eff}}^4. \tag{2.31}$$

The boundary conditions at the surface are governed by models of stellar atmospheres. These give us a relation between T and the optical depth τ, the so-called T–τ relationship. A simple and popular choice is the Eddington atmosphere, which states that

$$T^4 = \frac{3}{4} T_{\text{eff}}^4 \left(\tau + \frac{2}{3}\right). \tag{2.32}$$

The optical depth τ in the atmosphere can be written as

$$\tau = \int_R^\infty \kappa\rho \, dr = \bar{\kappa} \int_R^\infty \rho \, dr, \tag{2.33}$$

where $\bar{\kappa}$ is the average opacity in the atmosphere. Writing the equation of hydrostatic equilibrium in terms of radius, we have

$$\frac{dP}{dr} = -g\rho, \tag{2.34}$$

where g is the acceleration due to gravity. Thus we have

$$P_{r=R} = \int_R^\infty g\rho dr = g_0 \int_R^\infty \rho dr, \tag{2.35}$$

where g_0 is the surface value of g, and we have assumed that the atmosphere has negligible mass. Thus, using Eq. 2.33, we get

$$P_{r=R} = \frac{GM}{R^2}\frac{\tau_{\text{eff}}}{\bar{\kappa}}, \tag{2.36}$$

where τ_{eff} is the value of τ at $T = T_{\text{eff}}$. For the Eddington atmosphere, $\tau_{\text{eff}} = 2/3$.

Since the outer boundary conditions depend on T_{eff}, this usually means that the process of modeling begins by having to choose a value of L and R. Since models are constructed in essence by integrating from $m = 0$ outward, we have to choose a value of P_c and T_c to start the integrations, but we are not guaranteed that the luminosity and pressure at the surface will be consistent with the adopted surface values. For each value of T_c and P_c, we can define mismatch functions:

$$F_1(P_c, T_c) = L(M) - 4\pi\sigma_B R^2 T_{\text{eff}}^4, \tag{2.37}$$

and

$$F_2(P_c, T_c) = P(M) - \frac{GM}{R^2}\frac{\tau_{\text{eff}}}{\bar{\kappa}}. \tag{2.38}$$

The problem then reduces to finding a solution that gives

$$F_1(P_c, T_c) = F_2(P_c, T_c) = 0. \tag{2.39}$$

However, modern stellar evolution codes do not solve for stellar structure by integrating the equations. The most common way to determine the structure is to use the Henyey method, where the differential equations are discretized as a set of difference equations for every mass shell in the star. Starting from a guess solution, the equations are solved in an iterative manner. This method is very efficient on modern computers.

Since the equations need to be solved to determine how stars evolve in time, we also need initial conditions for each of the variables. The initial conditions needed to start evolving a star depend on where we start the evolution. If the evolution begins at the pre-main sequence phase (i.e., while the star is still collapsing), the initial structure is quite simple. Temperatures are low enough to make the star fully convective and hence chemically homogeneous. If evolution is begun at the zero-age main sequence (ZAMS), which is the point at which hydrogen fusion begins, a ZAMS model must be used.

2.5 INPUTS TO STELLAR MODELS

The equations of stellar structure and evolution cannot be solved without external inputs. Five external inputs are needed:

1. the equation of state,
2. radiative opacities,
3. nuclear reaction rates,
4. diffusion coefficients, and
5. an atmospheric model.

The first four inputs relate to fundamental properties of matter and are often referred to as the *microphysics* inputs.

The equation of state of ideal gases, $P/\rho = (\mathcal{R}/\mu)T$, μ being the mean molecular weight and $\mathcal{R}$ the gas constant, works well for simple models. However, it is not good enough to model real stars. Modern helioseismic data reveal significant deviations between the equation of state of solar matter and the ideal-gas equation of state. The ideal-gas law does not, by definition, include such effects as ionization caused by high temperatures and pressure, effects of electron degeneracy at high densities, radiation pressure, or the like, and all these processes affect the thermodynamic structure of a model. Modern equations of state are the result of complex numerical calculations and are usually given in a tabular form with important thermodynamic quantities, such as ∇_{ad} and C_P, listed as functions of T, P (or ρ), and composition. However, these tabulated equations of state suffer from the limitation that they are usually constructed for a fixed relative heavy-element abundance that cannot be changed by users and hence lack flexibility.

We need the opacity κ to calculate the radiative temperature gradient ∇_{rad} of stellar material as a function of temperature, density, and composition. Opacity is a measure of how opaque a material is to photons. Like modern equations of state, opacities are usually available in tabular form as a function of density, temperature, and composition. Unlike for equations of state, web-based interfaces are available to calculate opacities for any given mixture of elements. However, often tables are applicable only to the higher temperature conditions of stellar interiors, and in such cases, the tables have to be supplemented by specialized low-temperature opacities that also include opacity contributions from molecules.

Nuclear reaction rates are required to compute energy generation, neutrino fluxes, and composition changes. These rates are generally available in the form of so-called *astrophysical cross-sections*. The temperature and density dependences often need to be coded in.

Diffusion coefficients are seldom calculated by stellar modelers themselves and are usually used as inputs without further modification.

Model atmospheres are not microphysics inputs, but they are nonetheless used as external constraints to determine the outer boundary conditions of a star. Atmospheric models generally provide a relation between the temperature T in the atmosphere and the optical depth τ, and hence they are commonly referred to as T–τ relations. These can be the result of simple approximations (e.g., the aforementioned Eddington atmosphere) or of fairly sophisticated calculations. Sometimes semi-empirical models of the solar atmosphere are used to construct models of other stars.

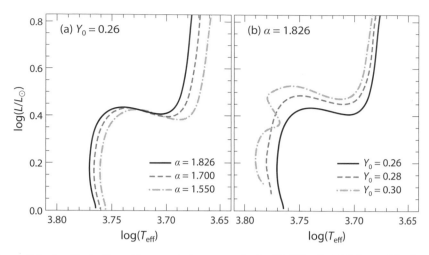

Figure 2.1. An illustration of how evolutionary tracks for [Fe/H] = 0, 1.12M$_\odot$ models change as we change α and Y_0. Note that increasing Y_0 to 0.30 changes the shape of the track and gives it the telltale shape of that of a star with a convective core. At low Y_0, the models have radiative cores.

2.6 MODELING A STAR

Let us assume we know the position of a star on the HR diagram, that is, we know its luminosity and effective temperature. How do we model this star?

Mass is the most fundamental input needed to model a star. Other important inputs include the initial heavy-element abundance Z_0 and the initial helium abundance Y_0. These quantities influence the equation of state and opacities and hence affect both the structure and the evolution of the star. Also required is the mixing-length parameter α. Usually the value of α is taken from solar models (see Section 2.7), but it is a parameter for which we have no independent constraints. Once these quantities are known (or chosen), models are evolved in time until they reach the observed temperature and luminosity. The initial guess of the mass may not result in a model with the required characteristics, and so a different mass may need to be chosen and the process repeated. Once a model that satisfies the observational constraints has been constructed, the model becomes a proxy for the star; the age of the star is assumed to be the age of the model, and the radius of the star is assumed to be that of the model.

Since the mixing-length parameter is essentially unconstrained, it is the largest source of uncertainty in low-mass stellar models. It controls the radius of a star at a given temperature; as a result, we cannot, strictly speaking, predict stellar radii. The uncertainty in the initial helium abundance Y_0 introduces a similar uncertainty. In Figure 2.1 we show how the evolutionary track of a stellar model changes as we change α and Y_0. Increasing α makes a model hotter at a given luminosity, implying that the radius of the model has to be smaller. However, the effects of α can only be seen on the main sequence and on the giant branch. Increasing Y_0 increases the luminosity at a given temperature at all stages of evolution.

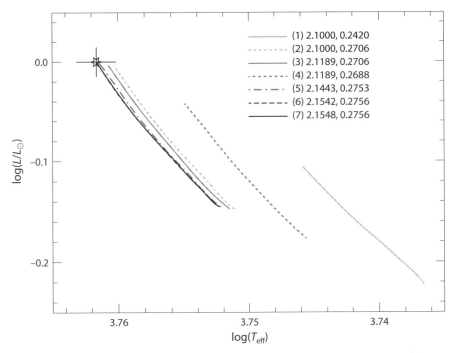

Figure 2.2. The process of constructing a solar model. The star marks the position of the Sun on the HR diagram. The pair of numbers in the legend denoting each track is the (α, Y_0) pair. The process starts with an initial guess of $\alpha = 2.10$ and $Y_0 = 0.242$ (track 1), and the iterative process ends with a solar model with $\alpha = 2.1548$, $Y_0 = 0.2756$ (track 7). The convergence criterion was set such that the iteration stopped when both $(L - L_\odot)/L_\odot$ and $(R - R_\odot)/R_\odot$ were less than 10^{-6}. At each step the model is evolved from the zero-age main sequence to the solar age.

2.7 CONSTRUCTING SOLAR MODELS

Solar models play a large role in stellar astrophysics, since we know more about the Sun than about any other star. Unlike other stars, we have independent estimates of the solar mass, radius, luminosity, and age. To be called a solar model, a $1M_\odot$ model must have the solar radius and luminosity at the solar age of 4.57 Gyr. Merely reproducing the position of the Sun on the HR diagram is not enough, unless we do so at the correct age. Thus solar models are constructed iteratively.

Once the microphysics of a stellar model is fixed, the set of equations has two free parameters, one being α, the other the initial helium abundance Y_0. As mentioned previously, α controls radius, and Y_0 controls luminosity. Thus to make a solar model, α and Y_0 are varied until one gets a model of $1L_\odot$ and $1R_\odot$ at the solar age. This has to be done in an iterative manner, since the equations of stellar structure are quite nonlinear. In addition to α and Y_0, the initial metal abundance Z_0 is also adjusted to reproduce the observed Z/X ratio in the solar envelope. The solar model obtained in this manner has no free parameters. When standard physics inputs are used, these models are called *standard solar models*, or SSMs. The convergence process is illustrated in Figure 2.2.

The value of α obtained for a solar model is often referred to as the *solar-calibrated* value of the mixing-length parameter and is also used to construct models

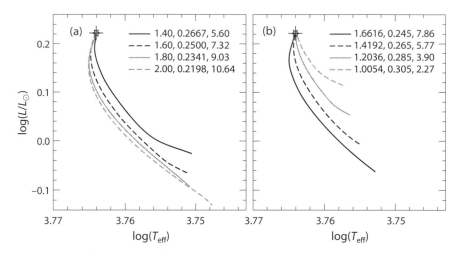

Figure 2.3. An illustration of how models can evolve to the same position in the HR diagram at different ages. Panel (a) shows Y_0 needed for different values of the mixing-length parameter α. Panel (b) shows how α changes for different values of Y_0. The trio of numbers that defines each track are (α, Y_0, age in Gyr). Note that for certain values of α the models require Y_0 to be lower than the primordial helium abundance, and hence those models may be considered unphysical.

of other stars. Unfortunately, the solar-calibrated value of α is not unique. It not only depends on the exact formulation of the mixing-length approximation used in the stellar evolution code, but also on input physics such as the atmospheric model used and whether or not the gravitational settling of helium and heavy elements is included. For example, for solar models constructed with the Yale Stellar Evolution Code (YREC) the value of α obtained for models with diffusion and Krishna Swamy atmospheres is 2.15, while α for the corresponding model with the Eddington T–τ relation is 1.87. If diffusion is not included, the calibrated α for a solar model with an Eddington atmosphere is 1.69.

2.8 CONSTRUCTING STELLAR MODELS OF A GIVEN MASS, RADIUS, AND T_{EFF}

In Chapter 1 we discussed how the average seismic parameters $\Delta\nu$ and ν_{max} are related to the mass, radius, and temperature of a star. This means that they can be used to estimate the mass and radius of a star. The effective temperature of a star can be determined from photometry or spectroscopy. Thus we have a situation where we have three of the four constraints that we have for the Sun and can model the star in an iterative fashion to get a model that satisfies all three constraints. The difference from the solar case is that we do not know the star's age. This means that we can either constrain α for a given value of Y_0 or constrain Y_0 for a given value of α. In each case, we will get a model that satisfies the global properties of the star, but they will have different ages and hence, different internal structures. This is illustrated in Figure 2.3. We cannot distinguish between the models without knowing the frequencies of individual modes.

2.9 FURTHER READING

The theory of stellar structure and evolution was covered in this chapter rather briefly. There is a very large body of literature on this topic. Readers interested in learning more details can consult the following books:

- Kippenhahn, R., and Weigert, A., 1990, *Stellar Structure and Evolution*, Springer-Verlag.
- Huang, R. Q., and Yu, K. N., 1998, *Stellar Astrophysics*, Springer-Verlag.
- Hansen, C. J., Kawaler, S. D., and Trimble, V., 2004, *Stellar Interiors: Physical Principles, Structure, and Evolution*, 2nd ed. Springer-Verlag.
- Weiss, A., Hillebrandt, W., Thomas, H.-C., and Ritter, H., 2004, *Cox and Giuli's Principles of Stellar Structure*, Cambridge Scientific Publishers.
- Maeder, A., 2009, *Physics, Formation and Evolution of Rotating Stars*, Springer-Verlag.

For practical details on how the equations of stellar structure and evolution are converted to difference equations and how the boundary conditions can be applied, see:

- Huang, R. Q., and Yu, K. N., 1998, *Stellar Astrophysics*, Springer-Verlag.

The details of the specific treatment of the mixing-length approximation described in this chapter is from:

- Kippenhahn, R., and Weigert, A., 1990, *Stellar Structure and Evolution*, Springer-Verlag.

Most stellar evolution codes are proprietary. Descriptions of the better-known codes, including the Yale Stellar Evolution Code (YREC) which is mentioned in this chapter, can be found in:

- 2008, *Astrophys. Sp. Sci.*, **316** (entire issue).

There is however, an open-source code called "Modules for Experiments in Stellar Astrophysics" or MESA, that is become increasingly popular. The code can be downloaded from http://mesa.sourceforge.net/. The description of the code can be found in:

- Paxton, B. et al., 2011, *Modules for Experiments in Stellar Astrophysics (MESA)*, *Astrophys. J. Suppl. Ser.*, **192**, 3.

A comprehensive account of how stellar models change with the input parameters can be found in:

- Salaris, M., and Cassisi, S., 2005, *Evolution of Stars and Stellar Populations*, John Wiley & Sons.

We have talked about microphysics inputs to stellar models. Descriptions of some of the most frequently used equations of state can be found in the following papers:

- The OPAL equation of state:
  - Rogers, F. J., Swenson, F. J., and Iglesias, C. A., 1996, *OPAL Equation-of-State Tables for Astrophysical Applications, Astrophys. J.*, **456**, 902.

- Rogers, F. J., and Nayfonov, A., 2002, *Updated and Expanded OPAL Equation-of-State Tables: Implications for Helioseismology, Astrophys. J.*, **576**, 1064.

- The MHD equation of state:

  - Däppen, W., Mihalas, D., Hummer, D. G., and Mihalas, B. W., 1988, *The Equation of State for Stellar Envelopes. III—Thermodynamic Quantities, Astrophys. J.*, **332**, 261.
  - Hummer, D. G., and Mihalas, D., 1988, *The Equation of State for Stellar Envelopes. I—An Occupation Probability Formalism for the Truncation of Internal Partition Functions, Astrophys. J.*, **331**, 794.
  - Mihalas, D., Däppen, W., and Hummer, D. G., 1988, *The Equation of State for Stellar Envelopes. II—Algorithm and Selected Results, Astrophys. J.*, **331**, 815.

- The SIREFF equation of state:

  - Guzik, J. A., and Swenson, F. J., 1997, *Seismological Comparisons of Solar Models with Element Diffusion Using the MHD, OPAL, and SIREFF Equations of State, Astrophys. J.*, **491**, 967.
  - Guzik, J. A., Willson, L. A., and Brunish, W. M., 1987, *A Comparison between Mass-Losing and Standard Solar Models, Astrophys. J.*, **319**, 957.

- FreeEOS:

  - http://freeeos.sourceforge.net/.
  - Cassisi, S., Salaris, M., and Irwin, A. W., 2003, *The Initial Helium Content of Galactic Globular Cluster Stars from the R-Parameter: Comparison with the Cosmic Microwave Background Constraint, Astron. Astrophys.*, **588**, 862.

Widely used sources of radiative opacities are:

- The OPAL opacities: Iglesias, C. A., and Rogers, F. J., 1996, *Updated Opal Opacities, Astrophys. J.*, **464**, 943.
- OP opacities: Badnell, N. R. et al., 2005, *Updated Opacities from the Opacity Project, Mon. Not. R. Astron. Soc.*, **360**, 458.
- OPAS opacities: Blancard, C., Cossé, P., and Faussurier, G., 2012, *Solar Mixture Opacity Calculations Using Detailed Configuration and Level Accounting Treatments, Astrophys. J.*, **745**, 10.

Sources of low-temperature opacities include:

- Kurucz, R. L., 1991, *New Opacity Calculations*, in *NATO ASIC Proc. 341: Stellar Atmospheres—Beyond Classical Models*, eds. Crivellari, L., Hubeny, I., and Hummer, D. G.
- Ferguson, J. W. et al., 2005, *Low-Temperature Opacities, Astrophys. J.*, **623**, 585.

Commonly used sources of nuclear reaction rates are:

- Adelberger, E. G. et al., 2011, *Solar Fusion Cross Sections. II. The pp Chain and CNO Cycles, Rev. Mod. Phys.*, **83**, 195.
- Angulo, C. et al., 1999, *A Compilation of Charged-Particle Induced Thermonuclear Reaction Rates, Nuclear Phys. A.*, **656**, 3.

The most used sources of diffusion coefficients are:

- Proffitt, C. R., and Michaud, G., 1991, *Gravitational Settling in Solar Models*, Astrophys. J., **380**, 238.
- Thoul, A. A., Bahcall, J. N., and Loeb, A., 1994, *Element Diffusion in the Solar Interior*, Astrophys. J., **421**, 828.

2.10 EXERCISES

1. In 1966, Robert Stein discussed some ways in which one could study stars using analytic models.[1] The following exercises are based on the so-called linear model. A *linear stellar model* is one for which we assume that density is a linear function of radius. In the following you may assume the star with mass M and radius R has a density that is a linear function of radius r, with $\rho = \rho_c$ at $r = 0$ and $\rho = 0$ at r = R.

(a) Show that mass varies as

$$m(r) = \frac{4\pi}{3} \rho_c r^3 \left(1 - \frac{3}{4} \frac{r}{R} \right) = M(4x^3 - 3x^4), \qquad (2.40)$$

where $x = r/R$.

(b) Also show that

$$\rho_c = \frac{3M}{\pi R^3}. \qquad (2.41)$$

(c) Using the expressions for ρ_c and $m(r)$ and the equation of hydrostatic equilibrium, show that

$$P = \frac{5}{4\pi} \frac{GM^2}{R^4} \left(1 - \frac{24}{5} x^2 + \frac{28}{5} x^3 - \frac{9}{5} x^4 \right), \qquad (2.42)$$

when it is assumed that $P = 0$ at $r = R$.

(d) Assuming an ideal gas, find an expression for temperature as a function of radius. Ideal gases follow the relation

$$\frac{P}{\rho} = \frac{\mathcal{R}}{\mu} T, \qquad (2.43)$$

where $\mathcal{R}$ is the gas constant, and μ is the mean molecular weight given by

$$\frac{1}{\mu} = 2X + \frac{3Y}{4} + \frac{Z}{2}, \qquad (2.44)$$

where X, Y, and Z are respectively the mass fractions of hydrogen, helium, and all other elements.

---

[1]Stein, R. F., 1966, *Stellar Evolution: A Survey with Analytic Models*, in *Stellar Evolution*, eds. Stein, R. F. and Cameron, A. G. W.

(e) Assuming that the energy generation rate varies as

$$\epsilon = \epsilon_0 \rho^k \left(\frac{T}{T_0} \right)^n,$$ (2.45)

find an expression for the total energy of the star, that is,

$$L = \int_0^R 4\pi r'^2 \rho(r') \epsilon \, dr'.$$ (2.46)

If you have trouble with the general integral, try the specific case of $k = 1$, $n = 4$.

2. This exercise continue on the theme of a linear model.

(a) Assume a star with the mass and radius of the Sun that has a linear density distribution as in Exercise 1. What is the central density? What is the central pressure?

(b) Assume that the star is chemically homogeneous, with $Z = 0.02$ and $X = 0.74$. What is the central temperature?

(c) In low-mass stars like the Sun, hydrogen fusion occurs predominantly through the so-called p-p chain. For the p-p chain, the nuclear energy generation can be approximated as

$$\epsilon = 50 \left(\frac{\rho}{100} \right) X^2 \left(\frac{T}{15 \times 10^6} \right)^4 \text{ erg s}^{-1} \text{ g}^{-1}.$$ (2.47)

Determine luminosity as a function of radius and plot the values.

3. Files `SolarModel1.txt` and `SolarModel2.txt` (available via the Princeton University Press website mentioned in the Preface) list stellar surface and core abundances as well as the radius of the base of the convection zone for models that satisfy the global properties of the Sun at the solar age. They differ in one crucial physical process. Use the data in the files to determine what the missing/added physical process is. Once you know what the process is, explain why it leads to the different rate of change of the convection-zone base.

3 Theoretical Underpinnings: Stellar Oscillations

To interpret seismic data, we need to know how seismic variables are related to the structure of a star. This not only requires a knowledge of the basics of the theory of stellar structure and evolution but also requires knowledge of the basics of the theory of stellar pulsations. In Chapter 2 we described the different equations and inputs needed to model a star. In this chapter we show how the oscillations of a star are related to its structure. We also describe some of the important properties of stellar oscillations that allow us to use seismic data to determine stellar properties.

Stellar oscillations are the response of a star to a mechanical disturbance, and thus the equations of stellar oscillations are derived by perturbing the equations of stellar structure. While a star may be spherically symmetric under equilibrium conditions, it need not retain spherical symmetry when it is perturbed, and as a result we can no longer avoid taking into account the three-dimensional nature of a star.

To a good approximation, solar-like oscillations can be described as being linear and adiabatic. For example, at the solar surface, modes of oscillation have velocity amplitudes of the order of tens of centimeters per second. This should be compared with the speed of sound in those layers, which is about 10 km s^{-1}. This puts the amplitudes comfortably in the linear regime. The oscillation timescale of minutes is much shorter than the heat-transport timescale in most layers, and hence adiabaticity is a good approximation. Thus we shall be deriving what is often called the LAWE: the *linear adiabatic wave equation*. The condition of adiabaticity does break down near the surface, where thermal timescales are short, and this changes the frequencies and adds to what we describe later as the *surface term* (Section 8.3).

3.1 THE EQUATIONS OF STELLAR OSCILLATIONS

To derive the equations of stellar oscillations, we begin with the basic equations of fluid dynamics, namely, the equation of continuity (i.e., the conservation of mass) and the equation describing the conservation of momentum. Since stars are self-gravitating objects, we use the Poisson equation to describe the gravitational field. We also need an equation to describe heat transport. The basic equations are therefore

$$\frac{\partial \rho}{\partial t} + \nabla \cdot (\rho \vec{v}) = 0, \tag{3.1}$$

$$\rho \left(\frac{\partial v}{\partial t} + \vec{v} \cdot \nabla \vec{v} \right) = -\nabla P + \rho \nabla \Phi, \quad \text{and} \tag{3.2}$$

$$\nabla^2 \Phi = 4 \pi G \rho, \tag{3.3}$$

where $\vec{v}$ is the velocity of the fluid element, Φ is the gravitational potential, and G is the gravitational constant. The heat equation is written in the form

$$\frac{dq}{dt} = \frac{1}{\rho(\Gamma_3 - 1)} \left(\frac{dP}{dt} - \frac{\Gamma_1 P}{\rho} \frac{d\rho}{dt} \right), \tag{3.4}$$

where q is heat, and Γ_1 and Γ_3 are adiabatic indices that are determined by the equation of state:

$$\Gamma_1 = \left(\frac{\partial \ln P}{\partial \ln \rho} \right)_{\text{ad}}, \quad \Gamma_2 = \left(\frac{\partial \ln T}{\partial \ln P} \right)_{\text{ad}}, \quad \text{and} \quad \Gamma_3 - 1 = \left(\frac{\partial \ln T}{\partial \ln \rho} \right)_{\text{ad}}. \tag{3.5}$$

For a monatomic ideal gas, $\Gamma_1 = \Gamma_2 = \Gamma_3 \equiv \gamma = 5/3$. In the adiabatic limit, $dq = 0$ and Eq. 3.4 reduces to

$$\frac{dP}{dt} = c_s^2 \frac{d\rho}{dt}, \tag{3.6}$$

where $c_s = \sqrt{\Gamma_1 P / \rho}$ is the speed of sound.

The equations that describe linear oscillations are a result of linear perturbations to the above fluid equations. For instance, we can express the linear perturbation in pressure as

$$P(\vec{r}, t) = P_0(\vec{r}) + P_1(\vec{r}, t), \tag{3.7}$$

where the subscript 0 denotes the equilibrium solution, the subscript 1 denotes the perturbation, and $|P_1/P_0| \ll 1$. The equilibrium pressure P_0 is spherically symmetric, and by definition does not depend on time. The perturbation, however, is not necessarily spherically symmetric, and it is also time dependent. Perturbations to other quantities can be written in the same way, thus

$$\rho(\vec{r}, t) = \rho_0(\vec{r}) + \rho_1(\vec{r}, t). \tag{3.8}$$

Note that Eqs. 3.7 and 3.8 are the *Eulerian perturbations* to pressure and density; that is, they are perturbations at a fixed point in space denoted by coordinates (r, θ, ϕ). In some cases, it is easier to use the *Lagrangian perturbation*: a perturbation seen by an observer moving with the fluid. The Lagrangian perturbation for density is

$$\delta\rho(r, t) = \rho(\vec{r} + \vec{\xi}(\vec{r}, t)) - \rho(\vec{r}) = \rho_1(\vec{r}, t) + \vec{\xi}(\vec{r}, t) \cdot \nabla \rho_0, \tag{3.9}$$

where $\vec{\xi}$ is the displacement from the equilibrium position. The perturbations to other quantities can be written in the same way.

The equilibrium state of a star is generally assumed to be static, and thus the velocity $\vec{v}$ in the fluid equations appears only after a perturbation, and then it is the rate of change of displacement of the fluid:

$$\vec{v} = d\vec{\xi}/dt. \tag{3.10}$$

Substituting the perturbed quantities into Eqs. 3.1–3.3 and keeping only linear terms in the perturbation, we get

$$\rho_1 + \nabla \cdot (\rho_0 \vec{\xi}) = 0, \tag{3.11}$$

$$\rho \frac{\partial^2 \vec{\xi}}{\partial t^2} = -\nabla P_1 + \rho_0 \nabla \Phi_1 + \rho_1 \nabla \Phi_0, \quad \text{and} \tag{3.12}$$

$$\nabla^2 \Phi_1 = 4\pi G \rho_1. \tag{3.13}$$

To perturb the heat equation in the adiabatic limit (Eq. 3.6), we first need to look at how to perturb a time derivative. The total time derivative of any quantity Q can be written as

$$\frac{dQ}{dt} = \frac{\partial Q}{\partial t} + \vec{v}(\vec{r}, t) \cdot \nabla Q. \tag{3.14}$$

Applying this to the pressure perturbation, we get

$$\frac{d(P_0 + P_1)}{dt} = \frac{\partial(P_0 + P_1)}{\partial t} + \vec{v} \cdot \nabla P_0 \tag{3.15}$$

when we only keep terms that are linear in perturbations. But P_0 is independent of time, and thus

$$\frac{d(P_0 + P_1)}{dt} = \frac{\partial P_1}{\partial t} + \vec{v} \cdot \nabla P_0 = \frac{\partial P_1 + \vec{\xi} \cdot \nabla P_0}{\partial t} = \frac{\partial \delta P}{\partial t}, \tag{3.16}$$

where δP is the Lagrangian perturbation to P. This means that the perturbed form of the adiabatic heat equation (Eq. 3.6) can be written as

$$\frac{\partial \delta P}{\partial t} = \frac{\Gamma_{1,0} P_0}{\rho_0} \frac{\partial \delta \rho}{\partial t}, \tag{3.17}$$

or

$$P_1 + \vec{\xi} \cdot \nabla P_0 = \frac{\Gamma_{1,0} P_0}{\rho_0} (\rho_1 + \vec{\xi} \cdot \nabla \rho_0). \tag{3.18}$$

The term $\Gamma_{1,0} P_0/\rho_0$ in Eqs. 3.17 and 3.18 is nothing but the squared, unperturbed sound speed $c_{s,0}^2$.

In the subsequent discussion, we drop the subscript 0 for the equilibrium quantities, and just keep the subscript 1 for the perturbations. Since stars are spherical, we cast the equations in spherical polar coordinates with the origin defined at the center of the star; thus r becomes the radial distance, θ the co-latitude, and ϕ the longitude.

We can decompose the different quantities into their radial and tangential components. For instance, the displacement $\vec{\xi}$ can be decomposed as

$$\vec{\xi} = \vec{\xi}_r + \vec{\xi}_t = \xi_r \hat{a}_r + \xi_t \hat{a}_t, \tag{3.19}$$

where $\hat{a}_r$ and $\hat{a}_t$ are the unit vectors in the radial and tangential directions, respectively; ξ_r is the radial component of the displacement vector; and ξ_t the transverse component. A big advantage of using spherical polar coordinates is the fact that tangential gradients of the equilibrium quantities are by definition zero. If they were not, they would give rise to horizontal motions that would wipe out the gradients. This simplifies the equations; for example, the heat equation (Eq. 3.18) becomes

$$\rho_1 = \frac{\rho}{\Gamma_1 P} P_1 + \rho \xi_r \left(\frac{1}{\Gamma_1 P} \frac{dP}{dr} - \frac{1}{\rho} \frac{d\rho}{dr} \right). \tag{3.20}$$

The tangential component of the equation of motion (Eq. 3.2) is

$$\rho \frac{\partial^2 \vec{\xi}_t}{\partial t^2} = -\nabla_t P_1 + \rho \nabla_t \Phi_1, \tag{3.21}$$

which, on taking the tangential divergence of both sides, becomes

$$\rho \frac{\partial^2}{\partial t^2} (\nabla_t \cdot \vec{\xi}_t) = -\nabla_t^2 P_1 + \rho \nabla_t^2 \Phi_1. \tag{3.22}$$

The term $\nabla_t \cdot \vec{\xi}_t$ can be eliminated from Eq. 3.22 by decomposing the continuity equation (Eq. 3.11) into its tangential and radial parts and using the tangential equation to determine $\nabla_t \cdot \vec{\xi}_t$. The elimination of $\nabla_t \cdot \vec{\xi}_t$ from Eq. 3.22 results in

$$-\frac{\partial^2}{\partial t^2} \left[\rho_1 + \frac{1}{r^2} \frac{\partial}{\partial r} (\rho r^2 \xi_r) \right] = -\nabla_t^2 P_1 + \rho \nabla_t^2 \Phi_1. \tag{3.23}$$

The radial component of the equation of motion gives

$$\rho \frac{\partial^2 \xi_r}{\partial t^2} = -\frac{\partial P_1}{\partial r} - \rho_1 g + \rho \frac{\partial \Phi_1}{\partial r}, \tag{3.24}$$

where we have used the fact that gravity acts in the negative r direction. Finally, the Poisson equation becomes

$$\frac{1}{r^2} \frac{\partial}{\partial r} \left(r^2 \frac{\partial \Phi_1}{\partial r} \right) + \nabla_t^2 \Phi_1 = -4\pi G \rho_1. \tag{3.25}$$

Equations 3.23–3.25 have two features in common. First, there are no mixed radial and tangential derivatives, and tangential gradients appear only as the tangential component of the Laplacian. Thus the tangential part of the perturbed quantities can be written in terms of eigenfunctions of the tangential Laplacian operator. Since we are dealing with spherical objects, using spherical harmonics (which are eigenfunctions of the tangential part of the Laplacian operator) makes the most sense. The second common feature is that time t does not appear explicitly in the coefficients of any of the derivatives. This implies that the time-dependent part of the solution can be separated from the spatial part. The time-dependent part of the solution can be written as $\exp(-i\omega t)$, where ω can be real (i.e., a solution that oscillates in time) or imaginary (a solution that grows or decays). Hence the perturbed quantities can be expressed as

$$\xi_r(r, \theta, \phi, t) \equiv \xi_r(r) Y_l^m(\theta, \phi) \exp(-i\omega t), \tag{3.26}$$

$$P_1(r, \theta, \phi, t) \equiv P_1(r) Y_l^m(\theta, \phi) \exp(-i\omega t), \quad \text{and} \tag{3.27}$$

$$\rho_1(r, \theta, \phi, t) \equiv \rho_1(r) Y_l^m(\theta, \phi) \exp(-i\omega t). \tag{3.28}$$

We now substitute these variables into Eqs. 3.20, 3.23, 3.24, and 3.25. Additionally, we eliminate ρ_1 from the equations using Eq. 3.20. Thus Eq. 3.23 becomes

$$\frac{d\xi_r}{dr} = -\left(\frac{2}{r} + \frac{1}{\Gamma_1 P}\frac{dP}{dr}\right)\xi_r + \frac{1}{\rho c_s^2}\left(\frac{S_l^2}{\omega^2} - 1\right)P_1 - \frac{l(l+1)}{\omega^2 r^2}\Phi_1, \tag{3.29}$$

where $c_s^2 = \Gamma_1 P/\rho$ is again the squared sound speed, and S_l^2 is the *Lamb frequency* defined by

$$S_l^2 = \frac{l(l+1)c_s^2}{r^2}. \tag{3.30}$$

Eq. 3.24 and the equation of hydrostatic equilibrium give

$$\frac{dP_1}{dr} = \rho(\omega^2 - N^2)\xi_r + \frac{1}{\Gamma_1 P}\frac{dP}{dr}P_1 + \rho\frac{d\Phi_1}{dr}, \tag{3.31}$$

where N is the *Brunt-Väisälä* or *buoyancy frequency*, defined as

$$N^2 = g\left(\frac{1}{\Gamma_1 P}\frac{dP}{dr} - \frac{1}{\rho}\frac{d\rho}{dr}\right). \tag{3.32}$$

This is the frequency with which a small element of fluid will oscillate when it is disturbed from its equilibrium position. When $N^2 < 0$, the fluid is unstable to convection. Finally, Eq. 3.25 becomes

$$\frac{1}{r^2}\frac{d}{dr}\left(r^2\frac{d\Phi_1}{dr}\right) = -4\pi G\left(\frac{P_1}{c_s^2} + \frac{\rho\xi_r}{g}N^2\right) + \frac{l(l+1)}{r^2}\Phi_1. \tag{3.33}$$

Equations 3.29, 3.31, and 3.33 form a set of fourth-order differential equations and constitute an eigenvalue problem with eigenvalue ω. We can determine the radial component ξ_r of the displacement eigenfunction as well as P_1, Φ_1, and $d\Phi_1/dr$ by solving the equations. Each eigenvalue is usually referred to as a *mode of oscillation*.

The transverse component of the displacement vector can be written in terms of $P_1(r)$ and $\Phi_1(r)$, and one can show that

$$\vec{\xi}_t(r, \theta, \phi, t) = \xi_t(r) \left(\frac{\partial Y_l^m}{\partial \theta} \hat{a}_\theta + \frac{1}{\sin \theta} \frac{\partial Y_l^m}{\partial \phi} \hat{a}_\phi \right) \exp(-i\omega t), \qquad (3.34)$$

where $\hat{a}_\theta$ and $\hat{a}_\phi$ are the unit vectors in the θ and ϕ directions, respectively, and

$$\xi_t(r) = \frac{1}{r\omega^2} \left(\frac{1}{\rho} P_1(r) - \Phi_1(r) \right). \qquad (3.35)$$

In Chapter 1 we said that modes are represented by three variables, l, n, and m, but as can be seen, the equations above have no n or m dependence. The different eigenvalues for a given value of l are given the label n. Conventionally, n can be any signed integer and can be positive, zero, or negative, depending on the type of the mode. In general $|n|$ represents the number of nodes that the radial eigenfunction has in the radial direction. As mentioned previously, values of $n > 0$ are used to specify acoustic modes, $n < 0$ label gravity modes, and $n = 0$ labels f modes. It is usual to denote the eigenfunctions with n and l; for example, the total displacement would be denoted as $\vec{\xi}_{nl}$, and the radial and transverse components as $\xi_{r,nl}$ and $\vec{\xi}_{t,nl}$, respectively.

The assumption of spherical symmetry explains the lack of an m dependence: m describes the latitudinal dependence of the eigenfunction, which means that we need an equator to define latitudes. This is not possible in a spherically symmetric system, where we could designate any arbitrary great circle as the equator and still the system would be unchanged. This means that all $2l + 1$ components of the mode must have the same frequency. Thus each mode is $(2l + 1)$-fold degenerate. Rotation and magnetic fields lift this degeneracy and introduce an m-dependence on the frequencies. The effect of slow rotation can be treated as a perturbation and is described in Section 3.5.

An important property of all modes is their mode inertia:

$$I_{nl} = \int_V \rho \vec{\xi}_{nl} \cdot \vec{\xi}_{nl} \, d^3\vec{r} = \int_0^R \rho [\xi_{r,nl}^2 + l(l + 1)\xi_{t,nl}^2] r^2 dr. \qquad (3.36)$$

Low-degree modes have higher mode inertia than do high-degree modes, and low-frequency modes have higher inertia than do high-frequency modes. As we shall see in Section 3.3, for a given perturbation in structure, frequencies of high-inertia modes change less than frequencies of low-inertia modes.

Since the normalization of eigenfunctions can be arbitrary, often I_{nl} is normalized explicitly as

$$I_{nl} = \frac{\int_0^R \rho [|\xi_{r,nl}|^2 + l(l + 1)|\xi_{t,nl}|^2] r^2 dr}{M \left[|\xi_{r,nl}(R)|^2 + l(l + 1)|\xi_{t,nl}(R)|^2 \right]}, \qquad (3.37)$$

where M is the total mass, and R is the total radius.

3.1.1 Boundary Conditions

The equations of stellar oscillations cannot be solved without boundary conditions. A complete solution of the equations requires four conditions. An examination of the equations around $r = 0$ shows that they have a singular point there; however, it is a singularity that allows regular (i.e., physically relevant) solutions.

As is usual, the central boundary conditions are obtained by expanding the solution around $r = 0$. This reveals that as $r \to 0$, $\xi_r \propto r$ for $l = 0$ modes, and $\xi_r \propto r^{l-1}$ for others. The quantities P_1 and Φ_1 vary as r^l.

The surface boundary conditions are complicated by the fact that the "surface" of a star is not well defined in terms of density, pressure, and so forth, but it is determined by the way the stellar atmosphere is treated. In fact, one can show that the calculated frequencies can change substantially, depending on where one assumes the outer boundary of the star lies. The surface boundary conditions are determined assuming that Φ_1 and $d\Phi_1/dr$ are continuous at the surface. Under the simple assumption that ρ_1 is zero at the surface, the Poisson equation can be solved to show that for Φ_1 to be zero at infinity, we must have

$$\Phi_1 = Ar^{-1-l}, \tag{3.38}$$

where A is a constant. It follows that

$$\frac{d\Phi_1}{dr} + \frac{l+1}{r}\Phi_1 = 0 \tag{3.39}$$

at the surface. As for pressure, one can assume that the pressure on the perturbed surface is zero; that is, the Lagrangian perturbation of pressure vanishes at the surface:

$$\delta P = P_1 + \xi_r \frac{dP}{dr} = 0 \tag{3.40}$$

at the surface. This condition implies that $\nabla \cdot \vec{\xi} \sim 0$ at the outer boundary. Combining this with Eq. 3.29, we get $\xi_r = P_1/g\rho$ at the surface. If Φ_1 is neglected (i.e., the so-called *Cowling approximation* is used), this reduces to

$$\xi_r = \xi_t R^3 \omega^2/(GM). \tag{3.41}$$

3.1.2 The Use of Dimensionless Variables

It is worth noting that the equations of stellar oscillations are generally solved after expressing all the physical quantities in terms of dimensionless variables. The dimensionless frequency σ is given by

$$\sigma^2 = \frac{R^3}{GM}\omega^2, \tag{3.42}$$

where R is the radius and M the mass of the star. Similarly, the dimensionless radius is $\hat{r} = r/R$; the dimensionless pressure is given by $\widehat{P} = (R^4/GM^2)P$; the dimensionless sound speed by $\hat{c}_s = \sqrt{R/(GM)}$; and the dimensionless density by $\hat{\rho} = (R^3/M)\rho$. These relations lead us to the dimensionless expressions for the perturbed quantities. It can be shown very easily that $\hat{\xi}_r$ (the dimensionless displacement eigenfunction), $\widehat{P}_1$ (the dimensionless pressure perturbation), and $\widehat{\Phi}_1$ (the dimensionless perturbation to the gravitational potential) can be written as

$$\hat{\xi}_r = \frac{\xi_r}{R}, \quad \widehat{P}_1 = \frac{R^4}{GM^2}P_1, \quad \text{and} \quad \widehat{\Phi}_1 = \frac{GM}{R}\Phi_1. \tag{3.43}$$

3.2 PROPERTIES OF STELLAR OSCILLATIONS

Equations 3.29, 3.31, and 3.33 are complicated enough that one cannot easily determine the properties of stellar oscillations from the equations without first simplifying them. We apply a few assumptions that help us distill out some of the more important properties.

The first assumption is that we can apply the *Cowling approximation*. This implies that the perturbation to the gravitational potential, Φ_1, can be ignored. The assumption is reasonably valid for modes with large values of n and also for modes of large l. This follows from the spherical harmonic expansion of the perturbation to the gravitational potential:

$$\Phi_1(r) = -\frac{4\pi G}{2l+1}\left[\frac{1}{r^{l+1}}\int_0^r \rho_1(r')r'^{l+2}dr' + r^l\int_r^R \frac{\rho_1(r')}{r'^{l-1}}dr'\right], \tag{3.44}$$

where r' is the dummy variable representing radius. At large l the $(r'/r)^{l+2}$ term in the first integral is small when $r' < r$; and when $r' > r$, the $(r/r')^{l-1}$ term in the second integral is small. When $|n|$ is large, ρ_1 in the two integrals is a rapidly varying function of r that alternates between positive and negative values, and therefore the value of the two integrals is small. The large n limit is what is most relevant to asteroseismic analyses.

The application of Cowling approximation makes Eq. 3.33 irrelevant, and it reduces Eqs. 3.29 and 3.31 to

$$\frac{d\xi_r}{dr} = -\left(\frac{2}{r} - \frac{1}{\Gamma_1}\frac{1}{H_P}\right)\xi_r + \frac{1}{\rho c_s^2}\left(\frac{S_l^2}{\omega^2} - 1\right)P_1, \tag{3.45}$$

and

$$\frac{dP_1}{dr} = \rho(\omega^2 - N^2)\xi_r - \frac{1}{\Gamma_1}\frac{1}{H_P}P_1, \tag{3.46}$$

where H_P is the pressure scale height.

Another assumption that can be used for modes of high $|n|$ is that the eigenfunctions vary much more rapidly than the equilibrium quantities. The implication of this assumption is that terms containing H_P^{-1} in Eqs. 3.45 and 3.46 can be neglected compared with the quantities on the left-hand sides of the two equations; we also assume that we are looking away from the center. This allows us to simplify Eqs. 3.45 and 3.46 to

$$\frac{d\xi_r}{dr} = \frac{1}{\rho c_s^2}\left(\frac{S_l^2}{\omega^2} - 1\right)P_1, \tag{3.47}$$

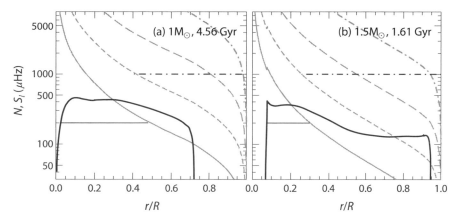

Figure 3.1. The propagation diagrams of two main sequence models. In both panels the heavy black line is the Brunt-Väisälä frequency N. The gray lines are the Lamb frequency, S, for $l = 1$ (solid), $l = 5$ (short-dashed line), $l = 20$ (long-dashed line), and $l = 100$ (dot-dashed line). The dot-dashed horizontal line marks the cavity for a p mode with frequency $1{,}000\ \mu$Hz, while the solid horizontal line marks the cavity for a $200\ \mu$Hz g mode.

and

$$\frac{\mathrm{d}P_1}{\mathrm{d}r} = \rho(\omega^2 - N^2)\xi_r. \tag{3.48}$$

These two equations can be combined to form one second-order differential equation:

$$\frac{\mathrm{d}^2 \xi_r}{\mathrm{d}r^2} = \frac{\omega^2}{c_s^2}\left(1 - \frac{N^2}{\omega^2}\right)\left(\frac{S_l^2}{\omega^2} - 1\right)\xi_r. \tag{3.49}$$

Eq. 3.49 is the simplest possible approximation to the equations of nonradial oscillation, but it suffices to illustrate some of the key properties. We can rewrite the equation as

$$\frac{\mathrm{d}^2 \xi_r}{\mathrm{d}r^2} = -K(r)\xi(r), \tag{3.50}$$

where

$$K(r) = \frac{\omega^2}{c_s^2}\left(\frac{N^2}{\omega^2} - 1\right)\left(\frac{S_l^2}{\omega^2} - 1\right), \tag{3.51}$$

which leads us to conclude that the equation does not always have an oscillatory solution. The solution is oscillatory when

1. $\omega^2 < S_l^2$, and $\omega^2 < N^2$, or
2. $\omega^2 > S_l^2$, and $\omega^2 > N^2$.

Figure 3.1 shows N and S_l plotted as a function of depth for a $1M_\odot$ model close to the age of the Sun (left panel) and a model of a main sequence $1.5M_\odot$ star (right panel). Such diagrams are often referred to as *propagation diagrams*, since they show where inside a star different modes can propagate. The figure shows that modes for which the first condition is true (i.e., $\omega^2 < S_l^2$ and $\omega^2 < N^2$) are trapped mainly in

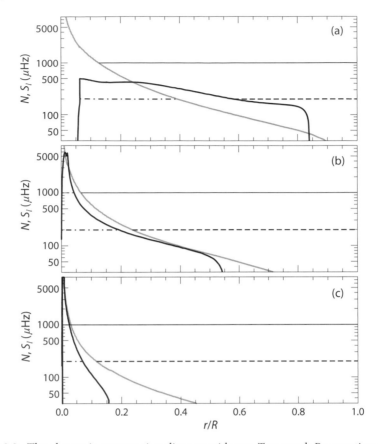

Figure 3.2. The change in propagation diagram with age. Top panel: Propagation diagram for a model of a $1.2M_\odot$ main sequence star. Middle panel: Propagation diagram for the same star but in the subgiant phase. Bottom panel: Model on the lower red-giant branch. The heavy black line is the buoyancy frequency, and the gray line is the Lamb frequency for $l = 1$. Note that the model has a convective core on the main sequence, and hence N drops sharply around $0.1R$ to become imaginary. The solid horizontal line marks the propagation cavity of an $l = 1$ p mode at 1 mHz, the dotted line marks the cavity of a 1 mHz g mode, the dashed line is that for a 0.2 mHz p mode, and the dot-dashed line is a 0.2 mHz g mode. Note that in the top panel, a 1 mHz g mode cannot exist. In the middle and bottom panels, the gap between the g- and p-mode cavities is so small that both 1 mHz and 0.2 mHz modes can have a g-mode-like character in the core and a p-mode-like one in the envelope, and are therefore mixed modes.

the inner regions. These are the *g modes*, whose restoring force is gravity. Modes that satisfy the second condition (i.e., $\omega^2 > S_l^2$ and $\omega^2 > N^2$) are oscillatory in the outer regions, though low-degree modes can penetrate right to the core. These are the *p modes*, whose restoring force is provided by the gradient of the pressure. The figure also shows that low-degree modes penetrate deeper than do high-degree modes. Thus modes of different degrees sample different layers of a star. For a given degree, modes of higher frequency penetrate deeper into a star.

The propagation diagram of a given star evolves with age (see Figure 3.2) as a result of the change in abundances due to nuclear reactions, and also because of

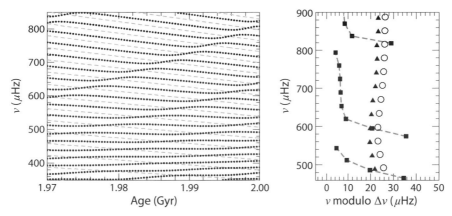

Figure 3.3. Left panel: The evolution with age of the $l = 1$ frequencies (points) of a $1.6\text{M}_\odot$ model. The gray dashed lines show the evolution of $l = 0$ modes, which change mainly because of the change in radius. Note that there are times when the $l = 1$ ridges come close together and then diverge again. These are the avoided crossings and are a result of mixed modes. As can be seen, the frequencies at which avoided crossings occur change rapidly with time. Right panel: The échelle diagram at 1.97 Gyr. The $l = 0$ and $l = 2$ modes are marked with the open circles and filled triangles, respectively. The filled squares are the $l = 1$ modes. Note that the $l = 1$ modes clearly show avoided crossings. Modes on either side of the avoided crossings have been joined by the dashed lines to guide the eye.

the change in the density profile as the core contracts in the subgiant and red-giant phases. This causes the buoyancy frequency N to become larger and larger as a star evolves until it is similar in magnitude to the Lamb frequency. In such cases one can get a *mixed* mode (i.e., a mode with g-mode-like characteristics near the center and p-mode-like characteristics near the surface); the region between the two cavities where the mode can decay is small enough that the g-mode cavity and the p-mode cavity can couple. The label n of mixed modes can be confusing. Since they have both p- and g-mode-like characteristics, they can be described with a value of n for the p-mode part, n_p, and one for the g-mode part, n_g. Often n for a mixed mode is defined as $n_p + n_g$. Recall that n_g is a negative integer, and thus for a mixed mode $|n| = |n_p| - |n_g|$. The frequencies of mixed modes evolve rapidly with time (Figure 3.3) and thus these are good diagnostics of age. Mixed modes give rise to avoided crossings in échelle diagrams of evolved stars.

The differences in the propagation diagrams between stars of different evolutionary states result in large differences in the oscillation spectra and hence in échelle diagrams. As a star evolves, more and more mixed modes appear in the spectrum. Mixed modes have larger mode inertia than p modes and hence are more difficult to excite than p modes; g modes are the most difficult modes to excite.

Figure 3.4 shows how mode inertia changes with mode frequency for models in different evolutionary phases. The differences are stark. For the main sequence star, we find that $l = 1$ and $l = 2$ modes have the same frequency-inertia relation as the $l = 0$ modes. As a result, all modes are seen easily. In the subgiant, a few modes have high inertia. These are the mixed modes, but there are only a small number of such modes. However, as the star evolves, the possible number of mixed modes increases and as we can see in the case of the red giants, there is a veritable jungle of possible

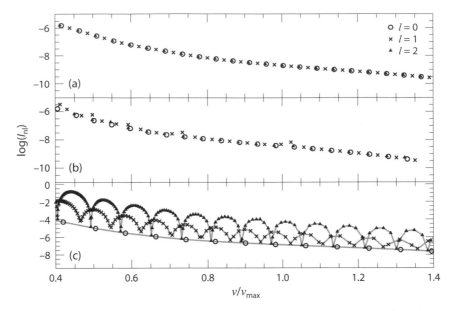

Figure 3.4. The relationship between mode frequencies and mode inertia of a (a) main sequence, (b) subgiant, and (c) red-giant model. The frequencies are normalized by ν_{max} to allow us to plot the modes on the same scale.

$l = 1$ and $l = 2$ modes; some of the nonradial modes have very large mode inertias. However, not all possible nonradial modes are excited; only the lowest inertia modes are. As a result, for each radial mode observed in a red giant, we observe only one $l = 2$ mode and not a multiplet. This is because the mode inertias of the $l = 2$ modes adjacent to the one with the lowest inertia are very high. As far as the $l = 1$ modes are concerned, we do indeed observe multiple $l = 1$ modes for each $l = 0$ mode.

The structure of the spectra is reflected in the échelle diagrams, where we see one ridge each for the $l = 0$ and $l = 2$ modes, and multiple modes of $l = 1$ for a given order. This is illustrated in Figure 3.5. Compared to power-spectra of red giants on the ascending part of the red-giant branch (i.e., stars with an inert helium core), stars in the red clump and on the horizontal branch (i.e., stars with helium fusion in the core) have more complicated spectra because of the larger number of $l = 1$ modes that are excited.

Red giants, particularly those on the ascending branch, are remarkably self-similar in their structure. They all have very dense cores with a thin hydrogen-burning shell around them, and they have very deep convective envelopes. The stars are close to being homologous and as a consequence, their frequencies form a pattern, sometimes referred to as the *universal pattern*. This is shown in Figure 3.6. One can see that when the frequencies are normalized by $\Delta\nu$, the $l = 0$ modes line up on the $(\nu/\Delta\nu)$–$\Delta\nu$ plane. The $l = 2$ modes do the same. The $l = 1$ modes show a small scatter since the relative spacing between the $l = 1$ frequencies of the same order depends on the details of the Brunt-Väisälä profile, which in turn depends on such details as whether overshoot was present, if and when the star had a convective core, whether there is mixing, and the like. The alignment of the radial and quadrupole modes (i.e., $l = 2$ modes) shows the universality of the oscillation pattern.

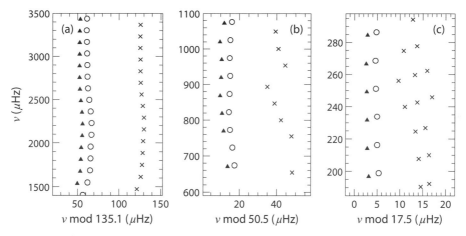

Figure 3.5. Échelle diagrams for (left panel) the Sun, (middle panel) the subgiant KIC 11026764, and (right panel) the red giant KIC 6448890 (Kepler 56). Circles denote $l = 0$ modes, triangles are $l = 2$ modes, and crosses are $l = 1$ modes.

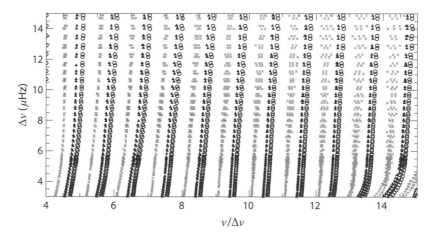

Figure 3.6. The universal pattern of red-giant frequencies. The circles, triangles, and crosses represent $l = 0$, $l = 2$, and $l = 1$ modes, respectively. Note how the $l = 0$ and $l = 2$ modes line up. For $l = 1$ modes we have only plotted the three lowest-inertia modes for each order. Stars evolve from the top of the diagram to the bottom. Frequencies of red giants obtained from *Kepler* and CoROT observations show the same pattern.

3.2.1 Asymptotic Properties of p and g Modes

We can use Eqs. 3.50 and 3.51 to derive some of the basic properties of p and g modes.

P Modes

These are modes with $\omega^2 > S_l^2$ and $\omega^2 > N^2$. The p modes are trapped between the surface and a lower or inner turning point r_t, given by $\omega^2 = S_l^2$:

$$\frac{c_s^2(r_t)}{r_t^2} = \frac{\omega^2}{l(l+1)}. \tag{3.52}$$

For p modes of high frequency (i.e., $\omega^2 \gg N^2$), we can approximate $K(r)$ in Eq. 3.51 as

$$K(r) \simeq \frac{\omega^2 - S_l^2(r)}{c_s^2(r)}. \tag{3.53}$$

This equation shows that the behavior of these p modes is basically governed by the sound-speed profile. This is not surprising, since p modes are pressure waves (i.e., sound waves).

We know that the dispersion relation of a sound wave is given by

$$\omega^2 = c_s^2 |\vec{k}|^2, \tag{3.54}$$

where $\vec{k}$ is the wavenumber that can be split into a radial part k_r and a horizontal part k_h, with $k^2 = k_r^2 + k_h^2$. We also know that the equation for a sound wave is simply

$$\frac{d^2 \xi_r}{dr^2} = -\frac{\omega^2}{c_s^2} \xi(r). \tag{3.55}$$

Thus we can identify $|\vec{k}|^2 = K(r)$. At the lower turning point, r_t, the wave has no radial component and hence the radial part of the wavenumber, k_r, vanishes, which leads to

$$k_r^2 = \frac{\omega^2 - S_l^2(r)}{c_s^2(r)}, \tag{3.56}$$

which, in turn, implies that

$$k_h^2 = \frac{l(l+1)}{r^2}. \tag{3.57}$$

Since the modes we observe are eigenmodes, there are other conditions on $K(r)$. A more complete analysis shows that the condition on $K(r)$ is

$$\int_{r_t}^{r_u} K(r)^{1/2} dr = \left(n - \frac{1}{2}\right) \pi, \tag{3.58}$$

where r_u is the upper turning point. We have an approximate expression for $K(r)$ (Eq. 3.53), but the analysis that led to it had no notion of an upper turning point. For the moment, we just assume that the upper turning point is at $r = R$. Thus the "reflection" at the upper turning point does not necessarily produce a phase shift of $\pi/2$ but some unknown additional shift, which we call $\alpha_p \pi$. In other words,

$$\int_{r_t}^{R} K(r)^{1/2} dr = \int_{r_t}^{R} (\omega^2 - S_l^2)^{1/2} dr = (n + \alpha_p)\pi. \tag{3.59}$$

Since ω does not depend on r, Eq. 3.59 can be rewritten as

$$\int_{r_t}^R \left(1 - \frac{L^2}{\omega^2} \frac{c_s^2}{r^2}\right)^{1/2} \frac{dr}{c_s} = \frac{(n + \alpha_p)\pi}{\omega}, \tag{3.60}$$

where $L = \sqrt{l(l+1)}$. The left-hand side of the equation is a function of $w \equiv \omega/L$, and the the equation is usually written as

$$F(w) = \frac{(n + \alpha_p)\pi}{\omega}, \tag{3.61}$$

where

$$F(w) = \int_{r_t}^R \left(1 - \frac{L^2}{\omega^2} \frac{c_s^2}{r^2}\right)^{1/2} \frac{dr}{c_s}. \tag{3.62}$$

Eq. 3.61 is usually referred to as the *Duvall Law* after Thomas Duvall Jr., who plotted $(n + \alpha_p)\pi/\omega$ as a function of $w = \omega/L$ and showed that the observed solar frequencies collapse into a single function of w. As we can see from Eq. 3.52, w is related to the lower turning point of a mode, since

$$w \equiv \frac{\omega}{\sqrt{l(l+1)}} = \frac{c_s(r_t)}{r_t}. \tag{3.63}$$

A version of the Duvall Law plot for modern data is shown in Figure 3.7, where we have used the fact that a better approximation to the oscillation equations gives $L = l + 1/2$.

Eq. 3.60 results in a simple relation for p-mode frequencies:

$$\nu_{nl} \simeq \left(n + \frac{l}{2} + \alpha_p\right) \Delta\nu, \tag{3.64}$$

where

$$\Delta\nu = \left[2 \int_0^R \frac{dr}{c_s}\right]^{-1} \tag{3.65}$$

is the inverse of the sound travel time from the surface to the center and back; it is the large frequency separation introduced in Chapter 1. This equation shows us that p modes of a given degree are uniformly spaced in frequency, and that modes with the same value of $n + l/2$ should have the same frequency (i.e., $\nu_{nl} - \nu_{n-1,l+2} \simeq 0$).

A higher-order mathematically rigorous asymptotic analysis of the equations shows that

$$\nu_{nl} \simeq \left(n + \frac{l}{2} + \frac{1}{4} + \alpha_p\right) \Delta\nu - (AL^2 - \delta)\frac{\Delta\nu^2}{\nu_{nl}}, \tag{3.66}$$

where δ is a constant, and Δ is given by

$$A = \frac{1}{4\pi^2\Delta\nu} \left[\frac{c_s(R)}{R} - \int_0^R \frac{dc_s}{dr} \frac{dr}{r}\right]. \tag{3.67}$$

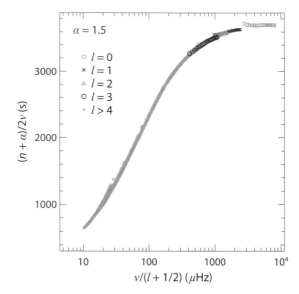

Figure 3.7. The Duvall Law demonstrated using a set of solar frequencies. The frequencies used in this figure are a combination of $l \leq 3$ mode frequencies obtained by the Birmingham Solar-Oscillations Network (BiSON) and frequencies of $4 \leq l \leq 200$ modes obtained by the Michelson Doppler Imager (MDI) on board the Solar and Heliospheric Observatory (SoHO). Note that all 2,000 modes fall on a narrow curve for $\alpha_p = 1.5$. The curve would have been narrower had we allowed α_p to be a function of frequency.

This immediately shows that when the term with the surface sound speed is neglected, we have

$$\nu_{nl} - \nu_{n-1,l+2} \equiv \delta\nu_{l,l+2}(n) \simeq -(4l+6)\frac{\Delta\nu}{4\pi^2\nu_{nl}}\int_0^R \frac{dc_s}{dr}\frac{dr}{r}. \tag{3.68}$$

$\delta\nu_{l,l+2}(n)$ is called the small frequency separation. As is clear from Eq. 3.68, it is sensitive to the gradient of the sound speed in the inner parts of a star. The sound-speed gradient changes with evolution as hydrogen is replaced by heavier helium, making the small separation a good diagnostic of the evolutionary stage of a star. Data on the small separations may be used to construct so-called asteroseismic HR diagrams (see Figure 3.8) and can used to find the age of a star if its metallicity is known. The small separation decreases with stellar age for stars on the main sequence. Unfortunately, it does not provide an age diagnostic for the more evolved phases of a star's life.

Often, observers may represent the frequency spectrum using the average value of $\Delta\nu$ obtained for $l=0$ modes and the average $\delta\nu_{02}$ between $l=0$ and $l=2$ modes:

$$\Delta\nu_0 = \langle\nu_{n+1,0} - \nu_{n,0}\rangle, \quad \text{and} \quad \delta\nu_{02} = \langle\nu_{nl} - \nu_{n-1,l+2}\rangle = (4l+6)D_0, \tag{3.69}$$

where

$$D_0 = \frac{\Delta\nu}{4\pi^2\nu_{nl}}\int_0^R \frac{dc_s}{dr}\frac{dr}{r}, \tag{3.70}$$

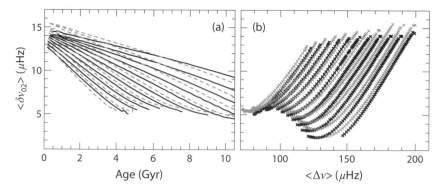

Figure 3.8. Left panel: The evolution of the average small separations with age on the main sequence. Black lines show the relation for models of [Fe/H] = 0.1, and gray dotted lines for [Fe/H] = −0.1. The different lines are for different masses at intervals of 0.04$M_\odot$ starting with 0.80$M_\odot$ at the top and 1.2$M_\odot$ at the bottom. Right panel: Plot showing the relation between the average large frequency separation and small frequency separation of the models shown in the left panel. The black and gray points denote metallicity as in the left panel. Masses of the models increase from right to left.

and Eq. 3.66 is presented in the form:

$$\nu_{nl} = \Delta\nu \left(n + \frac{l}{2} + \epsilon_p\right) - l(l+1)D_0. \tag{3.71}$$

In both equations $\epsilon_p = \alpha_p + 1/4$. The frequency dependence of α_p is often ignored. The term ϵ_p is the offset of the radial modes. The diagnostic potential of ϵ_p has been studied only recently. It can break the degeneracy of $\delta\nu_{l,l+2}$ at evolved stages. And as we shall see later, it can be used to differentiate between red giants with inert cores and those that are burning helium in the core.

Sometimes Eq. 3.71 appears in the form

$$\nu_{nl} = \Delta\nu \left(n + \frac{l}{2} + \epsilon_p\right) - \delta\nu_{0l}, \tag{3.72}$$

so that

$$\delta\nu_{0l} \equiv l(l+1)D_0. \tag{3.73}$$

This further suggests the possibility of using additional small frequency separations computed from individual modes (e.g., those involving the $l = 0$ and $l = 1$ modes), which we met in Section 1.3:

$$\delta\nu_{01}(n) = \nu_{n,0} - \frac{1}{2}\left(\nu_{n-1,1} + \nu_{n,1}\right),$$

$$\delta\nu_{10}(n) = \frac{1}{2}\left(\nu_{n,0} + \nu_{n+1,0}\right) - \nu_{n,1}, \tag{3.74}$$

which measure, respectively, deviations of the $l = 0$ modes from the exactly halfway frequencies of the adjacent $l = 1$ modes, and vice versa. In the asymptotic limit, we have $\delta\nu_{01}(n) = \delta\nu_{10}(n) = 2D_0$.

G modes

G modes are low-frequency modes with $\omega^2 < N^2$ and $\omega^2 < S_l^2$. The turning points of these modes are defined by $N = \omega$. Thus, in the case of the Sun, we would expect g modes to be trapped between the base of the convection zone and the core, while in the case of the 1.5$M_\odot$ star shown in Figure 3.1, the modes will be trapped between the edge of the envelope convection zone at 0.945 R and the core convection zone at 0.069 R.

For g modes $\omega^2 \ll S_l^2$, and thus Eq. 3.51 can be written as

$$K(r) \simeq \frac{1}{\omega^2}(N^2 - \omega^2)\frac{l(l+1)}{r^2}. \tag{3.75}$$

In other words, the properties of g modes are determined by the buoyancy frequency N.

An analysis similar to the one for p modes shows that for g modes, frequencies are determined by

$$\int_{r_1}^{r_2} L\left(\frac{N^2}{\omega^2} - 1\right)^{1/2}\frac{dr}{r} = \left(n - \frac{1}{2}\right)\pi, \tag{3.76}$$

where r_1 and r_2 mark the limits of the radiative zone. Thus we have

$$\int_{r_1}^{r_2} \left(\frac{N^2}{\omega^2} - 1\right)^{1/2}\frac{dr}{r} = \frac{(n - 1/2)\pi}{L} = G(\omega), \tag{3.77}$$

an expression similar to that for p modes (Eq. 3.61), but it shows how the buoyancy frequency, rather than the sound speed, matters for g modes.

A higher-order asymptotic analysis shows that the frequencies of high-order g modes can be approximated as

$$\omega = 2\pi\nu = \frac{L}{\pi(n + l/2 + \alpha_g)}\int_{r_1}^{r_2} N\frac{dr}{r}, \tag{3.78}$$

where α_g is a phase that varies slowly with frequency. To keep the same notation as in the p-mode case, often α_g is written as ϵ_g. In other words, while p modes are spaced equally in frequency, g modes are spaced equally in period, with the period spacing given by

$$\Delta\Pi_l = \frac{\Delta\Pi_0}{L}, \tag{3.79}$$

with

$$\Delta\Pi_0 = 2\pi^2\left(\int_{r1}^{r2} N\frac{dr}{r}\right)^{-1} \tag{3.80}$$

As we shall see in Section 9.3.3, $\Delta\Pi$ can be exploited to study red giants.

3.2.2 Mixed Modes and Avoided Crossings

Mixed modes are interesting entities from both a practical point of view and a mathematical one. Their dual nature—p-like close to the surface and g-like in the deeper layers—allows us to use them to study stellar cores; their fast evolution makes them very good diagnostics for determining ages. Mixed modes can be seen easily in échelle diagrams, where they prevent modes of a given l from lining up properly (see Figure 3.3). They cause *avoided crossings* between p modes of a given l but different adjacent n. We delve a bit more deeply into the issue of mixed modes in this section.

The presence of mixed modes was first reported in centrally condensed polytropes. Polytropes are simple models that satisfy the equations of stellar structure. They have a pressure-density relation given by $P \propto \rho^\gamma$, where γ, a constant, is the *polytropic exponent*. One can also define the *polytropic index* as $n = 1/(\gamma - 1)$. The structure of a polytrope can be determined easily, since the equations of stellar structure can be reduced to a single second-order differential equation, the Lane-Emden equation. Cold white dwarfs and neutron stars have polytropic equations of state. Fully convective normal stars can be described as polytropes under the assumption that they are composed of ideal gases. Other normal stars can generally be approximated as a combination of polytropes of different polytropic indices. This was a practical way of studying stars before computers became fast and readily available. As in the case of detailed stellar models, one can calculate the eigenfrequencies of polytropes.

As a polytrope is made denser in the core (this mimics what happens to a star on the subgiant and red-giant branches), the low-order p and g modes start behaving unusually, with the behavior characterized by the appearance of extra modes. These modes are of mixed character, g-mode like in the core and p-mode like toward the surface. This behavior was later seen in models of high-mass stars as well. As a star ages, the frequencies of g modes increase, while those of the f and p modes decrease. The frequencies of g modes increase until they approach the frequency of the f mode. This results in the change of both the f- and g-mode frequencies. The change in the f-mode frequency in turn changes the $n = 1$ p-mode frequency, which in turn affects the $n = 2$ p-mode frequency, and so on. This change in frequency of the f and p modes is known as *mode bumping*, since it effectively changes the value of n by unity.

One way to look at this phenomenon is the following: since g-mode frequencies increase with age, in the absence of a p-mode cavity, their frequencies would keep increasing and they would retain their pure g-like character. Such pure g modes are often called γ modes. Likewise, in the absence of a g-mode cavity, p modes would simply retain their p-like character; these are called π modes. When the increasing frequency of a γ mode approaches the decreasing frequency of a π mode, in a real star we see an avoided crossing.

The avoided crossings and the nature of mixed modes can be explained quite simply in terms of the coupling of the g- and p-mode cavities that are separated by a small attenuation region. Let us begin by assuming that a g mode couples with only one p mode. Then if $y_1(t)$ and $y_2(t)$ describe the oscillations, the coupled system can be described by

$$\frac{d^2 y_1(t)}{dt^2} = -\omega_1^2(\lambda)y_1 + \alpha\, y_2(t), \text{ and}$$

$$\frac{d^2 y_2(t)}{dt^2} = -\omega_2^2(\lambda)y_2 + \alpha\, y_1(t),$$

$$(3.81)$$

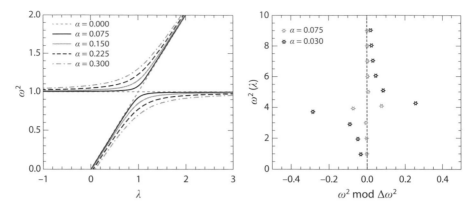

Figure 3.9. Left panel: Solutions for ω_+ (Eq. 3.83) for different values of the coupling parameter α, assuming $\omega_1^2 = 1$ and $\omega_2^2 = \lambda$. Note that the larger the coupling constant becomes, the larger is the zone of avoidance between the two solutions. Right panel: Échelle diagram of the solution for two different coupling constants. Increasing the value of the coupling constant increases the break in the frequencies.

where ω_1 and ω_2 are the frequencies of the uncoupled oscillations (i.e., of the γ and π modes), α is a coupling constant, and λ models the cavities as they evolve (in a real star, λ is age). If $\lambda = \lambda_0$ is the point where $\omega_1 = \omega_2$, then we would have $\omega_1 = \omega_2 = \omega_0$ if the equations were not coupled at that point. Of course the question is: what do the solutions look like when the equations are coupled? We want an oscillatory solution of course, that is, solutions of the form $y_1(t) = c_1 \exp(-i\omega t)$ and $y_2(t) = c_2 \exp(-i\omega t)$. Substituting these in Eq. 3.82, we get

$$\omega_1^2 c_1 = \omega_1^2 + \alpha c_2, \quad \text{and} \quad \omega_2^2 c_2 = \omega_2^2 + \alpha c_1. \tag{3.82}$$

There are two solutions for this system:

$$\omega_\pm = \frac{\omega_1^2 + \omega_2^2}{2} \pm \frac{1}{2}\left[(\omega_1^2 - \omega_2^2)^2 + 4\alpha^2\right]^{1/2}. \tag{3.83}$$

The solutions for the system for different values of the coupling parameter are shown in Figure 3.9. Note that if we set $\alpha = 0$ (i.e., we uncouple the equations), we recover the uncoupled solutions ω_1 and ω_2. When $\alpha \ll |\omega_1^2 - \omega_2^2|$, then the frequencies of the system are close to ω_1 and ω_2. If, however, $|\omega_1^2 - \omega_2^2| \ll \alpha$, then the frequencies can be approximated as $\omega_\pm = \omega_0^2 \pm \alpha$, that is, the two oscillators "avoid" the common frequency ω_0, and the frequency separation becomes larger as α becomes larger. In the stellar case α is a measure of the separation of the g- and p-mode cavities. The larger the separation becomes, the smaller will be the value of α.

On the subgiant branch in real stars, one g mode can couple with many different p modes. Assuming that the coupling constant can be assumed to be the same in each

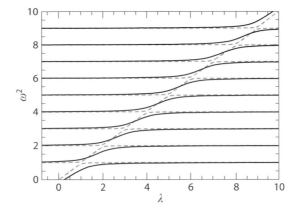

Figure 3.10. The solutions for ω for Eq. 3.84 for ten oscillators. The coupling constant is assumed to be the same in each case. The dashed lines are the solutions for a coupling constant of $\alpha = 0.075$, and the solid lines are for $\alpha = 0.30$. Note that the pattern of avoided crossings is very similar to the one in Figure 3.3.

case, we get n equations when $(n-1)$ p modes are coupled with one g mode:

$$\frac{d^2 y_1(t)}{dt^2} = -\omega_1^2(\lambda)y_1 + \alpha \; y_n(t),$$

$$\frac{d^2 y_2(t)}{dt^2} = -\omega_2^2(\lambda)y_2 + \alpha \; y_n(t),$$

$$\vdots \tag{3.84}$$

$$\frac{d^2 y_{n-1}(t)}{dt^2} = -\omega_{n-1}^2(\lambda)y_{n-1} + \alpha \; y_n(t),$$

$$\frac{d^2 y_n(t)}{dt^2} = -\omega_n^2(\lambda)y_n + \alpha \; y_1 + \alpha; y_2 + \cdots + \alpha \; y_{n-1}.$$

The solution for the case of ten oscillators is shown in Figure 3.10, where we have assumed that $\omega_n^2 = n$ for $n \leq 9$, and $\omega_{10}^2 = \lambda$.

Under the approximations that were used to obtain p- and g-mode asymptotics, it can be shown that the condition for mixed modes can be written in terms of a coefficient q that measures the strength of the coupling of the g- and p-mode cavities:

$$\cot \left(\int_{r_a}^{r_b} K(r)^{1/2} dr \right) \tan \left(\int_{r_c}^{R} K(r)^{1/2} dr \right) = q, \tag{3.85}$$

where r_a and r_b mark the limits of the g-mode cavity and r_c and R mark the limits of the p-mode cavity. The equation is often re-written as

$$\tan \theta_p = q \tan \theta_g. \tag{3.86}$$

Eq. 3.85 is a general condition that also holds for pure p and g modes when the coupling coefficient $q = 0$.

The coupling constant q is given by

$$q = \frac{1}{4} \exp \left[-2 \int_{r_b}^{r_c} |K(r)^{1/2}| dr \right] ; \qquad (3.87)$$

however, approximations made in deriving the asymptotic relation in Eq. 3.85 make this expression inexact. The coupling coefficient q is best determined by fitting mixed-mode frequencies.

For pure p modes, we can express the order n using Eq. 3.71 and thus the quantisation condition given in Eq. 3.59 can be written as

$$\theta_p = \pi \left(\frac{\nu}{\Delta \nu} - \frac{l}{2} - \epsilon_p \right). \qquad (3.88)$$

The g mode quantization condition and subsequent equations (Eqs. 3.75, 3.77, 3.78, and 3.80) give

$$\theta_g = \pi \left(\frac{1}{\nu \Delta \Pi_l} - \epsilon_g \right). \qquad (3.89)$$

In subgiants, usually one g mode couples with many p modes. In that case the relation in Eq. 3.85 can be used to express the frequency of a mixed mode in terms of the frequency of the closest g mode. In the case of mixed modes we can write $\theta_g = \theta_{g,\text{mixed}} - \theta_{g,\text{pure}}$, and thus we may write the g-mode phase as:

$$\theta_g = \frac{\pi}{\Delta \Pi_l} \left(1/\nu - 1/\nu_g \right) \qquad (3.90)$$

where ν_g is the pure g-mode frequency with the ϵ_g term absorbed into it. Setting $\tan \theta_p = q \tan \theta_g$ gives us the relation

$$\tan \left(\pi \left[\frac{\nu}{\Delta \nu} - \frac{l}{2} - \epsilon_p \right] \right) = q \tan \left(\frac{\pi}{\Delta \Pi_l} \left[\frac{1}{\nu} - \frac{1}{\nu_g} \right] \right). \qquad (3.91)$$

Rearranging the above to express ν in terms of perturbations about pure g-mode frequencies, we get

$$\frac{1}{\nu} = \frac{1}{\nu_g} + \frac{\Delta \Pi_l}{\pi} \arctan \left[\frac{1}{q} \tan \left(\pi \left[\frac{\nu}{\Delta \nu} - \frac{l}{2} - \epsilon_p \right] \right) \right]. \qquad (3.92)$$

In red giants, several g modes couple with a given p mode. As a result Eq. 3.85 is used to express the frequency of a mixed mode in terms of the frequency of the closest p mode. Since the majority of observed mixed modes are dipole in nature, in the subsequent discussion we will restrict ourselves to the $l = 1$ case. Since $q = 0$ implies pure p or g modes, in the case of mixed modes in red giants we can write

$$\theta_p = \theta_{p,\text{mixed}} - \theta_{p,\text{pure}}, \qquad (3.93)$$

or, in other words

$$\theta_p = \pi \left(\frac{\nu - \nu_p}{\Delta \nu} \right), \qquad (3.94)$$

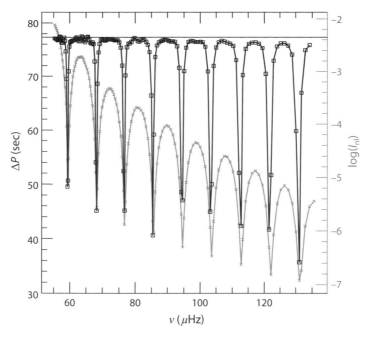

Figure 3.11. The period spacings between $l = 1$ modes for an ascending branch red-giant model plotted in black as a function of frequency. The black horizontal line is the asymptotic period spacing. In gray we plot the inertia of the modes. Note that modes with the lowest period spacings correspond to modes with the lowest inertia.

where ν_p is the pure p mode frequency with the terms $1/2$ and ϵ_p absorbed into it. Substituting Eq. 3.94 and Eq. 3.89 in Eq. 3.86 we get

$$\nu = \nu_p + \frac{\Delta\nu}{\pi} \arctan\left[q \tan\pi\left(\frac{1}{\nu\Delta\Pi_1} - \epsilon_g\right)\right].\qquad(3.95)$$

This equation is very useful for understanding the $l = 1$ mixed modes in red giants. The period spacings of a red giant model are shown in Figure 3.11. The period spacings between the pure g modes lie on a roughly horizontal line. These modes are not observed. The dips in the diagram correspond to the mixed modes that can be observed. Eq. 3.95 is often used to constrain fits to red-giant power spectra, as we shall see in Section 6.5.2.

3.2.3 What About the Upper Turning Point?

The analysis thus far assumes that the upper turning point of the modes is at $r = R$. This is not really the case. Additionally, the analysis does not yield f modes. While we will not be able to observe f modes in stars, solar f modes provide a lot of information about the Sun. The frequencies of f modes lie between those of g and p modes.

To see what happens close to the surface, we analyze the equations without making any assumptions about variation of the pressure scale height, but we keep the assumption that the curvature of the star (the $2/r$ term) can be neglected. One can show that under the Cowling approximation, the equations of adiabatic stellar

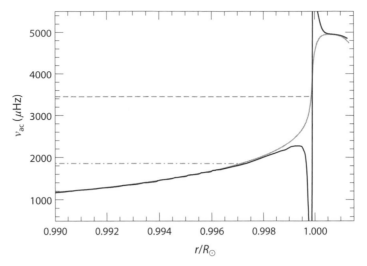

Figure 3.12. The profile of the acoustic cut-off frequency $\nu_{ac} = \omega_{ac}/(2\pi)$ of a solar model. The black line is the profile calculated as per Eq. 3.98. The thin gray line shows the profile when an isothermal atmosphere is assumed. The dot-dashed line is at a frequency of 1,860 μHz, and the point at which it meets the curves shows the upper limit of the cavity of a mode of this frequency. The dashed line is at a frequency of 3,450 μHz. Note that the upper turning point of the higher-frequency mode lies at a larger radius compared with that of the lower-frequency mode.

oscillations can be written as

$$\frac{d^2\Psi}{dr^2} + K^2(r)\Psi = 0, \tag{3.96}$$

where $\Psi = \rho^{1/2}c_s^2 \nabla \cdot \vec{\xi}$, and the wavenumber K is given by

$$K^2(r) = \frac{\omega^2 - \omega_{ac}^2}{c_s^2} + \frac{l(l+1)}{r^2}\left(\frac{N^2}{\omega^2} - 1\right), \tag{3.97}$$

with

$$\omega_{ac}^2 = \frac{c_s^2}{4H_\rho^2}\left(1 - 2\frac{dH_\rho}{dr}\right), \tag{3.98}$$

where H_ρ is the density scale height given by $H_\rho = -dr/d\ln\rho$. The quantity ω_{ac} is known as the *acoustic cut-off frequency*. The radius at which $\omega = \omega_{ac}$ defines the upper limit of the cavity for wave propagation, and that radius is usually called the upper turning point of a mode. For isothermal atmospheres, the acoustic cut-off frequency is simply $\omega_{ac} = c_s g\rho/(2P)$. The profile of the acoustic cut-off frequency for a solar model is shown in Figure 3.12, here in cyclic frequency (i.e., $\nu_{ac} = \omega_{ac}/(2\pi)$).

Eq. 3.96 can be solved for $\Psi = 0$. These are the f modes, and it can be shown that their dispersion relation is

$$\omega^2 \simeq gk, \tag{3.99}$$

k being the wavenumber, and thus f-mode frequencies are independent of the stratification of the star to a large extent, a fact that has been exploited for studies of the acoustic radius of the Sun.

3.2.4 Some Useful Scaling Relations

The two seismic parameters that are most easily observed from stellar oscillation power spectra are the large frequency separation $\Delta\nu$ and the frequency at which oscillation power is a maximum, ν_{max}. These two quantities can be related to the global parameters of a star.

Let us consider $\Delta\nu$ first. As per Eq. 3.65, it is related to the inverse of the sound-travel time in the star. A simple dimensional analysis of the equation leads to

$$(\Delta\nu)^2 \propto \frac{c_s^2}{R^2}. \tag{3.100}$$

In this simplified analysis we look at the sound speed at $r = R$. Since $c_s^2 = \Gamma_1 P/\rho$, and Γ_1 is of the order unity, this leads to

$$(\Delta\nu)^2 \propto \frac{P}{\rho R^2}. \tag{3.101}$$

An analysis of the stellar structure equations shows that pressure P scales as M^2/R^4, and ρ scales as M/R^3; thus

$$(\Delta\nu)^2 \propto \frac{M}{R^3} \propto \bar{\rho}, \tag{3.102}$$

where $\bar{\rho}$ is the mean density. This is usually written as

$$\Delta\nu = \Delta_{\nu,\odot}\sqrt{\frac{M/M_\odot}{(R/R_\odot)^3}}. \tag{3.103}$$

Unlike for $\Delta\nu$, there is no theory yet that fully explains the ν_{max} scaling relation. Since ν_{max} defines the maximum power in the oscillation spectrum, it should depend on the physical conditions in the near-surface layers where the modes are excited. Close to the surface, the behavior of the waves is influenced by the acoustic cut-off frequency, and as a result one can argue that ν_{max} will be related to this critical frequency. In the isothermal approximation, the acoustic cut-off frequency is simply $c_s g\rho/(2P)$, but $c_s = \sqrt{\Gamma_1 P/\rho}$, and for an ideal gas, $\rho/P = \mu/(\mathcal{R}T)$; thus we have

$$\nu_{ac} \propto \frac{g}{\sqrt{T}}. \tag{3.104}$$

If we consider conditions at the surface of a star, $g \propto M/R^2$ and $T = T_{eff}$, we have

$$\nu_{ac} \propto \frac{M}{R^2}\frac{1}{\sqrt{T_{eff}}}. \tag{3.105}$$

If $\nu_{\max}$ is assumed to be proportional to the cut-off frequency, we immediately have

$$\nu_{\max} = \nu_{\max,\odot} \frac{M/M_\odot}{(R/R_\odot)^2} \sqrt{\frac{T_{\text{eff},\odot}}{T_{\text{eff}}}}. \tag{3.106}$$

Eqs. 3.103 and Eq. 3.106 show that the two quantities $\Delta\nu$ and $\nu_{\max}$ are related. Eliminating radius R from the two equations gives us the relation

$$\Delta\nu \propto M^{-1/4} T_{\text{eff}}^{-3/8} \nu_{\max}^{3/4}. \tag{3.107}$$

Since the dependences on M and T_{eff} are small, they can be ignored to first order, and we get

$$\Delta\nu \propto \nu_{\max}^{3/4}. \tag{3.108}$$

This is very close to the observed dependence shown in Figure 1.3.

3.3 CHANGES IN FREQUENCIES IN RESPONSE TO CHANGES IN STELLAR STRUCTURE

While Eqs. 3.29, 3.31, and 3.33 are in forms that make it easy to calculate frequencies, they do not tell us much about how the frequencies of a star change if the structure is altered slightly. For that we need to cast the equations in a different form.

The starting point of the exercise is the perturbed momentum equation (Eq. 3.12):

$$\rho \frac{\partial^2 \vec{\xi}}{\partial t^2} = -\nabla P_1 + \rho_0 \vec{g}_1 + \rho_1 \vec{g}, \tag{3.109}$$

where $\vec{g}$ is the unperturbed acceleration due to gravity and $\vec{g}_1$ is the perturbed value. We already know that the time dependence of the perturbed quantities can be written as an oscillatory function, in other words, with a time dependence given by $\exp(-i\omega t)$. Substituting this in the equation, we get

$$-\omega^2 \rho \vec{\xi} = -\nabla P_1 + \rho \vec{g}_1 + \rho_1 \vec{g}. \tag{3.110}$$

Substituting for ρ_1 from the perturbed continuity equation (Eq. 3.11) and P_1 from Eq 3.18, we get

$$-\omega^2 \rho \vec{\xi} = \nabla(c_s^2 \rho \nabla \cdot \vec{\xi} + \nabla P \cdot \vec{\xi}) - \vec{g} \nabla \cdot (\rho \vec{\xi}) - G\rho \nabla \left(\int_V \frac{\nabla \cdot (\rho \vec{\xi}) d^3 r}{|\vec{r} - \vec{r}'|} \right), \tag{3.111}$$

where we have expressed the gravitational potential as an integral:

$$\vec{g}_1 = \nabla \Phi_1 = \nabla \int_V G \frac{\rho_1 d^3 r}{|\vec{r} - \vec{r}'|} = -\nabla \int_V \frac{G\nabla \cdot (\rho \vec{\xi}) d^3 r}{|\vec{r} - \vec{r}'|}. \tag{3.112}$$

Eq. 3.111 describes how the mode frequencies ω depend on the structure of the star. Note that $\xi \equiv \xi_{nl}$ and $\omega \equiv \omega_{nl}$, and thus there is one such equation for each mode described by the eigenfunction ξ_{nl}.

The first step in determining the frequency response of a star is to recognize that Eq. 3.111 is an eigenvalue equation of the form

$$\mathcal{L}(\vec{\xi}_{nl}) = -\omega_{nl}^2 \vec{\xi}_{nl}, \tag{3.113}$$

$\mathcal{L}$ being the differential operator in Eq. 3.111. It can be shown that under specific boundary conditions—namely, $\rho = P = 0$ at the outer boundary—the eigenvalue problem defined by Eq. 3.111 is a Hermitian one. This allows us to move ahead.

Hermitian operators have an important property: if $\mathcal{O}$ is a differential Hermitian operator, and x and y are two eigenfunctions of the operator, then

$$\int x^* \mathcal{O}(y) dV = \int y \mathcal{O}(x^*) dV, \tag{3.114}$$

where * indicates a complex conjugate. This is often written as

$$\langle x, \mathcal{O}(y) \rangle = \langle \mathcal{O}(x), y \rangle, \tag{3.115}$$

where the operation $\langle \cdot \rangle$ is usually called an inner product. The relevant inner product in the case of Eq. 3.111 is

$$\langle \vec{\xi}, \vec{\eta} \rangle = \int_V \rho \vec{\xi}^* \cdot \vec{\eta} d^3\vec{r} = 4\pi \int_0^R [\xi_r^*(r)\eta_r(r) + L^2 \xi_t^*(r)\eta_t(r)] r^2 \rho dr. \tag{3.116}$$

The eigenvalues of a Hermitian differential operator are real. Additionally, the variational principle can be applied, and this means that if a is the eigenvalue corresponding to the eigenfunction x, then

$$a = \frac{\langle x, \mathcal{O}(x) \rangle}{\langle x, x \rangle}. \tag{3.117}$$

Thus if the frequency ω_{nl} is the eigenvalue corresponding to the displacement eigenfunction $\vec{\xi}_{nl}$, then

$$-\omega_{nl}^2 = \frac{\int_V \rho \vec{\xi}_{nl}^* \cdot \mathcal{L}(\vec{\xi}_{nl}) d^3\vec{r}}{\int_V \rho \vec{\xi}_{nl}^* \cdot \vec{\xi}_{nl} d^3 r}. \tag{3.118}$$

Frequencies calculated in this manner are often referred to as *variational frequencies*.

The Hermitian nature of Eq. 3.111 allows us to use the variational principle to put the equation in a form that shows us how changing the internal structure will alter the frequencies. Eq. 3.113 is first linearized around a model, which is usually referred to as the *reference model*. Then if $\mathcal{L}$ is the operator of the reference model, the operator for the perturbed model can be written as $\mathcal{L} + \delta\mathcal{L}$. The corresponding displacement eigenfunctions for the reference model and the model after it is perturbed are $\vec{\xi}$ and $\vec{\xi} + \delta\vec{\xi}$, respectively, with corresponding frequencies ω and

$\omega + \delta\omega$. Thus we have

$$(\mathcal{L} + \delta\mathcal{L})(\vec{\xi} + \delta\vec{\xi}) = -(\omega + \delta\omega)^2(\vec{\xi} + \delta\vec{\xi}), \qquad (3.119)$$

which on expansion and retention of only linear terms gives

$$\int \vec{\xi}^* \mathcal{L}\vec{\xi}\,\mathrm{d}V + \int \vec{\xi}^* \delta\mathcal{L}\vec{\xi}\,\mathrm{d}V + \int \vec{\xi}^* \mathcal{L}\delta\vec{\xi}\,\mathrm{d}V + \int \vec{\xi}^* \delta\mathcal{L}\delta\vec{\xi}\,\mathrm{d}V = -\omega^2 \int \vec{\xi}^* \vec{\xi}\,\mathrm{d}V$$

$$- \omega^2 \int \vec{\xi}^* \delta\vec{\xi}\,\mathrm{d}V - (\delta\omega^2) \int \vec{\xi}^* \vec{\xi}\,\mathrm{d}V - 2\omega\delta\omega \int \vec{\xi}^* \vec{\xi}\,\mathrm{d}V. \qquad (3.120)$$

Because $\mathcal{L}$ is a Hermitian operator, all but two terms in Eq. 3.120 cancel out to give

$$\int \vec{\xi}^* \delta\mathcal{L}\vec{\xi}\,\mathrm{d}V = -2\omega\delta\omega \int \vec{\xi}^* \vec{\xi}\,\mathrm{d}V, \qquad (3.121)$$

or

$$\frac{\delta\omega}{\omega} = -\frac{\int_V \rho\vec{\xi} \cdot \delta\mathcal{L}\vec{\xi}\,\mathrm{d}^3\vec{r}}{2\omega^2 \int_V \rho\vec{\xi} \cdot \vec{\xi}\,\mathrm{d}^3\vec{r}} = -\frac{\int_V \rho\vec{\xi} \cdot \delta\mathcal{L}\vec{\xi}\,\mathrm{d}^3\vec{r}}{2\omega^2 I_{nl}}, \qquad (3.122)$$

where I_{nl} is the mode inertia, defined earlier in Eq. 3.36. This equation connects the change in frequencies to the change in structure represented as $\delta\mathcal{L}$. One such equation can be written for each mode. Note that the mode inertia appears in the denominator of Eq. 3.122, indicating that for a given perturbation, frequencies of modes with lower inertia will change more than those with higher inertia.

To determine what $\delta\mathcal{L}$ actually is, we have to return to Eq. 3.111, perturb it, and keep only terms linear in the perturbations. This gives us

$$\delta\mathcal{L}\vec{\xi} = \nabla(\delta c_s^2 \nabla \cdot \vec{\xi} + \delta\vec{g} \cdot \vec{\xi}) + \nabla\left(\frac{\delta\rho}{\rho}\right) c_s^2 \nabla \cdot \vec{\xi}$$

$$+ \frac{1}{\rho}\nabla\rho\delta c_s^2 \nabla \cdot \vec{\xi} + \delta\vec{g}\nabla \cdot \vec{\xi} - G\nabla \int_V \frac{\nabla \cdot (\delta\rho\vec{\xi})}{|\vec{r} - \vec{r}'|}\mathrm{d}^3\vec{r}', \qquad (3.123)$$

where δc_s^2, $\delta\vec{g}$, and $\delta\rho$ are the differences in sound speed, acceleration due to gravity, and density, respectively, caused by perturbing the model. Of course, the quantity $\delta\vec{g}$ can be expressed in terms of $\delta\rho$.

Substituting Eq. 3.123 in Eq. 3.122 and rearranging the terms, we get

$$\frac{\delta\omega}{\omega} = -\frac{1}{2\omega^2 I}(I_1 + I_2 + I_3 + I_4), \qquad (3.124)$$

where I is the mode inertia and

$$I_1 = -\int_0^R \rho(\nabla \cdot \vec{\xi})^2 \delta c_s^2 r^2 \mathrm{d}r, \qquad (3.125)$$

$$I_2 = \int_0^R \xi_r[\rho\nabla \cdot \vec{\xi} + \nabla \cdot (\rho\vec{\xi})]\delta g\, r^2 \mathrm{d}r, \qquad (3.126)$$

where

$$g(r) = \frac{4\pi G}{r^2} \int_0^r \delta\rho(s)s^2 ds, \qquad (3.127)$$

with s being a dummy variable representing radius. The other terms are

$$I_3 = \int_0^R \rho c^2 \xi_r \nabla \cdot \vec{\xi} \frac{d}{dr}\left(\frac{\delta\rho}{\rho}\right) r^2 dr, \qquad (3.128)$$

and

$$I_4 = \frac{4\pi G}{2l+1} \int_0^R \nabla \cdot (\rho\vec{\xi})r^2 \left[\frac{1}{r^{l+1}} \int_0^r s^{l+2}\left(\rho\nabla\cdot\vec{\xi} - \rho\frac{d\xi_r}{ds} - \frac{l+2}{s}\rho\xi_r\right)\frac{\delta\rho}{\rho} ds\right] dr, \qquad (3.129)$$

and again s is the dummy variable of integration.

One can rewrite Eq. 3.125 in terms of $\delta c_s^2/c_s^2$, the relative difference between the squared sound speed of the star and the reference model. Similarly Eqs. 3.126–3.129 can be rewritten in terms of the relative density difference $\delta\rho/\rho$ (after changing the the order of the integrals in some terms). Thus, Eq. 3.124 for each mode $i \equiv (n, l)$ can be written as

$$\frac{\delta\omega_i}{\omega_i} = \int K_{c_s^2,\rho}^i(r)\frac{\delta c_s^2}{c_s^2}(r)dr + \int K_{\rho,c_s^2}^i(r)\frac{\delta\rho}{\rho}(r)dr. \qquad (3.130)$$

The terms $K_{c_s^2,\rho}^i(r)$ and $K_{\rho,c_s^2}^i(r)$ are thus known functions of the unperturbed model that represent the change in frequency in response to changes in sound speed and density, respectively. These two functions are thus *kernels* that connect the change in structure to the changes in frequency.

We show a few sound-speed and density kernels in Figure 3.13. Note that kernels for $l = 3$ modes do not go as deep into the star as do kernels of $l = 0$; kernels of $l = 0$ reach all the way to the center. This was expected from our earlier discussion of how the lower turning points of modes of higher degree are shallower than those of lower-degree modes. From the figure we can also see that sound-speed kernels are positive at all radii, while density kernels oscillate between positive and negative values, reducing the contribution of the second term on the right-hand side of Eq. 3.130. This shows why we can use the asymptotic form of the oscillation equations, which assumes that p-mode frequencies can be explained by the sound speed alone (Eq. 3.60). The small discrepancies in the results obtained using the asymptotic relation are a consequence of ignoring the small contribution from density.

The kernels for sound speed and density can be converted to those of other quantities. It is relatively straightforward to change the (c_s^2, ρ) kernels to those for (Γ_1, ρ). Since $c_s^2 = \Gamma_1 P/\rho$, we have

$$\frac{\delta c_s^2}{c_s^2} = \frac{\delta\Gamma_1}{\Gamma_1} + \frac{\delta P}{P} - \frac{\delta\rho}{\rho}. \qquad (3.131)$$

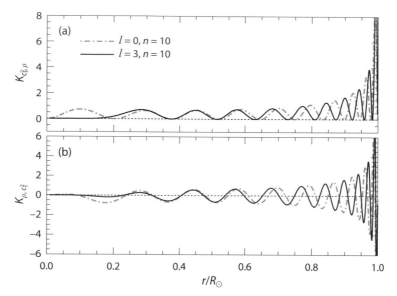

Figure 3.13. Sound-speed (top panel) and density (bottom panel) kernels for an $l = 0$ and an $l = 3$ mode of a solar model. Note that the sound-speed kernels are always ≥ 0, while the density kernels oscillate between positive and negative values. Also note that while the $l = 0$ kernels reach $r = 0$, the $l = 3$ do not, consistent with the fact that the Lamb frequency of $l = 3$ modes does not reach $r = 0$ at finite frequencies.

Substituting Eq. 3.131 in Eq. 3.130 gives

$$\frac{\delta\omega_i}{\omega_i} = \int K^i_{c_s^2,\rho}(r)\frac{\delta\Gamma_1}{\Gamma_1}\,dr + \int K^i_{c_s^2,\rho}\frac{\delta P}{P}dr + \int (K^i_{\rho,c_s^2} - K^i_{c_s^2,\rho})\frac{\delta\rho}{\rho}dr. \qquad (3.132)$$

Thus, $K_{\Gamma_1,\rho} = K_{c_s^2,\rho}$ and to obtain K_{ρ,Γ_1}, the equation of hydrostatic equilibrium, $dP/dr = -g\rho$, has to be used to write $\int K_{c_s^2,\rho}(\delta P/P)dr$ in terms of $\delta\rho/\rho$ though an integral.

Closed-form kernels are not possible for variable pairs other than (c_s^2, ρ) and (Γ_1, ρ). For example, going from (c_s^2, ρ) to (u, Γ_1), where $u \equiv P/\rho = c_s^2/\Gamma_1$, we find that $K_{\Gamma_1,u} \equiv K_{c_s^2,\rho}$ (from Eq 3.131), but the second kernel of the pair, K_{u,Γ_1}, does not have a closed-form solution. It can be written as

$$K_{u,\Gamma_1} = K_{c_s^2,\rho} - P\frac{d}{dr}\left(\frac{\psi}{P}\right), \qquad (3.133)$$

where ψ is a solution of the equation

$$\frac{d}{dr}\left(\frac{P}{r^2\rho}\frac{d\psi}{dr}\right) - \frac{d}{dr}\left(\frac{Gm}{r^2}\psi\right) + \frac{4\pi G\rho}{r^2}\psi = -\frac{d}{dr}\left(\frac{P}{r^2}K_{\rho,c_s^2}\right). \qquad (3.134)$$

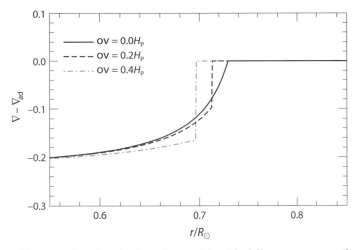

Figure 3.14. The quantity $\nabla - \nabla_{ad}$ for solar models with different amounts of overshoot. Note how the change in $\nabla - \nabla_{ad}$ becomes larger as overshoot is increased. This makes the acoustic glitch larger.

3.4 THE EFFECTS OF ACOUSTIC GLITCHES

There are regions in a star where variations in the structure occur over a length scale that is shorter than the radial wavelength of the modes. These are usually referred to as *acoustic glitches*. These leave very specific signatures on the mode frequencies.

Two important acoustic glitches occur in unevolved stars. The first is at the base of the outer convection zone, where the temperature gradient changes abruptly from the adiabatic value ∇_{ad} to the radiative one ∇_{rad}. This leaves a discontinuity in the second derivative of the speed of sound. The usual formulation of convective overshoot (or, rather, undershoot in this case) assumes that the overshooting region has adiabatic stratification, which abruptly changes to a radiative one (Figure 3.14); this not only shifts the position of the acoustic glitch (making it deeper) but also increases the amplitude of the glitch, making the signature larger.

The second glitch is the one due to the helium ionization zones. This is not a true discontinuity but a relatively rapid change in the sound speed because of changes in the adiabatic index Γ_1 as ionization takes place. The changes in Γ_1 for a few different models are illustrated in Figure 3.15. The quantity Γ_1 is essentially the ratio of the specific heats at constant pressure, C_P, and at constant volume, C_V. In ionization zones both C_P and C_V increase, because part of the energy goes into ionizing the gas rather than heating it. However, $C_P - C_V$ is nearly constant. To keep $C_P - C_V$ a constant while each quantity increases means that their ratio has to decrease. Since the sound speed varies as $\sqrt{\Gamma_1}$, sound speed decreases in that region, thereby affecting the frequencies. The amount of helium affects the decrease in Γ_1. The larger the amount of helium becomes, the more pronounced is the decrease and hence the larger the signal of the glitch.

Acoustic glitches introduce an oscillatory component to the frequencies with respect to the frequencies without such a glitch. In Figure 3.16 we show the observed signature of acoustic glitches in the Sun. For the range of masses over which acoustic

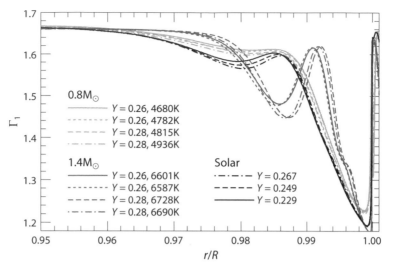

Figure 3.15. The change in the adiabatic index Γ_1 because of helium and hydrogen being ionized. The dip in Γ_1 close to the surface is a result of the ionization of HI and HeI, while the second, smaller dip is that due to HeII. We show three cases: The first is for solar models constructed with different helium abundances. The effect of the helium abundance is unambiguous in this case—the dip in the HeII ionization zone increases as the amount of helium increases. The second case is for $0.8M_\odot$ models with different helium abundances and different stages of evolution, as indicated by the effective temperatures listed in the plot. The last case is that of a set of $1.4M_\odot$ models. Note that the dip in Γ_1 changes as we change the mass and age.

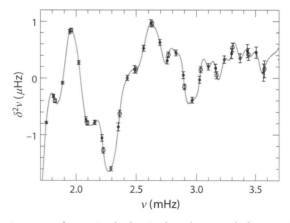

Figure 3.16. The signature of acoustic glitches in the solar p-mode frequencies. The different symbols denote modes of different degrees. We have amplified the signal by taking second frequency differences of the observed modes of a given l (i.e., $\delta^2 \nu = \nu_{l,n-1} - 2\nu_{l,n} + \nu_{l,n+1}$), which removes the predominant smooth trend of the frequencies as a function of n. We only show the results for $l = 0$ to $l = 2$, though for the Sun higher l modes are also observed. A spline has been drawn through the points to guide the eye. Note that there are two main "periods" seen in the curve, the longer period is the signature of the glitch at the HeII ionization zone, and the shorter period is that of the base of the convection zone.

glitches can be fitted, the convection zone is deeper than the helium ionization zone and leaves a higher-frequency signature that modulates the lower-frequency signature of the helium ionization zone.

The oscillatory component $\delta \nu$ can be expressed as

$$\delta \nu \propto \sin(4\pi \tau_g \nu_{nl} + \phi), \tag{3.135}$$

where ϕ is a phase factor and τ_g is the acoustic depth (i.e., the sound travel time of the glitch measured from the surface to the radial position of the glitch, r_g):

$$\tau_g = \int_{r_g}^{R} \frac{dr}{c_s}. \tag{3.136}$$

Each glitch will thus contribute an oscillatory signal to the frequencies with a period that is twice the acoustic depth. Thus a shallow glitch (e.g., the second helium ionization zone) will produce a signature with a longer period than that for a deeper glitch (e.g., the base of the convection zone). Both signatures can easily be seen in observed frequencies. The amplitude of the glitch is a function of the amount of change that takes place, that is, the extent to which the sound speed decreases in the helium ionization zone. The larger the change is, the larger the amplitude of the signature becomes.

The signatures of the acoustic glitches can be used to determine the acoustic depth of the surface convection zone in a star, which in turn can be used to constrain stellar models. They can also be used to determine the helium abundance in stars. We discuss this further in Chapter 8. An acoustic glitch also accurs at the boundary of convective cores. However, since the τ_g for these glitches is large, the signature is usually aliased at $T - \tau_g$, where T is the total acoustic radius of the star, given by $\int_0^R (1/c_s) dr$.

The frequency response of the glitches can, of course, be reproduced by the kernels introduced in Section 3.3. Figure 3.17 shows the response of the frequencies of a solar model to three sound-speed glitches: one close to the surface, one at $0.98 R_\odot$, and one at $0.71 R_\odot$.

The frequencies of red giants are affected by two types of glitches. The first is the acoustic glitch due to the helium ionization zone. This glitch affects the frequencies of the radial modes; their effects can also be seen in quadrupole modes. This feature can be used to determine the position of the HeII ionization zone as well as the amount of helium in the convection zone of the star.

The second type seen in red giants are glitches in the Brunt-Väisälä frequency (see Figure 3.18). These glitches affect both the frequencies and the mode inertias of mixed modes. The change in the frequency pattern can be seen as a variation in the underlying period spacing of pure g modes. The theoretical analysis of the glitches is complicated, but the effect can be written quite simply. If $\Delta\Pi_g$ is the period spacing in the presence of a glitch, and $\Delta\Pi$ is that without the glitch, then

$$\Delta\Pi_g = \Delta\Pi(1 + g), \tag{3.137}$$

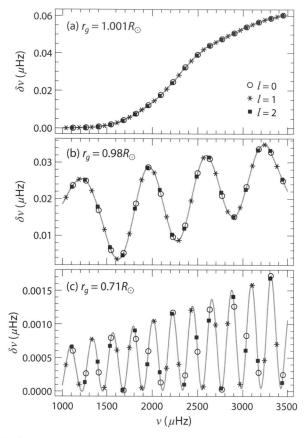

Figure 3.17. The frequency response of a solar model to a localized change in sound speed at radii corresponding to (top panel) the atmosphere, (middle panel) the second helium ionization zone, and (bottom panel) the base of the convection zone. The responses were calculated using Eq. 3.130 with $\delta c_s^2/c_s^2$ assumed to be a Gaussian centered at r_g and $\delta\rho/\rho = 0$. We only show $l = 0$, 1, and 2 modes. A spline has been drawn through the points to guide the eye.

where g is a function in $1/\nu$ (i.e., the period). The period of g is approximately

$$\Pi_g = \frac{\Delta\Pi_g}{\Delta\Pi} = \frac{\int_{r_1}^{r_2} N\frac{dr}{r}}{\int_{r_g}^{r_2} N\frac{dr}{r}}, \tag{3.138}$$

where r_1 and r_2 are, respectively, the inner and outer boundary of the radiative core, and r_g is the radius at which the glitch occurs.

3.5 THE EFFECTS OF ROTATION

Rotation causes a deviation from spherical symmetry and makes the frequencies m dependent. The effect of slow rotation can be treated as a linear perturbation.

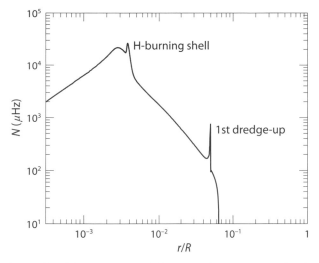

Figure 3.18. The profile of the Brunt-Väisälä frequency for a red giant. The glitches caused by the hydrogen-burning shell and the first dredge-up are marked.

A "slow" rotator is one for which the maximum centrifugal force at the surface is much smaller the the force of gravity. This approach works well for the Sun (rotation velocity $2 \, \mathrm{km \, s^{-1}}$), where the ratio of the centrifugal force to the gravitational force is of order 10^{-5}. However, even in the case of slow rotators, the perturbative treatment can only be applied to modes with frequencies much higher than the rotational frequency Ω.

To derive the oscillation equations in the spherically symmetric case, we had assumed that the velocity $\vec{v} = 0$ in the equilibrium state, and that $\vec{v} = \partial \vec{\xi} / \partial t$. The presence of rotation changes these conditions. If we assume that all layers of a star have the same axis of rotation, then even in the equilibrium state there is a velocity

$$\vec{v}_0 = \vec{\Omega} \times \vec{r} = \Omega r \sin \theta, \tag{3.139}$$

where Ω is the rotation rate. This changes the unperturbed continuity equations as well as the momentum equation. The full unperturbed continuity equation becomes

$$\frac{\mathrm{d}\rho_0}{\mathrm{d}t} + \nabla \cdot (\rho \vec{v}_0) = 0 \tag{3.140}$$

However, since we work under the assumption that the equilibrium density does not change with time, the equation reduces to

$$\nabla \cdot (\rho \vec{v}_0) = 0. \tag{3.141}$$

The momentum equation becomes

$$\rho_0 \frac{\mathrm{d}\vec{v}_0}{\mathrm{d}t} \equiv \rho_0 \frac{\partial \vec{v}_0}{\partial t} + \rho_0 \vec{v}_0 \cdot \nabla \vec{v}_0 = -\nabla P_0 + \rho_0 \vec{g}_0. \tag{3.142}$$

If we assume that the rotation rate does not change over the time scale of the oscillations, then $\partial \vec{v}_0 / \partial t = 0$. We have ignored the term of order $|v_0|^2$ in Eq. 3.142 since we are dealing with slow rotation.

The displacement $\vec{\xi}$ is related to the perturbed velocity $\vec{v}_0 + \vec{v}_1$ and

$$\frac{d\vec{\xi}}{dt} = \vec{v}_1 + (\vec{\xi} \cdot \nabla)\vec{v}_0. \tag{3.143}$$

Thus we have

$$\frac{d\vec{\xi}}{dt} = \frac{\partial \vec{\xi}}{\partial t} + (\vec{v}_0 \cdot \nabla)\vec{\xi}. \tag{3.144}$$

The perturbed momentum equation can now be written as

$$\rho_0 \frac{d^2\vec{\xi}}{dt^2} = -\nabla P_1 + \rho_1 \vec{g}_0 + \rho_0 \vec{g}_0, \tag{3.145}$$

Using Eq. 3.144 and neglecting terms in $(\vec{v}_0)^2$, Eq. 3.145 becomes

$$\rho_0 \frac{\partial^2 \vec{\xi}}{\partial t^2} + 2\rho_0(\vec{v}_0 \cdot \nabla)\left(\frac{\partial \vec{\xi}}{\partial t}\right) = -\nabla P_1 + \rho_1 \vec{g}_0 + \rho_0 \vec{g}_1. \tag{3.146}$$

At equilibrium, the structure is independent of time, and we can again write the time dependence as $\exp(-i\omega t)$; applying this to Eq. 3.146, we get

$$-\omega^2 \rho_0 \vec{\xi} - 2i\omega\rho_0(\vec{v}_0 \cdot \nabla)\vec{\xi} = -\nabla P_1 + \rho_1 \vec{g}_0 + \rho_0 \vec{g}_1. \tag{3.147}$$

If we assume that the $\vec{v}_0$ is small, we can compare the above equations to Eq. 3.110 and rewrite Eq. 3.147 as

$$-\omega^2 \vec{\xi} = \mathcal{L}(\vec{\xi}) + \delta\mathcal{L}(\vec{\xi}), \tag{3.148}$$

where $\mathcal{L}(\vec{\xi})$ is the differential operator on the right-hand side of Eq. 3.111, and the perturbation is

$$\delta\mathcal{L}(\vec{\xi}) = -2i\omega\rho_0(\vec{v}_0 \cdot \nabla)\vec{\xi}. \tag{3.149}$$

From the variational principle, we now see that the change in the frequency ω because of rotation can be expressed as

$$\delta\omega = -i\frac{\int_V \rho_0 \vec{\xi}^* \cdot (\vec{v}_0 \cdot \nabla)\vec{\xi}\,dV}{\int_V \rho_0 |\vec{\xi}|^2\,dV}. \tag{3.150}$$

The task of getting the desired expression for $\delta\omega$ reduces to computing $(\vec{v}_0 \cdot \nabla)\vec{\xi}$. For rotation that does not depend on θ, it can be shown that

$$\delta\omega_{nlm} = m\frac{\int_0^R \Omega(r)(\xi_r^2 + L^2\xi_t^2 - 2\xi_r\xi_t - \xi_t^2)r^2\rho\,dr}{\int_0^R (\xi_r^2 + L^2\xi_t^2)r^2\rho\,dr}. \tag{3.151}$$

Thus the frequency perturbation, or as it is usually called, the *frequency splitting*, is proportional to m.

It is conventional to write Eq. 3.151 in terms of a kernel:

$$\delta\omega_{nlm} = m\beta_{nl} \int_0^R \Omega(r) K_{nl}(r) \mathrm{d}r, \tag{3.152}$$

where the kernel $K_{nl}(r)$ is

$$K_{nl}(r) \equiv \frac{(\xi_r^2 + L^2\xi_t^2 - 2\xi_r\xi_t - \xi_t^2)r^2\rho}{\int_0^R (\xi_r^2 + L^2\xi_t^2 - 2\xi_r\xi_t - \xi_t^2)r^2\rho\,\mathrm{d}r}, \tag{3.153}$$

and

$$\beta_{nl} = \frac{\int_0^R (\xi_r^2 + L^2\xi_t^2 - 2\xi_r\xi_t - \xi_t^2)r^2\rho\,\mathrm{d}r}{\int_0^R (\xi_r^2 + L^2\xi_t^2)r^2\rho\,\mathrm{d}r}. \tag{3.154}$$

This definition of $K_{nl}(r)$ ensures that it is unimodular and that in the case of uniform rotation, when $\Omega(r) = \Omega_{\mathrm{u}}$,

$$\delta\omega_{nlm} = m\beta_{nl}\Omega_{\mathrm{u}}, \tag{3.155}$$

that is, the rotational splitting is simply proportional to the rotation and is governed only by β_{nl}.

Note from Eq. 3.151 that the frequency of the $m = 0$ component does not change. This is true only in the case of slow rotation; the situation is much more complicated when rotation is not slow. Fast rotation can distort the star and change the $m = 0$ frequencies too.

Very often the effect of uniform rotation is described in terms of the *Ledoux constant*, defined as

$$C_{nl} = 1 - \beta_{nl} = \frac{\int_0^R (2\xi_r\xi_t + \xi_t^2)r^2\rho\,\mathrm{d}r}{\int_0^R (\xi_r^2 + L^2\xi_t^2)r^2\rho\,\mathrm{d}r}, \tag{3.156}$$

and in the frame rotating with the star, the rotational splitting is simply $\delta\omega_{nlm} = -mC_{nl}\Omega_{\mathrm{u}}$.

An expression similar to Eq. 3.153 can be written when rotation is a function of both radius and latitude:

$$\delta\omega_{nlm} = m \int_0^R \int_0^\pi K_{nlm}(r,\theta)\Omega(r,\theta)\mathrm{d}r\,\mathrm{d}\theta. \tag{3.157}$$

The expression for the kernel is different, of course, and can be shown to be

$$K_{nlm}(r,\theta) = m\frac{R_{nlm}}{I_{nlm}}, \tag{3.158}$$

where

$$R_{nlm} = \left(|\xi_r(r)|^2 P_l^m(x)^2 + |\xi_t(r)|^2 \left[\left(\frac{\mathrm{d}P_l^m(x)}{\mathrm{d}\theta} \right)^2 + \frac{m^2}{1-x^2} P_l^m(x)^2 \right] \right.$$

$$- P_l^m(x)^2 [\xi_r^*(r)\xi_t(r) + \xi_r(r)\xi_t^*(r)]$$

$$\left. - 2 P_l^m(x) \frac{\mathrm{d}P_l^m}{\mathrm{d}\theta} \frac{x}{\sqrt{1-x^2}} |\xi_t(r)|^2 \right) \rho(r) r^2 \sin\theta, \tag{3.159}$$

where P_l^m are the associated Legendre polynomials, $x = \cos\theta$, and

$$I_{nlm} = \frac{2}{2l+1} \frac{(l+|m|)!}{(l-|m|)!} \int_0^R \left[|\xi_r|^2 + l(l+1)|\xi_t|^2 \right] \rho(r) r^2 \mathrm{d}r. \tag{3.160}$$

3.6 MODE EXCITATION

Most of the analysis techniques that are discussed in subsequent chapters depend on the analysis of mode frequencies. However, the oscillation power spectra also allow us to determine mode amplitudes and the widths of resonant peaks in the frequency domain. Interpreting mode amplitudes and widths requires us to know how modes are excited and damped, and that is by no means a resolved issue.

As mentioned in Chapter 1, there are essentially two types of oscillators. The first type are the so-called classical oscillators, such as Cepheids, RR Lyræ stars, δ Scuti and γ Dor stars, as well as β Cepheids and slowly pulsating B-type (SPB) stars. In these stars, modes are forced by a heat engine. Only certain modes—ones that are in resonance with the forcing—are excited in these stars, and the oscillations generally have large amplitudes. Oscillations in most of these stars are excited by the so-called κ-mechanism, where the variation of opacity with temperature in ionization zones plays the key role. Another excitation mechanism is that of *convective blocking*; this mechanism is believed to be responsible for pulsations in γ Dor stars. The second type of oscillators are the solar-like oscillators.

Solar-like oscillators, the subject of this book, are not excited by a heat engine; instead they are excited by the turbulent motion in the outer convection zones of these stars. Damping is a result of both radiative losses and the dissipation of energy by turbulence. The modes are not generally in resonance with the forcing, and as a result, mode amplitudes are small. However, unlike in the heat-engine case, all modes can be excited. Excitation is generally stochastic because of the stochastic nature of turbulence.

The processes of mode excitation and damping are nonadiabatic by definition: modes are receiving energy from the surroundings (excitation) and losing energy to the surroundings (damping). Often mode excitation and damping are treated as perturbations to the adiabatic equations of oscillations, that is, the modes are treated in a quasi-adiabatic manner.

3.6.1 Heat Engine Oscillations

For self-excited modes, such as those seen in classical pulsators, mode excitation and damping can be treated by rewriting the heat equation in the adiabatic case $(dq/dt = 0)$ as

$$\rho \frac{dq}{dt} = (\rho \epsilon - \nabla \cdot \vec{F}), \tag{3.161}$$

where $\rho \epsilon$ is the energy generated per unit mass, and $\vec{F}$ is the flux of energy. A linear perturbation to this equation yields

$$\rho \frac{\partial \delta q}{\partial t} = \delta(\rho \epsilon - \nabla \cdot \vec{F}) = (\rho \epsilon - \nabla \cdot \vec{F})_1, \tag{3.162}$$

where as usual, δq is the Lagrangian perturbation to q, and the subscript 1 refers to the perturbed quantity. In Eq. 3.162, we have used the fact that at equilibrium $\rho \epsilon = \nabla \cdot \vec{F}$. Using Eq. 3.162 now in the perturbed form of Eq. 3.4 and rearranging the terms, we get

$$\frac{P_1}{P} = \Gamma_1 \frac{\rho_1}{\rho} + \xi_r \left(\frac{d \ln P}{dr} - \Gamma_1 \frac{d \ln \rho}{dr} \right) + \frac{i}{\omega} \frac{\Gamma_3 - 1}{P} \left(\rho \epsilon - \nabla \cdot \vec{F} \right)_1$$

$$= \frac{P_{1,ad}}{P} + \frac{i}{\omega} \frac{\Gamma_3 - 1}{P} \left(\rho \epsilon - \nabla \cdot \vec{F} \right)_1, \tag{3.163}$$

where we have assumed the usual time dependence of the form $\exp(-i\omega t)$, and $P_{1,ad}$ is the pressure perturbation in the adiabatic case:

$$P_{1,ad} = P \Gamma_1 \frac{\rho_1}{\rho} + \xi_r \left(\frac{d \ln P}{dr} \right). \tag{3.164}$$

Thus instead of Eq. 3.111, we get

$$-\omega^2 \rho \vec{\xi} = -\nabla P_{1,ad} + \rho \vec{g}_1 + \rho_1 \vec{g} - \frac{i}{\omega} \nabla \left[(\Gamma_3 - 1)(\rho \epsilon - \nabla \cdot \vec{F})_1 \right]. \tag{3.165}$$

Hence we have

$$\omega^2 \xi_r = \mathcal{L}_{ad}(\vec{\xi}) + \delta \mathcal{L}(\vec{\xi}), \tag{3.166}$$

with

$$\mathcal{L}_{ad} = \frac{1}{\rho} \nabla P_{1,ad} - \vec{g}_1 - \frac{\rho_1}{\rho} \vec{g}, \tag{3.167}$$

and thus the nonadiabatic part of the frequency can be expressed as

$$\delta(\omega^2) = \frac{\frac{i}{\omega} \int_V \vec{\xi}^* \cdot \nabla \left[(\Gamma_3 - 1)(\rho \epsilon - \nabla \cdot \vec{F})_1 \right] dV}{\int_v \rho |\vec{\xi}|^2 dV}. \tag{3.168}$$

Whether modes are excited or damped depends on the sign of $\delta\omega$. For mode excitation to occur, it is necessary that $\delta\omega$ has the appropriate sign in the layers where the pulsation period is of the same order as the thermal relaxation time. The complexity in calculating Eq. 3.168 lies in the term $(\rho\epsilon - \nabla \cdot \vec{F})_1$.

3.6.2 Stochastic Oscillations

Solar-like pulsations are driven and damped in the super-adiabatic region that marks the transition between efficient convection in the interior of a star and radiation in the photosphere. Convection is inefficient in the super-adiabatic layers, and energy is transported by a combination of radiation and convection. This inefficiency leads to large convective velocities in this layer. Convective timescales reach a minimum, and the kinetic energy flux reaches a maximum. The complicated nature of the interaction of convection with pulsations leads to uncertainties in calculations of mode excitation and damping.

One way to look at mode excitation and damping is to consider the total mode energy,

$$E_{\mathrm{osc}}(t) = \int_V \rho |\vec{v}_{\mathrm{osc}}|^2 \mathrm{d}V, \tag{3.169}$$

where $\vec{v}_{\mathrm{osc}}$ is the mode velocity at position $\vec{r}$ and time t. If η represents the damping rate, and $\mathcal{E}$ the energy injected into the modes by all sources per unit time, then if $\mathcal{E}$ acts over a timescale that is less than $1/\eta$, the variation of $E_{\mathrm{osc}}(t)$ can be written as

$$\frac{\mathrm{d}E_{\mathrm{osc}}(t)}{\mathrm{d}t} = \mathcal{E} - 2\eta E_{\mathrm{osc}}(t). \tag{3.170}$$

Solar-like oscillations are stable modes; thus a time average of $\mathrm{d}E_{\mathrm{osc}}(t)/\mathrm{d}t$ should be zero:

$$\bar{E}_{\mathrm{osc}} = \frac{\bar{\mathcal{E}}}{2\eta}, \tag{3.171}$$

where $\bar{E}_{\mathrm{osc}}$ and $\bar{\mathcal{E}}$ are respectively the time averaged mode energy and excitation rate. Thus the mode energy is controlled by a balance between the energy given to the mode and the damping of the mode. Usually, mode excitation and damping are treated separately.

Mode excitation is modeled by adding a source term to the perturbed momentum equation (Eq. 3.12), which, assuming the Cowling approximation, becomes

$$\frac{\partial \rho \vec{v}}{\partial t} + \nabla : (\rho \vec{v}\vec{v}) + \nabla P_1 - \rho_1 \vec{g} = 0. \tag{3.172}$$

The gas velocity $\vec{v}$ now has two components: one the velocity of the pulsation $\vec{v}_{\mathrm{osc}}$, and the other the turbulent convective velocity $\vec{v}_t$. Differentiating Eq. 3.172, using the continuity equation while assuming incompressible turbulence ($\nabla \cdot \vec{v}_t = 0$), and neglecting nonlinear terms in $\vec{v}_{\mathrm{osc}}$, leads to the inhomogeneous oscillation equation

$$\rho \left(\frac{\partial^2}{\partial t^2} - \mathcal{L} \right) \vec{v}_{\mathrm{osc}} + \mathcal{D}\vec{v}_{\mathrm{osc}} = \frac{\partial \vec{S}}{\partial t}. \tag{3.173}$$

In the above equation, $\mathcal{L}$ is simply the linear, adiabatic operator discussed in Section 3.3. The source term $\vec{S}$ contains contributions from Reynold's stresses as well as entropy fluctuations. The operator $\mathcal{D}$ involves both the turbulent velocity $\vec{v}_{\mathrm{t}}$ and the oscillation velocity $\vec{v}_{\mathrm{osc}}$, and it contributes to the damping. In the absence of turbulence, one can show that we get back the original pulsation equation

$$\left(\frac{\partial^2}{\partial t^2} - \mathcal{L} \right) \vec{v}_{\mathrm{osc}} = 0, \tag{3.174}$$

and thus $\mathcal{L}(\vec{\xi}) = -\omega^2 \vec{\xi}$. In the presence of turbulence, the displacement eigenvector and the velocity are written in terms of the adiabatic solution and an instantaneous amplitude $A(t)$:

$$\vec{v}_{\mathrm{osc}} = \frac{\mathrm{d}\delta\vec{r}_{\mathrm{osc}}}{\mathrm{d}t} = \frac{1}{2}[-i\omega A(t)\vec{\xi}(\vec{r})\exp(-i\omega t) + cc], \tag{3.175}$$

where cc denotes the complex conjugate of the first term. The instantaneous amplitude $A(t)$ depends on both the driving and the damping forces. The mean mode energy can now be shown to be

$$\bar{E}_{\mathrm{osc}} = \frac{1}{2}|\bar{A}|^2 I \omega^2, \tag{3.176}$$

where the I is the mode inertia. If the period of the oscillation is assumed to be much shorter than its lifetime ($\sim 1/\eta$; η being the damping) such that $\omega \ll \eta$, then $|\mathrm{d}\ln A/\mathrm{d}t| \ll \omega$, and the second derivative of A can be ignored. In that case the equation for the amplitude reduces to

$$\frac{\mathrm{d}A}{\mathrm{d}t} + \eta A = \frac{1}{2\omega^2 I} \int_V \vec{\xi}^* \cdot \frac{\partial \vec{S}}{\partial t} \mathrm{d}V, \tag{3.177}$$

where $\vec{S}$ is the source term, and η is the damping rate. In the general case, η should contain all sources of damping and not just the small damping term that arises in the $\nabla : (\rho\vec{v}\vec{v})$ term in Eq. 3.172. The energy injection rate $\mathcal{E}$ is directly related to $\vec{S}$.

The assumption of adiabaticity has to be abandoned to calculate the mode damping rate. The difference between the heat engine case discussed in Section 3.6.1 and the stochastic oscillations discussed here is that the perturbation to the convective flux needs to be calculated. The formal solution of $A(t)$ is

$$A(t) = \frac{i \exp(-\eta t)}{2\omega I} \int_{-\infty}^{t} \mathrm{d}t' \int_V \mathrm{d}V \exp(\eta + i\omega)t' \vec{\xi}^* \cdot \vec{S}(\vec{r}, t'). \tag{3.178}$$

Given that the system is basically a damped oscillator, it is usual to write the equations simply as that for a damped harmonic oscillator with a forcing term:

$$\frac{\partial^2 \vec{\xi}}{\partial t^2} + 2\eta \frac{\partial \vec{\xi}}{\partial t} + \mathcal{L}\vec{\xi} = \mathcal{F}, \tag{3.179}$$

where $\mathcal{F}$ is the forcing term, and again $\mathcal{L}$ is the linear, adiabatic operator discussed in Section 3.3. The displacement eigenvector $\vec{\xi}$ is expressed as the oscillatory part of the time dependence and is considered to be the solution of the adiabatic wave equation: $\mathcal{L}\vec{\xi} = -\omega^2 \xi$. Equation 3.179 is used to explore the damping and forcing terms. One should note that we could also convert Eq. 3.179 to an equation for the amplitude, but it translates more easily into the amplitude of displacement rather than the amplitude of the change in velocity that we have in Eq. 3.177.

The biggest obstacle to calculating the excitation and damping rates of solar-like pulsators is the lack of a good description of convection. The mixing-length approach, or equivalent approximations, does not model the crucial super-adiabatic zone well. Nor does the approximation take into account the asymmetry between the upflows and downflows in a real convection zone or the pressure support provided by turbulence. Nonlocal formalisms, even those that include the anisotropy of flows, have been used, but all such formalisms have at least one (and sometimes more) free parameters and thus lack predictive power. As a result, it is becoming increasingly common to use simulations of stellar convection to perform calculations of mode excitation and damping. These simulations can be used to measure the required quantities directly. This has been applied successfully to the Sun and a few other stars. Given the difficulty in modeling convection analytically, we expect that using simulations of convection will become a common way to study mode excitation in solar-like oscillators.

The uncertainties in calculating mode excitation and damping have resulted in asteroseismic techniques being developed to exploit only oscillation frequencies. Mode widths and amplitudes are generally ignored at present when using asteroseismic data to infer stellar properties.

3.7 FURTHER READING

Details of the theory of stellar oscillations can be found in the following two excellent books:

- Cox, J. P., 1980, *Theory of Stellar Pulsation*, Princeton University Press.
- Unno, W., Osaki, Y., Ando, H., Saio, H. and Shibahashi, H., 1989, *Nonradial Oscillations of Stars*, University of Tokyo Press.

Other useful sources are:

- Aerts, C., Christensen-Dalsgaard, J., and Kurtz, D. W., 2011, *Asteroseismology*, Springer.
- Gough, D. O., 1993, *Linear Adiabatic Stellar Pulsation*, in *Astrophysical Fluid Dynamics—Les Houches 1987*, eds. Zahn, J.-P., and Zinn-Justin, J.

Early discussions of mixed modes and avoided crossings can be found in:

- Scuflaire, R., 1974, *The Non Radial Oscillations of Condensed Polytropes*, *Astron. Astrophys.*, **36**, 107.
- Osaki, Y., 1975, *Nonradial Oscillations of a 10 Solar Mass Star in the Main-Sequence Stage*, *Publ. Astron. Soc. Japan*, **27**, 237.
- Aizenman, M., Semeyers, P., and Weigert, A., 1977, *Avoided Crossing of Modes of Non-radial Stellar Oscillations*, *Astron. Astrophys.*, **58**, 41.

The Hermitian nature of Eq. 3.111 has been derived in:

- Chandrasekhar, S., 1964, *A General Variational Principle Governing the Radial and the Non-radial Oscillations of Gaseous Masses*, *Astrophys. J.*, **139**, 664.

Different versions of asteroseismic HR diagrams can be found in:

- Christensen-Dalsgaard, J., 1988, *A Hertzsprung-Russell Diagram for Stellar Oscillations*, in *Proc. IAU Symp.* 123, *Advances in Helio- and Asteroseismology*, eds. Christensen-Dalsgaard, J., and Frandsen, S.
- Mazumdar, A., 2005, *Asteroseismic Diagrams for Solar-Type Stars*, *Astron. Astrophys.*, **441**, 1079.
- White, T. R. *et al.*, 2011, *Calculating Asteroseismic Diagrams for Solar-Like Oscillations*, *Astrophys. J.*, **743**, 161.

The analysis of acoustic glitches and glitches in the Brunt-Väisälä frequency of red giants are discussed in the following papers:

- Gough, D. O. 1990, in *Comments on Helioseismic Inference*, *Lecture Notes in Physics*, **367**, 283, Springer-Verlag.
- Gough, D. O., and Sekii, T., 1993, *On the Detection of Convective Overshoot*, in *Astron. Soc. Pac. Conf. Ser.*, 42, 177, ed. Brown, T. M., Astronomical Society of the Pacific.
- Roxburgh, I. W, and Vorontsov, S. V., 1994, *Seismology of the Solar Envelope— The Base of the Convective Zone as Seen in the Phase Shift of Acoustic Waves*, *Mon. Not. R. Astron. Soc.*, **268**, 880.
- Miglio, A., Montalbán, Noels, A., and Eggenberger, P., 2008, *Probing the Properties of Convective Cores through g Modes: High-Order g Modes in SPB and γ Doradus Stars*, *Mon. Not. R. Astron. Soc.*, **386**, 1487.
- Cunha, M. S., Stello, D., Avelino, P. P., Christensen-Dalsgaard, J., and Townsend, R.H.D., 2015, *Structural Glitches Near the Cores of Red Giants Revealed by Oscillations in g-mode Period Spacings from Stellar Models*, *Astrophys. J.*, **805**, 127.
- Mosser, M., Vrard, M., Belkacem, K., Deheuvels, S., and Goupil, M. J., 2015, *Period Spacings in Red Giants. I. Disentangling Rotation and Revealing Core Structure Discontinuities*, *Astron. Astrophys.*, **584**, 50.

We have used the variational principle to derive rotational splittings under very specific conditions. The derivation of a general variational principle for rotating bodies can be found in:

- Lynden-Bell, D., and Ostriker, J. P., 1967, *On the Stability of Differentially Rotating Bodies*, *Mon. Not. R. Astron. Soc.*, **136**, 293.

For a description of how differential rotation may be measured with asteroseismology, see:

- Gizon, L., and Solanki, S. K., 2004, *Measuring Stellar Differential Rotation with Asteroseismology*, Solar Phys., **220**, 169.

For a taste of what fast rotation can do, see the following article and references in it:

- Reese, D. R., K. B. MacGregor, S. Jackson, A. Skumanich, and Metcalfe, T. S., 2009, *Pulsation Modes in Rapidly Rotating Stellar Models Based on the Self-Consistent Field Method*, Astron. Astrophys., **506**, 189.

The study of mode excitation and damping is an active field of research, and what we have done here is just a taste of how the problem is approached. The books by Unno et al., and Cox mentioned at the start of this reading list do describe excitation and damping; however, those interested in studying the field more thoroughly could start with two excellent reviews:

- Gautschy, A., and Saio, H., 1995, *Stellar Pulsations Across the HR Diagram: Part 1*, Ann. Rev. Astron. Astrophys., **33**, 75.
- Gautschy, A. and Saio, H., 1996, *Stellar Pulsations Across the HR Diagram: Part 2*, Ann. Rev. Astron. Astrophys., **34**, 551.

To study solar-like oscillations, one could start with the seminal paper:

- Goldreich, P., and Keeley, D., *Solar Seismology. II—The Stochastic Excitation of the Solar p-Modes by Turbulent Convection*, 1977, Astrophys. J., **212**, 243. Perhaps a gentler introduction can be found in the following:
- Houdek, G., and Dupret, M.-A., 2015, *Interaction Between Convection and Pulsation*, Living Rev. Solar Phys., **12**, 2.
- Samadi, R., 2011, *Stochastic Excitation of Acoustic Modes in Stars*, Lecture Notes in Physics, **832**, 305, Springer-Verlag.
- Belkacem, K., and Samadi, R., 2013, *Connections Between Stellar Oscillations and Turbulent Convection*, in *Lecture Notes in Physics*, **865**, 179, Springer-Verlag.
- Stein, R. F., et al. 2004, *Excitation of Radial P-Modes in the Sun and Stars*, Solar Phys., **220**, 229.
- Houdek, G., et al. 1999, *Amplitudes of Stochastically Excited Oscillations in Main-Sequence Stars*, Astron. Astrophys., **351**, 582.

3.8 EXERCISES

1. Files BV1.txt, BV2.txt, and BV3.txt (available via the Princeton University Press Web site mentioned in the Preface) list the Brunt-Väisälä frequency and Lamb frequency for $l = 1$ modes for three stars.

   (a) In each case plot the Lamb frequency for the $l = 2$ and $l = 3$ modes. Also plot the Lamb frequency for the $l = 5$ and $l = 20$ modes.

   (b) Assume that you have a 500 μHz mode. Can the mode exist as a mixed mode in any of the three stars? Which ones?

2. File `SolarFreq_MDI.txt` lists solar oscillation frequencies observed by the Michelson Doppler Imager on board the Solar and Heliospheric Observatory (SoHO). Use the solar p mode frequencies (i.e., modes with $n > 0$) to show that the Duvall Law holds by plotting $(n + \alpha)/\nu$ as a function of ν/L. What value of α do you need to use to collapse the points onto one curve? Note that while the simple asymptotic relation we derived gave $L^2 = l(l + 1)$, a second-order asymptotic derivation shows that it is more correct to use $L = (l + 1/2)$ in the asymptotic expression.

3. File `solarmodel_gs98_diffusion.txt` lists the internal properties of a solar model. Assume that the sound-speed profile of this model is the same as that of the Sun (a reasonable assumption, since the sound speed of the model matches that of the Sun to within fractions of a percent).

   (a) Use the sound-speed profile to calculate the lower turning points of the solar p modes used in Exercise 2.
   (b) What is the maximum depth that you can probe if you only have modes with $l = 20$ to $l = 30$?
   (c) Calculate l for modes that have turning points between 0.9 and $0.95 R_\odot$.

4. File `Kernels_gs98_c2_rho.txt` lists the $l = 0$, 1, and 2 sound-speed kernels for the solar model with internal properties in file `solarmodel_gs98_diffusion.txt`. The corresponding density kernels are in file `Kernels_gs98_rho_c2.txt` The mode frequencies are in file `freqs_gs98_diffusion.txt`.

   (a) Calculate the frequency response to a sound-speed difference $\delta c_s^2/c_s^2$ of the form

   $$A \exp\left[-\left(\frac{B-r}{C}\right)^2\right],$$

   with $A = 0.0001$, $B = 0.71$, and $C = 0.005$. Assume that the density difference $\delta\rho/\rho = 0$. Note that r is the fractional radius.
   (b) Calculate the frequency response to a density difference of the form

   $$A \exp\left[-\left(\frac{B-r}{C}\right)^2\right]$$

   with A, B, and C as given in Exercise 4(a), and the sound-speed difference $\delta c_s^2/c_s^2 = 0$.
   (c) Examine the consequences of shifting the peak of the Gaussian-shaped perturbations to different radii.

5. Files `rotker.m1.20_2.78Gyr.txt` and `rotker.m1.5_2.5Gyr.txt` contain the rotation kernels $\beta K(r)$ for a $1.20 M_\odot$ main sequence model and a $1.5 M_\odot$ evolved model, respectively. For each model calculate rotational splittings for the following rotation rates:

   (a) $\Omega = 400$ nHz for $0 \le r/R \le 1$.
   (b) $\Omega = 450$ nHz for $0 \le r/R \le 0.3$, and $\Omega = 350$ nHz for $0.3 < r/R \le 1$.
   (c) $\Omega = 350$ nHz for $0 \le r/R \le 0.7$, and $\Omega = 450$ nHz for $0.7 < r/R \le 1$.

4 Observational Data: Overview and Fundamentals

The basic datasets from which the asteroseismic and other intrinsic stellar parameters are extracted are usually lightcurves of photometric observations or time series of Doppler velocity observations. In this chapter we consider some of the fundamentals associated with these data, in particular how the observational technique affects the amplitudes of the observed oscillations. We also introduce the other intrinsic stellar signals that manifest in the data, specifically those due to granulation (signatures of near-surface convection) and magnetic activity. Our aim is to familiarize the reader with the basic content of the typical data. In addition, we lay some important groundwork for the detailed presentations that follow in Chapter 5, which covers the appearance and characteristics of the data in the frequency domain (where the extraction of the asteroseismic parameters is usually performed), and in Chapter 6, which covers frequency-domain analysis in detail.

Let us start by considering how the observational technique affects the measured perturbations caused by the modes, beginning with how the spatial response of the observations affects the visibility of modes of different angular degrees l.

4.1 SPATIAL RESPONSE OF THE OBSERVATIONS

When we observe a star's intensity or Doppler velocity, perturbations due to oscillations are integrated over the visible disc of the star. For solar-like oscillations, which give rise to small-amplitude variations about the equilibrium state, the perturbations may be described in terms of spherical harmonic functions. The integration inherent in giving the disc-averaged signal applies a strong spatial filter, so that the observations are sensitive only to modes of low angular degree l. The amplitudes due to modes of medium or high l average to undetectably small values, due to net cancellation of the signal across the visible disc of the star. In spite of this limitation, nature has been kind: it is the low-l modes that provide information on the deep layers inside stars and hence their evolutionary states.

The weights that perturbations on different parts of the visible disc receive during the integration depend not only on the spherical harmonic component (i.e., the mode we are considering) but also on the method of observation, and in turn (though usually to a lesser extent) on the intrinsic stellar properties. For example, observations in Doppler velocity of rapidly rotating stars can provide better sensitivity to modes of higher l than would be the case for slowly rotating stars (so-called *Doppler imaging*, a point we return to later in this section).

We begin by considering perturbations in intensity. Let the position- and time-dependent intensity perturbation due to the oscillations, written in co-latitude θ and longitude ϕ, be $I(t; \theta, \phi)$. The integrated signal $I(t)$ is given by integration over the area a of the visible disc:

$$I(t) = \int_a I(t; \theta, \phi) W(\cos \theta) \mathrm{d}a, \tag{4.1}$$

where $W(\cos \theta)$ is a weighting factor that is assumed to depend only on the angular distance from the rotation axis of the star. For photometric observations the dominant contribution describes the change in intensity due to limb darkening, $g(\cos \theta)$, that is, we have $W(\cos \theta) \equiv g(\cos \theta)$. Examples of commonly adopted limb-darkening laws with coefficients c_k are a simple power law

$$g(\cos \theta) = I(\theta)/I(0) = 1 - \sum_{k=1}^{N} c_k \left[1 - \cos \theta\right]^k, \tag{4.2}$$

which is usually expanded up to $N = 2$ (quadratic), or a nonlinear law of the form

$$g(\cos \theta) = I(\theta)/I(0) = 1 - \sum_{k=1}^{N} c_k \left[1 - (\cos \theta)^{k/2}\right]. \tag{4.3}$$

Coefficients may be calculated from stellar atmosphere models for different stellar effective temperatures and surface gravities.

Returning to Eq. 4.1, for $\mathrm{d}a = \mathrm{d}\phi \cos \theta \sin \theta \mathrm{d}\theta$, we have

$$I(t) = \int_0^{2\pi} \mathrm{d}\phi \int_0^{\pi/2} \mathrm{d}\theta \, I(t; \theta, \phi) W(\cos \theta) \cos \theta \sin \theta. \tag{4.4}$$

The contribution of each mode to the observed disc-averaged intensity depends on the relevant spatial properties. These properties are described by spherical harmonics

$$Y_l^m(\theta, \phi) = (-1)^m c_{lm} P_l^m(\cos \theta) \exp(im\phi), \tag{4.5}$$

with $P_l^m(\cos \theta)$ being the so-called *associated* Legendre polynomials, with

$$c_{lm}^2 = \frac{(2l + 1)(l - m)!}{4\pi (l + m)!}. \tag{4.6}$$

We assume that the intensity perturbation of a mode of a given l, m, and radial order n, which has an amplitude A_{nlm} and angular frequency ω_{nlm}, may be written in terms of the real part of the eigenfunction at the surface:

$$I_{nlm}(t; \theta, \phi) = \mathrm{Re}\{\sqrt{4\pi}\, Y_l^m(\theta, \phi) A_{nlm} \exp{(-i\omega_{nlm}t)}\}. \tag{4.7}$$

We take the real part of the complex functions on the right-hand side to get the observed scalar quantity. Substituting for $Y_l^m(\theta, \phi)$ gives

$$I_{nlm}(t; \theta, \phi) = \sqrt{4\pi}(-1)^m c_{lm} P_l^m(\cos\theta) A_{nlm} \cos(m\phi - \omega_{nlm}t). \tag{4.8}$$

To estimate the relative contribution of each mode, we consider the case where the rotation axis of the star points along the line-of-sight direction (i.e., a stellar angle of inclination i_s of zero degrees). Under these circumstances only components with $m = 0$ contribute to the disc-averaged signal; net cancellation of all $m \neq 0$ results. We shall address the sensitivity in m a little later in this section.

The associated Legendre polynomials $P_l^0(\cos\theta)$ then correspond to the simpler Legendre polynomials $P_l(\cos\theta)$. We may then write the disc-integrated, signal in the form

$$I_{nl}(t) = S_l A_{nl} \cos(\omega_{nl}t), \tag{4.9}$$

where we have grouped several terms together to give the *visibility function*:

$$S_l = 2\pi\sqrt{2l+1} \int\limits_0^{\pi/2} P_l(\cos\theta) W(\cos\theta) \cos\theta \sin\theta \, \mathrm{d}\theta. \tag{4.10}$$

The visibility is assumed to be independent of n, which is a good assumption for the high-order modes typically observed in solar-like oscillators. This assumption becomes problematic at low radial orders, where it can no longer be assumed that the perturbations due to the modes are dominated by radial motion; the horizontal component becomes progressively more important, affecting the mode visibilities. In the Cowling approximation—where perturbations of the gravitational potential by the modes are neglected—the ratio of the amplitudes of the horizontal and radial motions (δ_h/δ_r) depends only on frequency and may be written as

$$\delta_h/\delta_r = \sqrt{l(l+1)}\left(\frac{GM}{\omega^2 R^3}\right)$$

$$\equiv \sqrt{l(l+1)}\left(\frac{G\bar\rho}{3\pi \nu^2}\right). \tag{4.11}$$

Here, $M, R,$ and $\bar\rho$ are the stellar mass, radius, and mean density, respectively, and the mode frequency is given in angular (ω) and temporal (ν) units. For modes typically observed in cool main sequence and sub giant stars, this ratio lies significantly below unity. Only for evolved red giants may the ratio reach levels of $\simeq 10\%$ in some of the detectable modes.

Figure 4.1 shows values of the estimated visibilities for photometric observations, where we follow usual practice in showing sensitivities relative to the radial

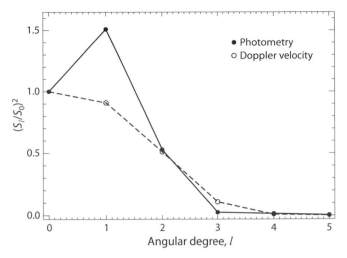

Figure 4.1. Estimated squared, normalized sensitivities $(S_l/S_0)^2$ for photometric and Doppler velocity observations with a quadratic limb-darkening law having coefficients $c_1 = 0.44$ and $c_2 = 0.23$ (see Eq. 4.2).

mode sensitivity. Moreover, we plot the ratio squared, $(S_l/S_0)^2$, since the analysis of the data is usually performed in the frequency domain on a power (as opposed to amplitude) spectrum. We applied Eq. 4.2 to describe the limb-darkening, using the quadratic form with coefficients appropriate to a Sun-like star (specifically, $c_1 = 0.44$ and $c_2 = 0.23$). These calculations of course neglect any consideration of nonadiabatic effects, and it is therefore not surprising that ratios in photometric data do in practice show some (albeit modest) departures from these simple predictions.

What about the sensitivities in Doppler velocity? Here, the main modification to the formulas above is the introduction of a factor of $\cos\theta$ to account for the projection of the velocity perturbation onto the line of sight. The modified form for the visibility is then

$$S_l = 2\pi\sqrt{2l+1}\int\limits_0^{\pi/2} P_l(\cos\theta)W(\cos\theta)\cos^2\theta\sin\theta\,d\theta. \tag{4.12}$$

Figure 4.1 also plots estimated values of $(S_l/S_0)^2$ for the Doppler velocity observations, using the same limb-darkening law as for the photometric predictions. One can see that the Doppler observations are more sensitive to modes of higher l than are the photometric observations. This is a result of the extra factor of $\cos\theta$, which acts to reduce the area of the visible disc over which the perturbations are effectively sampled. This result can be thought of in terms of an uncertainty principle: a reduction in area (on the disc) allows one to sample a larger range of values in spatial frequency (i.e., in angular degree l).

Doppler imaging can further accentuate this effect. If the width (in velocity units) of the instrumental response and the spectral line being used are less than the surface rotational velocity, then the rotation will significantly affect the sensitivity across the disc. The tendency is again to reduce the effective area on the surface over which the observations are sensitive to the modes (because rotation acts to broaden the observed line profile). The impact of Doppler imaging is usually not a major issue for observations of solar-like oscillators, where in the vast majority of cases the rotation rates are fairly modest.

Finally in this section we turn to the issue of the dependence of the visibilities on the azimuthal order m. To proceed, we make two important assumptions. First, we assume that contributions to the observed mode signal across the visible stellar disc depend only on the angular distance from the disc center (i.e., on $\cos\theta$). This is a good assumption for photometric observations, where limb darkening controls the weighting, but it is potentially more problematic for Doppler velocity observations. Second, we assume that energy is equipartitioned among the different m components. As mentioned in Chapters 1 and 3, solar-like oscillations are stochastically excited and intrinsically damped, and so they have finite lifetimes (as we shall see later, beginning in Section 5.3). In the limit of the observations spanning a large number of lifetimes, the observed amplitudes will approach their the underlying values. This assumption will break down in very rapid rotators, because the rotation can then affect convection, and it is convection in turn that excites and damps the modes. In spite of this caveat, the effects are still rather modest, even for rotation periods as short as a few days (departures from energy equipartition are then at the $\simeq 1$–2% level).

Invoking the above assumptions, we may then write the amplitudes A_{nlm} as

$$A_{nlm} = |A_{nl}| \exp(i\delta_{nlm}), \tag{4.13}$$

where δ_{nlm} is a phase factor. Using this and Eq. 4.7, it is possible to derive the dependence of the visibilities on l, m and the angle of inclination of the star, i_s. The visibilities, in power, are found to be

$$\mathcal{E}_{lm}(i_s) = \frac{(l-|m|)!}{(l+|m|)!} \left[P_l^{|m|}(\cos i_s) \right]^2, \tag{4.14}$$

where $P_l^{|m|}$ are again associated Legendre polynomials, and the visibilities are normalized so that

$$\sum_{m=-l}^{m=l} \mathcal{E}_{lm}(i_s) = 1. \tag{4.15}$$

Expanding Eq. 4.14, the visibilities of the $l=1$ components are

$$\mathcal{E}_{10}(i_s) = \cos^2 i_s, \tag{4.16}$$

and

$$\mathcal{E}_{1\pm1}(i_s) = \frac{1}{2} \sin^2 i_s, \tag{4.17}$$

while those of the $l=2$ components are

$$\mathcal{E}_{20}(i_s) = \frac{1}{4} \left(3\cos^2 i_s - 1\right)^2, \tag{4.18}$$

$$\mathcal{E}_{2\pm1}(i_s) = \frac{3}{8} \sin^2(2i_s), \tag{4.19}$$

and

$$\mathcal{E}_{2\pm2}(i_s) = \frac{3}{8} \sin^4 i_s. \tag{4.20}$$

Measuring the relative power of the azimuthal components of different $|m|$ in a nonradial multiplet therefore provides a direct estimate of the stellar angle of inclination i_s, or more properly $|i_s|$, since symmetries inherent in Eq. 4.14 mean we cannot discriminate between i_s and $-i_s$, and $\pi - i_s$ and $\pi + i_s$. In Chapter 6 we discuss the practical challenges of making such a measurement.

4.2 PHOTOMETRIC AND DOPPLER VELOCITY OBSERVATIONS

Next let us consider how the observational technique affects the detected amplitudes of the oscillation signal, starting with observations made in intensity.

4.2.1 Photometric Observations

The following discussion is pertinent to radial ($l = 0$) modes, for which induced perturbations are uniform over the surface of the star. Our discussion is predicated on the assumption that variations in temperature dominate the observed photometric changes in flux arising from oscillations. The response of the obser-vations to intrinsic fluctuations in temperature is then dependent on the wavelength response of the observations. As we shall see below, the effects are more pronounced in narrow-band than in wide-band wavelength data.

We use the Planck function to describe the stellar spectrum as a function of wavelength λ:

$$B(\lambda, T) = \frac{2hc^2}{\lambda^5} \left[\exp\left(\frac{hc}{\lambda k_B T}\right) - 1 \right]^{-1}, \tag{4.21}$$

where k_B is the Boltzmann constant, and h is Planck's constant, c is the speed of light, and the temperature T may be regarded as being the effective temperature T_{eff}. The relative flux variation for narrow-band observations made at a wavelength λ is then given by

$$\left(\frac{\delta F_\lambda}{F_\lambda}\right) = \delta B(\lambda, T_{eff})/B(\lambda, T_{eff}) \simeq \delta \ln B(\lambda, T_{eff}). \tag{4.22}$$

Since

$$\ln B(\lambda, T_{eff}) \simeq \ln(2hc^2) - 5\ln\lambda - \left(\frac{hc}{\lambda k_B T_{eff}}\right), \tag{4.23}$$

we may write

$$\delta \ln B(\lambda, T_{eff}) \simeq \left(\frac{hc}{\lambda k_B}\right)\left(\frac{\delta T}{T_{eff}^2}\right). \tag{4.24}$$

For a given fluctuation in temperature, it then follows that

$$\left(\frac{\delta F_\lambda}{F_\lambda}\right) \propto \lambda^{-1}. \tag{4.25}$$

The simple prediction is, therefore, a linear dependence of the observed oscillation amplitude on the wavelength at which narrow-band photometric observations are made. Shorter-wavelength observations give higher measured amplitudes. It is remarkable that narrow-band observations of solar oscillations, made at three different wavelengths (blue, green, and red) by the SPM/VIRGO package on the ESA/NASA SoHO spacecraft, show ratios in amplitude that are not far from the above simple linear dependence.

The wavelength dependence of the observations must clearly be taken into account when interpreting results from real data, in particular when data are available from several instruments with different response functions. Converting these wavelength-dependent amplitudes to the expected bolometric amplitude provides a suitable reference for comparison. The required conversion is straightforward, using $F_{\text{bol}} \propto T_{\text{eff}}^4$. From Eq. 4.24, we then have

$$\left(\frac{\delta F_{\text{bol}}}{F_{\text{bol}}}\right) = 4\left(\frac{\delta T}{T_{\text{eff}}}\right) \equiv 4\lambda T_{\text{eff}}\left(\frac{k_B}{hc}\right)\left(\frac{\delta F_\lambda}{F_\lambda}\right). \tag{4.26}$$

We may write the above in a slightly more convenient form, with the numerical value of the ratio $(4k_B/hc)$ absorbed in terms given in the solar effective temperature $T_{\text{eff},\odot} = 5{,}777$ K and the wavelength of the observations:

$$\left(\frac{\delta F_{\text{bol}}}{F_{\text{bol}}}\right) \simeq \left(\frac{\lambda}{623\,\text{nm}}\right)\left(\frac{T_{\text{eff}}}{5{,}777\,\text{K}}\right)\left(\frac{\delta F_\lambda}{F_\lambda}\right). \tag{4.27}$$

In contrast, photometric observations made by *Kepler* and CoRoT, and those planned for TESS and PLATO,, have extended wavelength responses. Figure 4.2 shows the spectral responses $\epsilon(\lambda)$ (i.e., efficiencies as a function of wavelength) of *Kepler*, CoRoT and TESS. We must now convolve the intrinsic fluctuations with the relevant spectral response to calculate the expected amplitudes. Provided the observed fluctuations are sufficiently small so that they may be treated as linear perturbations, we may write the expected change in measured flux $(\delta F/F)$ for a given instrumental response, $\epsilon(\lambda)$, as

$$\left(\frac{\delta F}{F}\right) = \frac{\int_\lambda \mathcal{T}'(\lambda)\partial[B(\lambda, T_{\text{eff}})/\partial T_{\text{eff}}]\delta T\,d\lambda}{\int_\lambda \mathcal{T}'(\lambda)B(\lambda, T_{\text{eff}})\,d\lambda}, \tag{4.28}$$

where

$$\mathcal{T}'(\lambda) = \epsilon(\lambda)\left(\frac{hc}{\lambda}\right)^{-1}. \tag{4.29}$$

Although *Kepler* and CoRoT have quite similar wavelength responses, the response of CoRoT extends slightly further into the red. As a result, the amplitudes of intrinsic variability for Sun-like stars are about 5% lower for CoRoT than for *Kepler*. TESS has

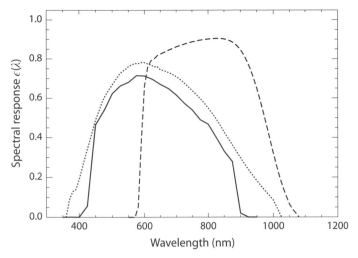

Figure 4.2. Spectral wavelength responses, that is, efficiencies $\epsilon(\lambda)$, of *Kepler* (solid line), CoRoT (dotted line), and TESS (dashed line).

an even redder response, and so TESS amplitudes are about 15% lower than those observed by *Kepler*.

We may also convert the extended flux amplitudes ($\delta F/F$) to the expected bolometric amplitudes using

$$\left(\frac{\delta F_{\mathrm{bol}}}{F_{\mathrm{bol}}}\right) = \mathcal{C}_{\mathrm{bol}}\left(\frac{\delta F}{F}\right), \tag{4.30}$$

where

$$\mathcal{C}_{\mathrm{bol}} = \left(\frac{4}{T_{\mathrm{eff}}}\right)\left(\frac{\int_\lambda \mathcal{T}'(\lambda)B(\lambda, T_{\mathrm{eff}})\mathrm{d}\lambda}{\int_\lambda \mathcal{T}'(\lambda)\partial[B(\lambda, T_{\mathrm{eff}})/\partial T_{\mathrm{eff}}]\mathrm{d}\lambda}\right). \tag{4.31}$$

Finally, we may also fold the instrumental wavelength response into the weighting functions $W(\cos\theta)$ for photometric observations that we introduced in Section 4.1 (see Eq. 4.1). The weighting factor is given by

$$W(\cos\theta) = \frac{\int_\lambda \mathcal{T}'(\lambda)T_{\mathrm{eff}}\partial[B(\lambda, T_{\mathrm{eff}})/\partial T_{\mathrm{eff}}]H_\lambda g_\lambda(\cos\theta)\mathrm{d}\lambda}{\int_\lambda \mathcal{T}'(\lambda)B(\lambda, T_{\mathrm{eff}})H_\lambda G_\lambda \mathrm{d}\lambda}, \tag{4.32}$$

where

$$G_\lambda = \int_0^1 g_\lambda(\cos\theta)\cos\theta\, \mathrm{d}(\cos\theta), \tag{4.33}$$

and

$$H_\lambda = \left(\int_0^1 g_\lambda(\cos\theta)\,\mathrm{d}(\cos\theta)\right)^{-1}, \tag{4.34}$$

and $g_\lambda(\cos\theta)$ is a suitable limb-darkening law (e.g., like those introduced in Equations 4.2 and 4.3).

4.2.2 Doppler Velocity Observations

Unpicking the impact of the detailed nature of the observations on the measured oscillation amplitudes is rather more complicated where Doppler velocity data are concerned. However, the fundamental principle underlying those changes is conceptually straightforward: different parts of a given spectral line, and different lines, are formed across a range of different heights in stellar atmospheres. This is in contrast to photometric observations, which at visible and infrared wavelengths show very little height dependence (for both narrow-band observations in the continuum or wide-band observations).

Let us assume that the energy density of the oscillations is conserved at different heights in the stellar atmosphere, implying that there are no significant losses. (Note that it is usually the case that the p modes are evanescent at the photospheric level and above.) The velocity amplitude of the p modes will have an exponential dependence with height z through the atmosphere of the form

$$v(z) = v(z_0) \exp\left(\frac{z - z_0}{H_v}\right), \tag{4.35}$$

where $v(z_0)$ is the velocity amplitude at some fiducial reference height z_0, and H_v is the velocity amplitude scale height

$$\frac{1}{H_v} = \left[\frac{1}{2}\left(\frac{1}{H_\rho} - \frac{1}{H_E}\right)\right], \tag{4.36}$$

with H_ρ the atmospheric density scale height and H_E the energy density scale height. Radiative losses or other forms of dissipation will decrease the size of H_E, making H_v longer.

In the isothermal atmosphere approximation, the density scale height H_ρ and the pressure scale height H_P are equivalent:

$$H_\rho \equiv H_P = \left(\frac{\mathcal{R}T}{\mu g}\right), \tag{4.37}$$

where $\mathcal{R}$ is the gas constant, and μ is the mean molecular weight.

Sun-as-a-star observations of low-l oscillations imply that $H_v \simeq 600\,\mathrm{km}$ in the solar atmosphere. If two different solar observations use different spectral lines—or different parts of the same line—having an effective height of formation difference of 100 km, we would then expect the observed signal amplitudes to differ by approximately 15 %.

The devil here is in the details, specifically, in the estimation of the effective height of the observational response. Spectral lines are formed over—and therefore map to—a range of heights in the atmosphere. The impact of this mapping on the measured oscillation amplitudes depends not only on the lines that are used, but also on the way the spectral information is used. For example, one may use the entire line—or more commonly a set of lines—by cross-correlating observed line profiles with standard reference or synthetic spectra. One is then in effect averaging, in some sense, the impact of the height dependence both across each line and between different lines. At the other extreme, Doppler velocity observations can be made by measuring the relative intensities in narrow passbands in the blue and red wavelength wings of a single spectral line. Any sensitivity calculation should formally convolve the resulting effective height response with the exponential variation of the oscillation amplitudes described by Eq. 4.35.

Heights of formation can vary quite significantly across spectral lines, usually by a few hundred kilometers. Some lines may even have a significant response at chromospheric heights (e.g., the sodium D1 and D2 lines at 589.6 nm and 589 nm). Lines due to potassium (at 770 nm), nickel (676 nm), and sodium (see above) have all been used extensively in observations of solar oscillations (by, respectively, BiSON, MDI/SoHO, and GONG, and GOLF/SoHO). The median heights of formation of these lines differ by as much as 200 km or more. The lines most commonly used by spectrographs making observations of stars are those due to neutral transitions of iron. The median formation heights are slightly lower in the atmosphere than those of potassium and sodium, meaning, for example, that stellar spectrographs measure slightly lower "Sun-as-a-star" oscillation amplitudes than do the dedicated BiSON or GOLF Sun-as-a-star instruments.

4.2.3 Comparison of Mode Amplitudes in Photometry and Doppler Velocity

Early empirical results on p-mode oscillations, including data on classical pulsators, suggested a scaling of the observed mode amplitudes in Doppler velocity of the form

$$v \propto (L/M), \tag{4.38}$$

where L is the luminosity, and M the mass of the star. More recent results, including input from theoretical calculations, have suggested a scaling of the form $(L/M)^s$, with s in the range of $\simeq 0.7$–0.8. Calibrating such a relation against the measured solar amplitude in velocity therefore provides a simple way to make an approximate Doppler velocity amplitude prediction for other stars:

$$v = v_\odot \left(\frac{L/L_\odot}{M/M_\odot} \right)^s \tag{4.39}$$

We may also express the ratio L/M as a ratio of atmospheric or near-surface properties (i.e., in proximity to where p modes are excited and damped), allowing us to write

$$v = v_\odot \left(\frac{T_{\mathrm{eff}}^4 / T_{\mathrm{eff},\odot}^4}{g/g_\odot} \right)^s \tag{4.40}$$

The measured amplitudes of the strongest radial solar p modes depend on the instrument being used to make the observations (see Section 4.2.2). For example, BiSON observations give a maximum radial p-mode amplitude of $v_\odot \simeq 20\,\mathrm{cm\,s^{-1}}$ ($\simeq 14\,\mathrm{cm\,s^{-1}}$ RMS); while for spectrographs designed to make observations of stars, the amplitude is slightly lower: $v_\odot \simeq 18.5\,\mathrm{cm\,s^{-1}}$ ($\simeq 13\,\mathrm{cm\,s^{-1}}$ RMS).

Given an observed or predicted amplitude in Doppler velocity, can we predict the expected amplitude for photometric observations (or vice versa)? Detailed computations of the expected mode amplitudes depend on a complicated nonadiabatic treatment of the interactions of the oscillations with the near-surface convection. We may nevertheless use some basic physics and simplifying approximations to make estimates for p modes that turn out to be quite reasonable when compared to real observations. These simple calculations are very useful, because they can be used when planning observations and selecting targets for asteroseismic studies. In Section 5.3 we shall introduce further expressions relating to the excitation and damping of solar-like oscillations.

We begin here with the simplifying approximation that the oscillations are adiabatic throughout the star. Under this approximation, assuming an ideal gas, we can approximate the structure to be a polytrope,

$$P \propto \rho^\gamma, \tag{4.41}$$

where $\gamma = \Gamma_1$ and has been defined in Eq. 3.5. The ideal gas equation of state gives us

$$P \propto \rho T, \tag{4.42}$$

where T is the temperature of the gas. Combining these two equations implies that

$$T \propto \rho^{\gamma-1}, \tag{4.43}$$

and it therefore follows that

$$\frac{\delta T}{T} \propto (\gamma - 1) \frac{\delta \rho}{\rho}. \tag{4.44}$$

For p-mode oscillations, it is compression of the gas that dominates. A change in density from compression of the gas may be equated to the corresponding velocity perturbation arising from the compression via

$$\left(\frac{\delta \rho}{\rho} \right) = \frac{v}{c_s}, \tag{4.45}$$

where the adiabatic sound speed in the medium is c_s, and

$$c_s^2 \propto T/\mu, \tag{4.46}$$

μ being the mean molecular weight of the gas. The relevant temperature for the observations is the surface temperature, which is close to $T_{\rm eff}$. If we assume that the adiabatic index close to the surface does not change, it then follows that

$$\frac{\delta T}{T_{\rm eff}} \propto v\, T_{\rm eff}^{-1/2}. \tag{4.47}$$

We know from Section 4.2.1 that relating a change in temperature to an expected change in observed flux, $(\delta F/F)$, depends on the wavelength response of the photometric observations. Taking the three cases considered—bolometric, narrow-band, and extended-wavelength observations—we have

$$\left(\frac{\delta F_{\text{bol}}}{F_{\text{bol}}}\right) \propto \left(\frac{\delta T}{T_{\text{eff}}}\right) \propto v\, T_{\text{eff}}^{-1/2}, \tag{4.48}$$

$$\left(\frac{\delta F_\lambda}{F_\lambda}\right) \propto (\lambda\, T_{\text{eff}})^{-1} \left(\frac{\delta F_{\text{bol}}}{F_{\text{bol}}}\right) \propto v\lambda^{-1} T_{\text{eff}}^{-3/2}, \tag{4.49}$$

and

$$\left(\frac{\delta F}{F}\right) \propto (\mathcal{C}_{\text{bol}})^{-1} \left(\frac{\delta F_{\text{bol}}}{F_{\text{bol}}}\right) \propto v\, (\mathcal{C}_{\text{bol}})^{-1}\, T_{\text{eff}}^{-1/2}. \tag{4.50}$$

Equations 4.48–4.50 may be calibrated by, for example, using the amplitudes of the solar oscillations observed in Doppler velocity and photometry. Taking the bolometric case as an example, and following rearrangement to make the ratio of the amplitudes the subject of the relation, we find that

$$\left(\frac{\delta F_{\text{bol}}}{F_{\text{bol}}}\right) v^{-1} \simeq 20 \left(\frac{T_{\text{eff}}}{5{,}777\,\text{K}}\right)^{-1/2} \text{ ppm per m s}^{-1}. \tag{4.51}$$

In practice, a negative temperature exponent closer to unity provides a better match to the observations. Moreover, detailed comparisons with nonadiabatic computations show additional discrepancies that are also frequency dependent. Equation 4.51 does nevertheless provide a useful back-of-the-envelope estimate.

We shall come back to the scaling of the mode amplitudes later in Section 5.4.

4.3 OTHER CONTRIBUTIONS TO THE OBSERVED DATA

In addition to signatures from detectable oscillations, the observational data will also have contributions from other forms of intrinsic stellar variability, as well as from instrumental and photon-counting noise.

First, there is a contribution from the surface signatures of convection, the so-called pattern of *granulation*. These signatures manifest as a honeycomb-like pattern of individual convective cells or *granules*, comprised of bright (hot) upward moving gas at their centers, surrounded by darker (cooler) downward moving gas in the intergranular lanes. Granulation leads to fluctuations in photometric and Doppler velocity data that are stochastic (random) in nature. The underlying strength of the variations and the characteristic timescales on which those variations occur vary, depending on the physical properties of the outer layers of the stars. As we shall see in Section 4.31, it is possible to make approximate, and very useful, predictions for the granulation parameters based on the fundamental stellar properties (subject to several key simplifying assumptions). Because variations due to the granulation are governed by a characteristic timescale, discrete measurements of fluctuations in brightness or Doppler velocity will be correlated in time. As shown in Chapter 5, the frequency spectrum of the granulation is therefore not flat, but increases in power toward low frequencies. In this chapter we make predictions of the characteristic amplitude and timescale of the granulation for cool stars, and we leave discussion of the expected shape of the frequency spectrum to Chapter 5.

The second significant stellar contribution is from surface manifestations of magnetic activity. Large features, such as starspots or active regions, are dark relative to the surrounding quieter photosphere, and when on the visible disc they can give rise to significant signatures not only in photometric observations but also in Doppler velocity data. Their signals can have a periodic component: the rotation of the star moves spots onto, across, and then off the visible disc (assuming a suitably favorable inclination of the rotation axis with respect to the observer). The amplitude and time-dependent form of these signatures will depend not only on the properties, size, and surface distributions of the spots and on the surface rotation, but also on the orientation of the rotation axis of the star relative to the line of sight.

There will be contributions from instrumental noise—for example, from instrumental drift or jitter—and a contribution from the finite rate at which the stellar photons are detected. This type of noise—which is a result of measuring a phenomenon comprising discrete, independent events—is called *Poisson* or *shot* noise (the latter a term borrowed from electronics and descriptions of the noise arising from discrete electric charges in electronic devices). When the photon counting rate is high, as is the case for the observations we are concerned with, the Poisson distribution approaches the normal distribution. We may therefore assume photon noise contributes a Gaussian noise source in the time domain, which has a flat or *white* spectrum in frequency. The size of the shot-noise contribution can often be the limiting factor in determining whether we can detect solar-like oscillations. The shot noise may be predicted from the apparent magnitude of the target star and the known instrumental performance properties (e.g., see near the end of Section 6.2.1 and the discussion accompanying Eq. 6.48).

Finally, there may also be periodic signatures in the data due to exoplanets or binary companions. These signatures of course carry rich information on the underlying phenomena; but they can also present complications for the analysis of the asteroseismic data, since they manifest in the frequency spectrum as complicated, harmonic signatures. We leave further discussion of these signatures to Section 5.6.

In what remains of this chapter we introduce the intrinsic stellar signals due to granulation and activity. What follows is based on simple physics and some important assumptions. Nevertheless, the resulting predictions are informative and provide very useful prior information on the amplitudes and timescales we expect for the signals, which then folds into target-selection planning and preparing the asteroseismic analysis itself (where these background signatures must be fully accounted for). Let us begin with the granulation signal.

4.3.1 Signal Due to Granulation

Our goal in this section is to predict the characteristic amplitude and timescale of the granulation. We are guided by the simplifying assumptions inherent in mixing-length theory, which we met in Section 2.2. As noted previously, this theory aims to provide a description of the average behavior of convective motions by reducing the problem to a description of convection on some characteristic length scale, the so-called mixing length. Let us start with the expected signatures in photometric observations, using the Sun as a baseline reference, before we go on to make predictions for other stars.

Signatures of Granulation in Photometry

Let $(\delta I/I)_c$ be the typical relative intensity fluctuation due to an individual granule (convective cell). A typical solar value for the relative fluctuation at visible wavelengths is about 5%. The observed amplitudes of the fluctuations are wavelength dependent, as we might expect, given our discussion in Section 4.2.1. The observed dependence differs from a simple λ^{-1} prediction (see Eq. 4.25), an obvious reason being that one must also account for changes to the observed intensity fluctuation with $\cos\theta$.

If we make the simplifying approximation that all granules behave in a statistically independent manner, the disc-averaged RMS intensity fluctuation will then be given approximately by

$$\sigma_{c,I} \simeq (\delta I/I)_c / \sqrt{n} , \tag{4.52}$$

where

$$n \simeq \left(\frac{R}{r_c}\right)^2 \tag{4.53}$$

is the number of granules on the visible disc of the star, with r_c the typical convective granule size and R the stellar radius. We have ignored the geometric effects of foreshortening and changes in intensity contrast with $\cos\theta$. On the Sun, the peak contribution to the observed intensity fluctuations occurs at a granule size scale of approximately 1,000 km. Substituting this value into Eq. 4.52 gives an estimate for $\sigma_{c,I}$ of several tens of ppm, fairly close to the observed value.

To estimate the photometric signal for other cool main sequence, sub giant or red-giant stars, we assume that intensity fluctuations due to individual granules are the same from star to star. Moreover, we also assume that the characteristic granule size scales with the mixing length l_c. In Section 2.2 we saw that the mixing length is some multiple α (i.e., the mixing-length parameter) of the pressure scale height H_P, that is, $l_c = \alpha H_P$ (Eq. 2.17). Here we ignore possible variations in α when scaling to other stars.

For a simple isothermal atmosphere (where we also note that the pressure and density scale heights are equal, i.e., $H_P = H_\rho$), we have $H_P \propto T_{\mathrm{eff}}/g$. It then follows that

$$\sigma_{c,I} \propto \left(\frac{T_{\mathrm{eff}}}{gR}\right) \propto \left(\frac{T_{\mathrm{eff}}R}{M}\right). \tag{4.54}$$

We may then, for example, express the above in the form of a solar-normalized scaling relation:

$$\sigma_{c,I} \simeq \sigma_{c,I,\odot} \left(\frac{T_{\mathrm{eff}}}{T_{\mathrm{eff},\odot}}\right) \left(\frac{R}{R_\odot}\right) \left(\frac{M_\odot}{M}\right). \tag{4.55}$$

This may be simplified further for cool main sequence stars. If the exponent in the power-law relation $\Delta\nu \propto \nu_{\max}^{\beta}$ is fixed to $\beta = 0.75$—a reasonable match for solar-like oscillators on the main sequence—use of the scaling relations for $\Delta\nu$ and $\nu_{\max}$ (see Chapter 3) then implies $M \propto T_{\mathrm{eff}}^{1.5}$. Substitution into Eq. 4.54 then allows us to rewrite the dependence of the granulation amplitude as

$$\sigma_{\mathrm{c},I} \propto g^{-1/2} T_{\mathrm{eff}}^{1/4} \propto \nu_{\max}^{-1/2}, \tag{4.56}$$

and to recast the calibrated relation to the simpler form

$$\sigma_{\mathrm{c},I} \approx \sigma_{\mathrm{c},I,\odot} \left(\frac{\nu_{\max,\odot}}{\nu_{\max}}\right)^{1/2}. \tag{4.57}$$

The solar granulation has an amplitude of $\sigma_{\mathrm{c},I,\odot} \simeq 40\,\mathrm{ppm}$ in the *Kepler* bandpass.

Real stars follow the above relations to a fairly reasonable approximation (extending even into the subgiant and red-giant regime). As noted above, this helps in target selection: if we have some a priori information on the fundamental properties of a potential target, we can predict its granulation amplitude, which then feeds into predictions of whether we can detect oscillations against the combined background of granulation and shot noise. We introduce some simple models in Chapter 5 to describe the underlying spectrum of the granulation signal in the frequency domain, where the detailed analysis is performed. We then introduce some statistical detection tests in Chapter 6.

The other parameter that fixes the form of the frequency spectrum is the characteristic timescale of the granulation. We assume that the timescale τ_{c} is given approximately by

$$\tau_{\mathrm{c}} \simeq l_{\mathrm{c}}/v_{\mathrm{c}} \equiv \alpha H_P / v_{\mathrm{c}}, \tag{4.58}$$

where v_{c} is the velocity associated with the convective motions. If one assumes that the convective velocity scales with the sound speed in the near-surface layers (i.e., $v_{\mathrm{c}} \propto c_{\mathrm{s}} \propto T_{\mathrm{eff}}^{1/2}$), substitution into Eq. 4.58 implies that

$$\tau_{\mathrm{c}} \propto H_P \, T_{\mathrm{eff}}^{-1/2} \propto g^{-1} T_{\mathrm{eff}}^{1/2} \propto \nu_{\mathrm{ac}}^{-1} \propto \nu_{\max}^{-1}. \tag{4.59}$$

Eq. 4.59 may be transformed into a scaling relation calibrated against the observed solar timescale:

$$\tau_{\mathrm{c}} \simeq \tau_{\mathrm{c},\odot} \left(\frac{g_{\odot}}{g}\right) \left(\frac{T_{\mathrm{eff}}}{T_{\mathrm{eff},\odot}}\right)^{1/2} \equiv \tau_{\mathrm{c},\odot} \left(\frac{\nu_{\max,\odot}}{\nu_{\max}}\right), \tag{4.60}$$

again providing a potentially useful a priori constraint.

Analysis of data on the Sun suggests a dominant timescale of the order of 200 sec. Although fits to photometric data on real main sequence, subgiant, and red-giant stars show a reasonable adherence to predictions based on Eq. 4.60, there are small but significant differences. One obvious problem with the above derivation for the timescale is the assumption that the Mach number $\mathcal{M}_c$ associated with the near-surface turbulence, which is defined as

$$\mathcal{M}_c = v_c/c_s, \tag{4.61}$$

is the same from star to star. In what follows, we therefore derive a slightly more complicated relation for the granulation timescale, using an estimate of v_c from mixing-length theory that does allow for such variations. Here we set aside any discussion of the potentially bigger issue of the general applicability of simple mixing length theory over the range of stellar properties of interest.

Parcels of gas in the convection zone are assumed to behave adiabatically and to maintain pressure equilibrium with their surroundings. If the gas is also assumed to behave in an ideal manner, it then follows that the fractional deficit (or excess) in density compared to the surroundings is related to the corresponding excess (or deficit) in temperature according to

$$\Delta\rho/\rho = -\Delta T/T, \tag{4.62}$$

where we have ignored any changes in the mean molecular weight due to ionization. The acceleration arising from the convective buoyancy is therefore just

$$a = -g\left(\frac{\Delta\rho}{\rho}\right) = g\left(\frac{\Delta T}{T}\right). \tag{4.63}$$

Now, if over a time t the parcel is accelerated a distance of one mixing length, l_c, we have $l_c = at^2/2$, from which it follows that the average velocity of the parcel will be

$$v_c = l_c/t \equiv \left(\frac{1}{2}l_c a\right)^{1/2} \tag{4.64}$$

Substituting for the acceleration using Eq. 4.63 therefore gives

$$v_c = \left(\frac{gl_c\Delta T}{2T}\right)^{1/2}. \tag{4.65}$$

To progress to a scaling relation in fundamental stellar properties, we must determine the dependence of the temperature difference ΔT on those properties. A useful relation for the temperature difference follows from the expression for the energy flux carried by the convection, which is

$$\mathcal{F}_c = \rho v_c C_P \Delta T, \tag{4.66}$$

where C_P is the specific heat per unit mass at constant pressure. At the top of the subsurface convection zone, we assume that convection carries all of the stellar flux. Since the surface flux is just equal to $\sigma_B T_{\text{eff}}^4$, where σ_B is the Stefan-Boltzmann constant and we have taken $T \simeq T_{\text{eff}}$, we may write

$$\mathcal{F}_c \simeq \sigma_B T_{\text{eff}}^4. \tag{4.67}$$

Equating Eqs. 4.66 and 4.67, the temperature difference at the surface is given by:

$$\Delta T \simeq \left(\frac{\sigma_B T_{\mathrm{eff}}^4}{\rho v_c C_P} \right) \tag{4.68}$$

We may now combine Eqs. 4.65 and 4.68 to eliminate ΔT and write an expression for the convective velocity in the form

$$v_c \simeq \left(\frac{g\alpha H_P \sigma_B T_{\mathrm{eff}}^3}{2\rho C_P} \right)^{1/3}. \tag{4.69}$$

Now, we know the dependence of H_P on fundamental stellar properties and we may also assume that g is just the surface gravity. That just leaves the tricky issue of estimating how the surface density ρ changes from star to star. Unfortunately, this scales poorly with the mean stellar density $\bar{\rho} \propto M/R^3$. We therefore use the following line of reasoning.

In the T_{eff} range of interest to us, the near-surface opacity in stars is dominated by the tendency of hydrogen atoms to acquire an extra electron. The resulting H^- configuration is very unstable and so acts as a copious source of opacity. The absorption coefficient κ for this transition may be shown to scale like

$$\kappa \propto \rho^{1/2} T^9, \tag{4.70}$$

where ρ is now the density in the near-surface layers, and we may again assume that $T \simeq T_{\mathrm{eff}}$. We can translate this absorption coefficient to an equivalent optical depth across a granule. This is equal approximately to $\kappa \rho \alpha H_P$, where we have assumed that the parameters do not alter significantly over the associated change in depth (thereby simplifying the integration over radius to a product of average parameter values). At the photospheric level, the disc-averaged optical depth is equal by definition to 2/3, implying that

$$\kappa \rho \alpha H_P \simeq 2/3 \equiv \text{constant}. \tag{4.71}$$

Combining Eqs. 4.70 and 4.71, we have

$$\rho \propto (\kappa \alpha H_P)^{-1} \propto T_{\mathrm{eff}}^{-20/3} g^{2/3}. \tag{4.72}$$

We may now go back to Eq. 4.69, and find the dependence of v_c on stellar properties. This gives

$$v_c \propto T_{\mathrm{eff}}^{32/9} g^{-2/9}. \tag{4.73}$$

Substituting into Eq. 4.58, we arrive at the sought-for dependence of the granulation timescale:

$$\tau_c \propto T_{\mathrm{eff}}^{-23/9} g^{-7/9}. \tag{4.74}$$

As per the simpler treatment above (see Eqs. 4.59 and 4.60), we may recast Eq. 4.74 as a scaling relation normalized by the solar timescale:

$$\tau_c \simeq \tau_{c,\odot} \left(\frac{g_\odot}{g} \right)^{7/9} \left(\frac{T_{\text{eff},\odot}}{T_{\text{eff}}} \right)^{23/9}. \qquad (4.75)$$

For red giants, the spread in T_{eff} is quite modest, hence it is the change in surface gravity that dominates the prediction above. In the cool main sequence and subgiant regimes, things are a little more complicated given the spread in gravity and effective temperature.

Finally in this section, we can rewrite the dependence of τ_c on stellar properties in a form that includes the Mach number $\mathcal{M}_c$ explicitly. Starting from Eq. 4.58, we may reexpress the timescale in terms of the Mach number and the acoustic cut-off frequency:

$$\tau_c \propto H_P/v_c \propto H_P/(\mathcal{M}_c c_s) \propto 1/(\mathcal{M}_c \nu_{\text{ac}}). \qquad (4.76)$$

It then remains to derive the dependence of $\mathcal{M}_c$ on stellar properties for each of the two scenarios discussed above. Recall that in the first scenario, it was assumed that the Mach number is constant. In the second, we relied on mixing-length theory to derive v_c; the dependence of the Mach number under this scenario follows from using Eqs. 4.73 and 4.74. Putting this all together, we may summarize as follows:

$$\tau_c \propto 1/(\mathcal{M}_c \nu_{\text{ac}}) \begin{cases} \mathcal{M}_c = \text{constant}, \quad v_c \propto c_s \\ \mathcal{M}_c \propto T_{\text{eff}}^{3.06} g^{-0.22}, \ v_c = \mathcal{M}_c c_s. \end{cases} \qquad (4.77)$$

Signatures of Granulation in Doppler Velocity

In Doppler velocity, a basic prediction of the disc-averaged signal is given approximately by

$$\sigma_{c,v} \simeq (2/\pi) \frac{f_c v_c}{\sqrt{n}}, \qquad (4.78)$$

where the factor $2/\pi$ is to allow for weighting of the line-of-sight signal on $\cos\theta$, and f_c is a factor to correct for partial cancellation of the signals due to hot (bright) upward moving gas (blue-shifted signal) and cooler (darker) sinking gas (red-shifted signal). The net signal is blue-shifted, due to the bigger relative contribution of the hot, brighter elements, and for the Sun it amounts to a few hundred meters per second. Taken together, the above implies a solar value of $\sigma_{c,v}$ of a few tens of centimeters per second.

To scale $\sigma_{c,v}$ to other stars we must include the dependence of n, v_c, and f_c on stellar properties. The dependence of n follows straightforwardly from Eq. 4.53, while that of v_c was explored above in the context of the granulation timescale. We then make the crude assumption that f_c is the same from star to star. Putting this together allows us to make the following predictions:

$$
\sigma_{c,v} \propto \begin{cases} T_{\text{eff}}^{3/2} R M^{-1}, & v_c \propto c_s \\[2ex] T_{\text{eff}}^{41/9} R^{13/9} M^{-11/9}, & v_c = \mathcal{M}_c c_s. \end{cases} \tag{4.79}
$$

The most striking difference between the two scenarios is the much stronger dependence on temperature given by the mixing-length predictions. This strong dependence is also in marked contrast to the photometric prediction for $\sigma_{c,I}$, which showed only a linear dependence on effective temperature.

Finally in this section, note that we have assumed that the timescales of the granulation signatures in Doppler velocity are the same as those in intensity.

Relative Granulation Amplitudes in Photometry and Doppler Velocity

How do the amplitudes of the granulation signals compare for photometry and Doppler velocity? First, the ratio of the amplitudes for the Sun is $\simeq 100$ ppm per m s^{-1}, which is notably higher than the ratio for the oscillations (see Eq. 4.51 in Section 4.2.3). This tells us that the solar granulation signal is much more prominent relative to the solar oscillation signal when observations are made in intensity as opposed to in Doppler velocity. This has the effect of making the lower-frequency, weaker-amplitude p modes much harder to detect in photometry.

Can we make a rough prediction of the ratio for other stars? From the preceding sections, we may write

$$
\sigma_{c,I}/\sigma_{c,v} \simeq \left(\frac{\pi}{2 f_c} \right) (\delta I/I)_c \, v_c^{-1}. \tag{4.80}
$$

If we make the simplifying assumption that $(\delta I/I)_c$ and f_c are the same from star to star, the above reduces to

$$
\sigma_{c,I}/\sigma_{c,v} \propto v_c^{-1}. \tag{4.81}
$$

From earlier in this section, we therefore have

$$
\sigma_{c,I}/\sigma_{c,v} \propto \begin{cases} T_{\text{eff}}^{-1/2}, & v_c \propto c_s \\[2ex] T_{\text{eff}}^{-32/9} g^{2/9}, & v_c = \mathcal{M}_c c_s. \end{cases} \tag{4.82}
$$

Finally, if we calibrate to the solar values, we have

$$
\sigma_{c,I}/\sigma_{c,v} \simeq \begin{cases} 100 \left(\dfrac{T_{\text{eff}}}{5{,}777\,\text{K}} \right)^{-1/2} \text{ppm per m s}^{-1}, & v_c \propto c_s \\[3ex] 100 \left(\dfrac{T_{\text{eff}}}{5{,}777\,\text{K}} \right)^{-32/9} \left(\dfrac{g}{g_\odot} \right)^{2/9} \text{ppm per m s}^{-1}, & v_c = \mathcal{M}_c c_s. \end{cases} \tag{4.83}
$$

4.3.2 Signal Due to Stellar Activity

Let us now consider the signal arising from signatures of stellar activity. We expect the signal to have a periodic component due to the rotational modulation of starspots and active regions, as well as a stochastic component due to the evolution and decay of spots and active regions.

First consider the photometric signal. We assume that, to first order, it is signal from the spots that dominates. On the Sun, the temperature contrast between a sunspot and the surrounding, quiet photosphere is about 1,500 K, implying a brightness contrast $(\delta I/I)_{\mathrm{act}}$ up to 70%. In practice, measurements of sunspot contrast, spanning the range of spot sizes, return values between about 10 and 80% with larger spots giving a higher contrast.

The fractional spot coverage in area $(\delta a/a)$ over recent solar cycles has typically peaked at around 0.3% (although the maximum of the current, comparatively weak cycle 24 is significantly lower, at less than 0.2%). These values of course reflect the additive contributions from all spots lying on the visible disc at epochs of high stellar activity. Multiplying the spot coverage by a suitable average contrast value implies a peak disc-averaged signal amplitude,

$$\sigma_{\mathrm{act},I} \simeq \left(\frac{\delta I}{I}\right)_{\mathrm{act}} \left(\frac{\delta a}{a}\right), \qquad (4.84)$$

of roughly $\approx 1,000$ ppm. Translating this figure to other stars requires estimates of the two basic inputs to Eq. 4.84, in addition to information on the angle of inclination of the star's rotation axis, as well as the latitudinal distribution of the spots. If the angle is low, so that the star is observed nearly pole-on, and moreover, we assume a similar spot and active-region architecture to that shown on the Sun, we would expect a much weaker modulation of the spot signal.

The sensitivity to Doppler velocity arises because a spot acts almost like an opaque disc covering a patch on the star (just like a transiting exoplanet). If the rotation axis of the star is close to the plane of the sky, a spot on the approaching (blue-shifted) half of the visible disc will give rise to a small red-shifted perturbation relative to the signal expected in the absence of the spot; similarly, once the spot has crossed to the receding (red-shifted) half of the disc, it will give a small blue-shifted perturbation. As per our discussion above, if the star is observed close to pole-on, the observed signal will be much weaker.

An approximate estimate of the signal in Doppler velocity may be made using

$$\sigma_{\mathrm{act},v} \approx \left(\frac{\delta I}{I}\right)_{\mathrm{act}} \left(\frac{\delta a}{a}\right) v \sin i_{\mathrm{s}}, \qquad (4.85)$$

where $v \sin i_{\mathrm{s}}$ is the projected rotational velocity at the surface of the star. For the Sun, which has $v \sin i_{\mathrm{s}} \simeq 1.9 \, \mathrm{km \, s^{-1}}$, we expect a disc-averaged signal $\sigma_{\mathrm{act},v}$ of a few meters per second.

The stochastic signal due to the evolution and decay of starspots and active regions is less straightforward to predict. However, even though the amplitudes of the photometric and Doppler signals can be significantly higher than those of the corresponding signals due to oscillation modes and granulation, it is important to note the separation of associated timescales: the activity signal is most prominent at much longer timescales (i.e., much lower frequencies) and so does not usually interfere with analysis of oscillation spectra.

The same is of course also notionally true for the abovementioned rotational modulation signals. However, there is the potential in very active stars for strong harmonics of the rotational signal to ring into higher-frequency parts of the power spectrum and into the regime where fitting of granulation and oscillations is performed. We look at this issue in the final part of Section 5.6.

4.4 FURTHER READING

The two papers below cover in detail the calculations needed to convert photometric amplitudes for different instrumental response functions to equivalent bolometric amplitudes:

- Michel, E., et al., 2009, *Intrinsic Photometric Characterisation of Stellar Oscillations and Granulation. Solar Reference Values and CoRoT Response Functions*, Astron. Astrophys., **495**, 979.
- Ballot, J., Barban, C., and van't Veer-Menneret, C., 2011, *Visibilities and Bolometric Corrections for Stellar Oscillation Modes Observed by* Kepler, Astron. Astrophys., **531**, 124.

The first of the next two references is one of the foundation papers of the field. It presents extensive predictions for amplitudes of stellar oscillations based on simple physics, predictions that were made before detections of solar-like oscillations had been confirmed in a star other than the Sun. The second paper, written by the same authors some 15 years later, is an update on the first paper and utilizes the knowledge gained from detections made using ground-based Doppler velocity data, and photometric data from CoRoT and *Kepler*. It also includes scaling relations for granulation.

- Kjeldsen, H., and Bedding, T. R., 1995, *Amplitudes of Solar Oscillations: The Implications for Asteroseismology*, Astron. Astrophys., **193**, 87.
- Kjeldsen, H., and Bedding, T. R., 2011, *Amplitudes of Solar Oscillations: A New Scaling Relation*, Astron. Astrophys., **529**, 8.

These papers discuss in detail scaling relations for granulation and inclusion of an explicit dependence on the Mach number:

- Samadi, R., Belkacem, K., and Ludwig, H.-G., 2013, *Stellar Granulation as Seen in Disk-Integrated Intensity. I. Simplified Theoretical Modelling*, Astron. Astrophys., **559**, 39.
- Samadi, R., et al., 2013, *Stellar Granulation as Seen in Disk-Integrated Intensity. II. Theoretical Scaling Relations Compared with Observations*, Astron. Astrophys., **559**, 40.

4.5 EXERCISES

1. Derive the visibility ratio in power at $l = 1$ (i.e., $(S_1/S_0)^2$), for photometric observations with limb darkening, assuming a simple linear limb-darkening law (Eq. 4.2) with $c_1 = 3/5$.

2. Derive the visibility ratio in power at $l = 3$, (i.e., $(S_3/S_0)^2$), again for photometric observations, but this time without limb darkening. Comment on your result.

3. The maximum bolometric amplitude of radial solar p modes is about $A = 3.5$ ppm. What fractional change in radius or surface temperature would be needed to give this signal? What would the maximum amplitude be in narrow band-pass filters at 500 nm and 800 nm?

4. Make use of the scaling relations used in this chapter to estimate the ratio of the maximum radial p mode amplitude to the granulation amplitude for a $1.2 M_\odot$ star at two different stages in its evolution. You should compute ratios of the amplitudes in both bolometric intensity (i.e., $(\delta F_{bol}/F_{bol})/\sigma_{c,I}$) and Doppler velocity (i.e., $v/\sigma_{c,v}$) when the star is:

> (a) on the main sequence, at an evolutionary epoch when $T_{eff} = 6{,}295$ K, and $R = 1.27 R_\odot$; and
> (b) on the low-luminosity part of the red-giant branch, at an evolutionary epoch when $T_{eff} = 4{,}775$ K, and $R = 4.67 R_\odot$.

Take the equivalent solar values to be $(\delta F_{bol}/F_{bol})_\odot \simeq 2.4$ ppm RMS, $v_\odot \simeq 13$ cm s^{-1} for the maximum RMS radial p mode amplitudes, and $\sigma_{c,I,\odot} \simeq 40$ ppm and $\sigma_{c,v,\odot} \simeq 39$ cm s^{-1} for the granulation amplitudes. Comment on your results.

5. The file oneMsolar.txt contains fundamental properties for a $1 M_\odot$ stellar evolutionary track. Use the scaling relations presented in this chapter to predict the granulation amplitude in Doppler velocity, $\sigma_{c,v}$, and the corresponding spot-modulation (activity) amplitude $\sigma_{act,v}$, from the zero age main sequence (ZAMS) to the terminal age main sequence (TAMS).

You may assume that the mean surface rotation period follows the Skumanich Law (i.e., $P \propto t^{1/2}$). Also make the simplifying assumption that the intensity contrast and surface area coverage due to starspots remain unchanged over the evolutionary timescales considered.

5 Observational Data: Detailed Characteristics in the Frequency Domain

In this chapter we consider in detail the frequency-domain properties of observational data on stars. It is from the analysis of the frequency spectrum of the observations that the asteroseismic parameters are usually extracted.

Because they are intrinsically damped, most detectable solar-like oscillations are not coherent over the length of the observations. The phase is not maintained— the resonant peaks due to the oscillations therefore have width in the frequency spectrum—and so we usually perform the analysis on the frequency-*power* spectrum, since this leads to no loss of information. For the very longest-lived, effectively coherent modes, it does however remain advantageous to use the phase information. The frequency-domain appearance of the modes will be discussed at length below, including the impact of the finite lifetimes.

The peaks due to the oscillations are superimposed on a slowly varying, broadband background. This broad-band background will have several contributions. There are the intrinsic stellar signatures due to granulation and activity that we met in Section 4.3. Both have stochastic signatures that show increasing power toward lower frequencies. There is then a flat (white) contribution due to photon shot noise; we also see power from instrumental contributions, such as drifts, which are usually not flat in frequency. Narrow-band signatures are also present, for example, those arising from periodic signals due to the rotational modulation of starspots (the aforementioned activity), exoplanets, or binary companions. More often than not we also have instrumental signatures arising from operational functions (e.g., from regular angular momentum dumps on a satellite, or on-board processing of data). Finally, there may be significant gaps in the data (e.g., due to breaks in ground-based coverage), which can introduce significant artifacts in the frequency domain.

In the sections that follow we look at each of the intrinsic stellar contributions in turn, always with an eye to how each must be handled in the asteroseismic analysis (which we describe in detail in Chapter 6). We also look in this chapter at the impact on the analysis of gaps in the data.

As a first step in the analysis, we must apply frequency spectrum estimation techniques to the time-domain data. We therefore begin by covering practical fundamentals associated with estimation of the frequency spectra, which underpin everything that follows. We assume that the reader is already familiar with Fourier transforms.

5.1 POWER SPECTRUM ESTIMATION

We have seen that the basic observations are usually a lightcurve of photometric data or a time series of Doppler velocity data. Let us denote the time-domain data by the generic $x(t)$. Fourier-based methods may be used to compute the frequency spectrum of the observations. The standard Fourier transform $X(\omega)$ of the function $x(t)$ may be written as

$$X(\omega) = \frac{1}{\sqrt{2\pi}} \int_{-\infty}^{+\infty} x(t) \exp(-i\omega t) dt. \tag{5.1}$$

In reality we deal of course with discrete data. Let us consider a set of N such observations, each denoted by x_j. Various numerical methods are available to compute the discrete transform X_k, for example, the fast Fourier transform is but one (which assumes the data are presented on a regular grid in time). The discrete transform may be written as

$$X_k = \frac{1}{\sqrt{N}} \sum_{j=0}^{N-1} x_j \exp(-i2\pi kj/N), \tag{5.2}$$

where the normalization factor is a matter of convention; here, the factor $1/\sqrt{N}$ means the discrete transform is unitary. The power spectrum is then just

$$P(\omega) \equiv P_k = |X_k|^2. \tag{5.3}$$

The analysis of data is usually performed in cyclic as opposed to angular frequency units, and so in what follows we shall usually express equations and terms in ν rather than in ω.

5.1.1 Frequency Resolution

The new generation of space telescopes usually collect data on a regular sampling cadence, which we denote here by Δt. If there are N such observations, so that the length of the dataset is

$$T = N\Delta t, \tag{5.4}$$

then cyclic frequencies in the discrete transform will be uncorrelated at separations of $1/T$. We call this the *natural frequency resolution* of the dataset. If the discrete transform is computed at this natural resolution, then the *binwidth* in cyclic frequency will be

$$\Delta_T = T^{-1} \equiv (N\Delta t)^{-1}. \tag{5.5}$$

In what follows we shall assume that the sampling is close to regular.

It can be advantageous to oversample the frequency spectrum, meaning that the transform is computed at a resolution that is finer than Δ_T (e.g., to separate features that are very closely spaced in frequency and therefore barely resolvable). However, if models are fitted to the oscillation peaks or other statistical analysis of the spectrum is to be performed (see Chapter 6), such analyses should be made at the natural frequency resolution since they are predicated on the assumption that the individual bins are statistically independent. Otherwise, explicit allowance should be made for the resulting correlations in the oversampled spectrum.

5.1.2 Nyquist Frequency

For any given periodic signal, we need to have at least two samples Δt per period to completely determine the signal, so that there is no ambiguity in the measured frequency. The highest frequency in the transform that satisfies this requirement is called the *Nyquist* frequency, ν_{Nyq}, which is defined by

$$\nu_{\mathrm{Nyq}} \overset{\text{def}}{=} (2\Delta t)^{-1}. \tag{5.6}$$

At frequencies $\nu \leq \nu_{\mathrm{Nyq}}$, we say the signal is oversampled. What happens if, instead, the signal is undersampled, so that $\nu > \nu_{\mathrm{Nyq}}$? We will see a peak at the true frequency, ν, which will then lie in what we call the *super-Nyquist* regime of the transform; that peak will also be reflected, or aliased, back into the frequency region below ν_{Nyq}. Writing the true frequency as $\nu = \nu_{\mathrm{Nyq}} + \nu'$, peaks will occur at $\nu_{\mathrm{Nyq}} + \nu'$ and $\nu_{\mathrm{Nyq}} - \nu'$. The estimated frequency is ambiguous—we will not know which peak is associated with the real frequency—and we are reliant on using other knowledge to pick the true frequency (see Section 5.7).

If the sampling of the data in the time domain is irregular, there is no longer a hard definition for the Nyquist frequency (i.e., the spectrum will no longer comprise exact, reflected copies about multiples of ν_{Nyq}). However, a notional estimate of the Nyquist frequency based on the *median* sampling time provides an approximate guide to the frequency around which features of the spectrum are reflected.

5.1.3 Signal Attenuation Due to Sampling Integration

One other aspect of the sampling affects the appearance of the transform: the integration time. If a high fraction of each cadence Δt is used to collect data (i.e., to integrate photons) from the star, we may significantly underestimate the true amplitude of the oscillations, because each datum will average the time-varying signal.

If the integration time per cadence is $\Delta t'$, a signal of frequency ν will have its amplitude attenuated by the factor

$$\eta(\nu) = \mathrm{sinc}\left[\pi\left(\nu\Delta t'\right)\right]. \tag{5.7}$$

Recall that the sinc function of some quantity x is defined according to

$$\mathrm{sinc}(x) = \frac{\sin x}{x}. \tag{5.8}$$

When the fraction of each cadence given over to integration is unity, so that $\Delta t' = \Delta t$, the attenuation may be written as

$$\eta(\nu) = \text{sinc}\,[\pi\,(\nu\Delta t)] = \text{sinc}\left[\frac{\pi}{2}\left(\frac{\nu}{\nu_{\text{Nyq}}}\right)\right]. \tag{5.9}$$

The attenuation in power is given by the square of the sinc function, (i.e., by η^2). Even when the integration duty cycle is close to 100%, Eq. 5.9 indicates that one still has sensitivity to signals in the super-Nyquist regime ($\nu > \nu_{\text{Nyq}}$), since the first zero of the sinc function does not occur until $\nu = 2\nu_{\text{Nyq}}$. This is something that is not widely appreciated, and it opens the possibility to perform super-Nyquist asteroseismology, as we shall discuss in Section 5.7.

5.1.4 Parseval's Theorem and Calibration of the Transform

Calibration of the power in the discrete transform follows from the application of Parseval's theorem. Simply put, the total power in the transform must equal the mean-square power of the data in the time domain. Although this might seem a rather trivial matter, there are actually a few practical choices one must make when calibrating the transform. If these choices are not clearly stated, the potential to cause a good deal of confusion arises (e.g., if one wishes to compare results of one research study with those of another).

The first choice we make is whether to use a single- or double-sided definition for the discrete transform. A double-sided transform is determined at positive *and* negative frequencies. However, since our data comprise real values (e.g., intensities or velocities) the negative-frequency transform will be just a mirror image of the positive-frequency transform. We therefore adopt a single-sided calibration, in which all power from the time domain appears in the positive-frequency transform, and we disregard the negative-frequency side.

Taking the N cadences that have recorded data in the time domain, a statement of Parseval's Theorem, which our calibrated transform must satisfy, is then

$$\sum_{k=0}^{N/2} P_k = \frac{1}{N}\sum_{j=1}^{N} x_j^2. \tag{5.10}$$

Notice how, with a single-sided calibration, the power is spread over $N/2 + 1$ bins in the transform (left-hand side), the extra bin being the zero-frequency bin. If the average of the input data is zero, no power appears in this bin. The right-hand side is the *mean-square* power of the time-domain signal.

If P_k' are the raw values of the power spectrum (power per bin) given by our Fourier transform code prior to calibration, then multiplication of those raw values by a numerical calibration coefficient c_P will ensure we have a correctly calibrated power spectrum, P_k:

$$P_k = c_P P_k'. \tag{5.11}$$

From Eq. 5.10 it follows that

$$c_P = \left(\frac{1}{N} \sum_{j=1}^{N} x_j^2 \right) \Big/ \left(\sum_{k=0}^{N/2} P_k' \right). \tag{5.12}$$

Equation 5.12 gives calibrated power spectral densities (PSD) in units of power per frequency bin. We may also express the power spectrum in units of power per Hertz (i.e., power per fixed frequency interval) by dividing the correctly calibrated values by the natural binwidth $\Delta_T = T^{-1}$ (Eq. 5.5). We come back in a few paragraphs to discuss an obvious advantage of adopting these units. Here, we may summarize the required calibrations as follows:

$$P_k = \begin{cases} c_P P_k', & \text{power per bin,} \\ c_P P_k' T \equiv c_P P_k' / \Delta_T, & \text{power per Hz.} \end{cases} \tag{5.13}$$

We might also wish to calculate an oversampled spectrum, one for which the resolution exceeds that defined by the natural resolution. If we choose to oversample by a numerical factor m, so that there are $mN/2 + 1$ bins in the frequency domain, the calibration coefficient becomes

$$c_{Pm} = m \times \left(\frac{1}{N} \sum_{j=1}^{N} x_j^2 \right) \Big/ \left(\sum_{k=0}^{mN/2} P_k' \right). \tag{5.14}$$

A summary for the oversampled case, including the conversion to power per Hertz, is

$$P_k = \begin{cases} c_{Pm} P_k', & \text{power per bin} \\ c_{Pm} P_k' T \equiv c_{Pm} P_k' / (m\Delta_{Tm}), & \text{power per Hz,} \end{cases} \tag{5.15}$$

where $\Delta_{Tm} = \Delta_T / m$ is the binwidth of the oversampled spectrum.

When there are gaps in the dataset—due to, for example, diurnal (daily) gaps in ground-based observations—extra care may be needed when applying the above calibrations depending on how the time-domain data are presented. We discuss how to handle gapped data in Section 5.1.5.

Belt-and-braces checks of the calibrations may be made by considering limiting dataset cases comprising of either white noise or commensurate sinusoids. It is extremely useful to be able to make a quick back-of-the-envelope prediction of the expected power spectral density for the white-noise case. This provides an estimate of the shot noise background we would expect given known photon-counting rates. We consider this belt-and-braces case first, assuming that there are no gaps (missing cadences) in the data.

A normally distributed (Gaussian) noise source of zero mean, characterized by a sample standard deviation σ (assumed stationary in time) will have a flat (white) spectrum in the frequency transform. The mean-square power in the time domain is just

$$\frac{1}{N} \sum_{j=1}^{N} x_j^2 = \sigma^2. \tag{5.16}$$

This power will be spread evenly across $N/2$ bins (there will be no power in the zero-frequency bin), and so it is a straightforward matter to compute the average power per bin we should expect to see for a correctly calibrated power spectrum: we will have

$$\langle P \rangle = 2\sigma^2/N, \tag{5.17}$$

where the units are power per frequency bin. The required calibration coefficient is given by

$$c_P = \sigma^2 \left(\sum_{k=0}^{N/2} P_k' \right)^{-1}. \tag{5.18}$$

One drawback here is that even though the noise is stationary, the average power per bin depends on N: the longer the dataset, the lower is the average power. This is not ideal if we wish to compare, for example, the intrinsic noise quality of one dataset with another, when the lengths of those datasets differ. We may instead convert values to power per Hertz. Under this scaling, a stationary noise source will give the same underlying average power per Hertz irrespective of the length of the dataset.

To summarize, the average power levels for the two different scalings should be

$$\langle P \rangle = \begin{cases} 2\sigma^2/N, & \text{power per bin,} \\ 2\sigma^2 T/N = 2\sigma^2 \Delta t, & \text{power per Hz,} \end{cases} \tag{5.19}$$

where we have used $T = N\Delta t$ in the power-per-Hertz conversion.

Next, consider tests with a sinusoidal signal

$$x_j = A \sin(2\pi \nu_0 t_j) \tag{5.20}$$

of amplitude A, where the frequency ν_0 is chosen to be commensurate with the dataset length T (i.e., to give an exact integer number of periods in the dataset). The mean-square power in the time domain is now

$$\frac{1}{N} \sum_{j=1}^{N} x_j^2 = A^2/2. \tag{5.21}$$

We therefore expect a peak $P(\nu_0)$ at the frequency of the sinusoid having a height equal to

$$P(\nu_0) = \begin{cases} A^2/2, & \text{power per bin} \\ A^2 T/2 \equiv A^2 N\Delta t/2, & \text{power per Hz.} \end{cases} \tag{5.22}$$

It is worth stressing that our calibration recovers the mean-square power $A_{\text{rms}}^2 \equiv A^2/2$, and *not* A^2. The required calibration coefficient in, for example, the power-per-bin scaling is

$$c_P = \frac{A^2}{2} \left(\sum_{k=0}^{N/2} P_k' \right)^{-1}. \tag{5.23}$$

The approach of replacing all data in the lightcurve or time series by a sinusoidal signal sampled at the same times as the original data is commonly used as a way of testing the calibration, usually with a signal of unit amplitude. One may also circumvent the strict requirement that the signal be commensurate by averaging the spectra obtained from a pure sinusoidal signal and a pure cosinusoidal signal, each of the same frequency and amplitude. This approach is particularly instructive when the data have gaps, allowing one to see the more complicated structure that the oscillation signal will then have in the frequency spectrum, which we now discuss.

5.1.5 The Window Function: The Effect of Gaps in the Data

An obvious desirable attribute for the time-domain data is that there should be as few gaps (i.e., cadences with no data) as possible. Observations made from satellites often provide excellent continuity, with only a few cadences being lost (e.g., to regular spacecraft operations, such as communication repointings or angular momentum dumps). Maintaining continuity of observations can be a major logistical challenge for ground-based observations, demanding either a dedicated network of observatories (e.g., the BiSON and GONG helioseismology programs) or a well coordinated observing campaign comprising telescopes spreadout in longitude.

Impact of the Window Function

The *window function* defines the times or epochs at which we have data. We may formally define the window function $w(t)$ to be

$$w(t) = \begin{cases} 1, & \text{Data available} \\ 0, & \text{Otherwise.} \end{cases} \tag{5.24}$$

The inclusion of zeros in the above is strictly relevant for lightcurves or time series that are prepared for a fast Fourier transform algorithm. Such algorithms require that the input data be placed on a regular timing grid: if some cadences have no data, then zeros must be explicitly included at the corresponding cadences. Other Fourier transform tools, which can handle nonregular time sampling, require presentation only of the cadences where data are available. In such cases the window function is specified by replacing the data at each available cadence by the value unity, that is, $w(t) = 1$.

When data are missing in the time domain—be that missing cadences from regular sampling or simply periods of time with no data for irregular cadences—the appearance of the power spectrum will be affected. That is because power will be *aliased* from the actual signal frequencies to other frequencies, with the exact nature of the aliasing depending on the properties of the window function.

A regular pattern of gaps in the time domain gives rise to additional peaks surrounding the actual signal frequencies. Astronomical observations can be beset by such difficulties, the most common problem being diurnal gaps in ground-based observations. Consider, for example, ground-based observations from a single site. Suppose we begin and end our observations at the same times, night after night. When we compute the frequency transform of the entire dataset—after the data for all nights have been joined together in a coherent manner, respecting the times at which data were collected—we will find not only strong peaks at the frequencies of any oscillation signals but also aliased peaks. The spacing in frequency of these aliased peaks depends on the timing of the gaps. Here, the dominant period given by the day-night cycle is one day. This means that frequencies that fit either one extra or one fewer oscillation cycle into one day than the true signal frequency will also provide a good fit to the data. If the gaps are big, frequencies fitting more (or fewer) multiple cycles into a day may show significant power. Each oscillation peak will therefore be surrounded by several additional peaks, the frequency spacing between adjacent peaks being one over a day (i.e., $11.57\,\mu$Hz). In some cases the spacing may unfortunately coincide with the separations shown by different modes in the oscillation spectrum of the star, making extraction of the mode parameters more challenging. Gaps showing a more random structure, such as those arising from instrumental problems or weather, will give broadband noise, potentially decreasing the height-to-background levels of the oscillation peaks in the frequency domain.

Let us consider the impact of the window function in more detail by first defining $x(t)$ to be the data we would have in the absence of gaps. The actual dataset available to us is then $x(t) \times w(t)$. We may write the expected frequency transform of this composite dataset by making use of the *convolution theorem*. If $X(\nu)$ and $W(\nu)$ are the frequency transforms of $x(t)$ and $w(t)$, respectively, then with $\mathcal{F}$ denoting the Fourier transform operation, we may write

$$x(t) \times w(t) \overset{\mathcal{F}}{\longleftrightarrow} X(\nu) * W(\nu), \tag{5.25}$$

where the asterisk symbol denotes the convolution operation. The frequency transform of the signal multiplied by the window function is just the convolution of the frequency transforms of each of the two time-domain components. The convolution may be written for discrete data as

$$(X * W)_k = \sum_{m=-\infty}^{\infty} X_m W_{k-m} = \sum_{m=-\infty}^{\infty} X_{k-m} W_m, \tag{5.26}$$

with k tagging each bin of the convolution, and m being a dummy integer.

If we know $X(\nu)$ and $W(\nu)$, we are therefore in a position to predict the exact form of the observed power spectrum. Before we even consider the impact of gaps in the dataset, the above allows us to make an important point regarding the impact of the *finite length* of any real dataset. If we have continuous observations (no internal gaps) spanning a duration T, the window function $w(t)$ is just the normalized boxcar (or rectangular) function:

$$\Pi_T(t) \stackrel{\text{def}}{=} \begin{cases} 1, & 0 \le t \le T \\ 0, & \text{Otherwise.} \end{cases} \tag{5.27}$$

The Fourier transform of Eq. 5.27 is the function $\text{sinc}(\pi \nu T)$. From Eq. 5.25, it then follows that the observed transform of the *finite* dataset will be the Fourier transform of the signal $x(t)$ convolved with the above sinc function.

Let us take a simple example for $x(t)$ of a sinusoidal signal of frequency ν_0 (see Section 5.1.4). Its frequency spectrum is a single peak of height $A^2/2$ at frequency ν_0. The convolved spectrum in power at the discrete frequencies ν_k of the transform is therefore

$$P_k = \frac{A^2}{2} \text{sinc}^2 \left[\pi (\nu_k - \nu_0) T \right]. \tag{5.28}$$

Notice that when the transform is sampled at its natural resolution, $1/T$, the zeros of the sinc function lie exactly at the sampled frequencies of the transform. We see only the power at the frequency ν_0. In contrast, if the signal is not exactly commensurate, or we oversample the frequency transform, we will partially or fully trace out the sinc-squared profile in power. The appearance of the spectrum is no longer that of just a peak (or spike) in a single frequency bin.

Let us now introduce gaps in the full duration T of the observations. Taking the single-site ground-based case, suppose we collect T_D of data each day:

$$\Pi_{T_D}(t) \stackrel{\text{def}}{=} \begin{cases} 1, & 0 \le t \le T_D \\ 0, & \text{Otherwise.} \end{cases} \tag{5.29}$$

These T_D-long segments recur daily, that is, they are repeated at intervals D equal to the length of a day (24 hours). This grid is the Dirac comb:

$$\text{III}_D(t) \stackrel{\text{def}}{=} \sum_{k=-\infty}^{\infty} \delta(t - kD). \tag{5.30}$$

A description of the window function is therefore

$$w(t) = \left[\Pi_{T_D}(t) * \text{III}_D(t) \right] \times \Pi_T(t). \tag{5.31}$$

We have the boxcar $\Pi_{T_D}(t)$, which describes each segment of data, convolved with the Dirac comb describing the daily frequency of the observations. This convolved part is multiplied by another boxcar, $\Pi_T(t)$, which describes the finite length of the observations. The frequency transform of the window function is then

$$W(\nu) = \left[\text{sinc}(\pi \nu T_D) \times \frac{1}{D} \text{III}_{1/D}(\nu) \right] * \text{sinc}(\pi \nu T), \tag{5.32}$$

where the transform of the Dirac comb is another (inverse spaced) Dirac comb:

$$\mathrm{III}_D(t) \overset{\mathcal{F}}{\longleftrightarrow} \frac{1}{D} \mathrm{III}_{1/D}(\nu) = \sum_{m=-\infty}^{\infty} \exp(-i2\pi\nu m D). \qquad (5.33)$$

The transform $W(\nu)$ of the window function will be convolved with the frequency transform $X(\nu)$ of the signal. Let us assume that the time-domain signal $x(t)$ gives a sharp peak in $X(\nu)$. The first two terms on the right-hand side of Eq. 5.32 (those in square braces) are the most important in describing the extra structure we will see around the true signal frequency. The final term on the right-hand side is just the sinc profile due to the finite length of the dataset.

The inverse Dirac comb $\mathrm{III}_{1/D}(\nu)$ describes the sideband peaks, that is, the aliased peaks separated by $1/D = 11.57\,\mu\mathrm{Hz}$ in frequency. The amplitudes of these peaks relative to the signal peak are controlled or modulated by the $\mathrm{sinc}(\pi\nu T_D)$ term, which depends on how much data T_D are collected each day. The shorter T_D (i.e., the lower the duty cycle of the observations) the more slowly the sinc function falls off in frequency, meaning that the sidebands will have significant amplitudes. As T_D approaches D—implying high duty-cycle observations—the sinc function falls off much more rapidly in frequency, meaning that the sidebands have much lower amplitudes relative to the signal peak.

Figure 5.1 shows some example *spectral window functions*, power spectra of sinusoidal signals $x(t)$ injected through different time-domain window functions, $w(t)$, of length 10 years. The window functions come from ground-based observations made by BiSON. We constructed three different window functions, comprised of data from one, three, and six telescopes, respectively, in the network. The resulting duty cycles are marked on each panel.

As more stations are added, and the overall duty cycle increases, so power in the diurnal sidebands is significantly reduced. The insets show zooms in frequency of the respective spectral windows. The inset to the top panel reveals structure arising from seasonal variations in the length of the observing day, which not only spreads power but also gives rise to yearly sidebands (note the peaks at $\pm0.032\,\mu\mathrm{Hz}$).

Impact of Gaps on the Calibration of the Power Spectrum

We now turn to the second important consequence of gaps in the data: their implications for the correct calibration of the frequency spectrum. The devil is again in the details, specifically, the way in which we present data for the Fourier transform code.

If we only need to present the cadences that have data, then direct application of Eqs. 5.12 and 5.14, as specified in Section 5.1.4, will still correctly calibrate the spectrum. However, if we are using a fast Fourier transform—implying regular cadences, where we must also include those cadences that have no data (as zero-valued entries)—extra care is needed. We can either compute the mean-square power in the time domain using *only* the nonzero cadences, or alternatively we can correct the mean-square power of the full lightcurve or time series, which includes the zero-valued cadences, by the fractional duty cycle of the data.

Computation of the duty cycle d is trivial for a dataset having a regular cadence. Formally, it is then just the fraction of nonzero cadences divided by the total number

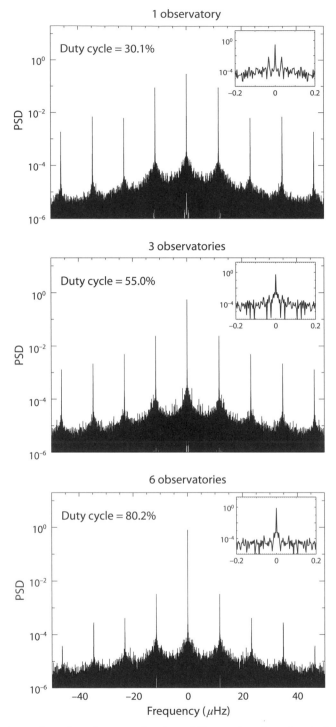

Figure 5.1. Spectral window functions for three different time-domain window functions, comprised of one (top panel), three (middle panel) and six (bottom panel) ground-based telescopes in BiSON. Insets: Zoom at zero frequency of the respective spectral windows.

of cadences in the dataset (including all zero-valued data). Eq. 5.12 then becomes

$$c_P = \frac{1}{d} \left(\frac{1}{N} \sum_{j=1}^{N} x_j^2 \right) \Bigg/ \left(\sum_{k=0}^{N/2} P_k' \right), \tag{5.34}$$

where we again stress that N now runs over *all* entries in the lightcurve or time series, including the zero-valued entries. Implicit in the above approaches is the assumption that the mean-square power of the available data is representative of that for the missing data.

With the calibration taken care of, there is, however, one last catch: we need to make an additional correction if we wish to measure the power of a mode in the frequency spectrum. This is because power from the signal is now present not only at the main signal peak but also at several diurnal sidebands (and also between the sidebands if the gap structure is irregular). It is the sum over all these components that equals the mean-square power of the signal in the time domain, *not* the power over just the main signal peak. An estimate of the signal power based on measurement of the main signal peak only (using techniques we shall discuss in Chapter 6) will therefore now underestimate the true power.

The extra correction is trivial if we have a dataset sampled on a regular cadence. We simply divide the measured power by another factor of the duty cycle (i.e., by d). For a dataset with irregular cadences, we can inject a commensurate sinusoid into the window function and find the extra multiplicative factor that is needed to recover the correct mean-square power $A^2/2$. This factor can then be applied to our analysis of the real oscillation peaks.

5.2 STATISTICS OF THE POWER SPECTRUM

We shall see in Chapter 6 that key stages of the asteroseismic analysis involve fitting multiparameter models describing the various components of the observed frequency-power spectrum. This demands not only the choice of an appropriate multicomponent model to describe the observed spectrum, but also, crucially, the fitting procedures must incorporate the correct noise statistics for the spectrum. In this section we shall introduce the basic components of the models that are used to describe all observed features of the power spectrum. However, first we make some introductory remarks on the statistics, since they are fundamental to understanding the appearance of a real frequency spectrum of data, noise and all. To be more specific, we seek to describe how the real, observed power spectrum is distributed about the true, underlying *limit* power spectrum. The latter is the spectrum we would obtain in the limit of averaging an infinite number of different (independent) noise realizations of the data. The main goal of the analysis is of course to find a best-fitting model that matches the actual limit spectrum as closely as possible.

Let us start with a simple example. Consider a dataset composed of normally distributed (Gaussian) random noise. This noise is assumed to be stationary—its properties do not alter over time—and so each datum is statistically independent and derived from the same underlying normal distribution. The complex (real and imaginary) parts of the discrete Fourier transform will also be normally distributed,

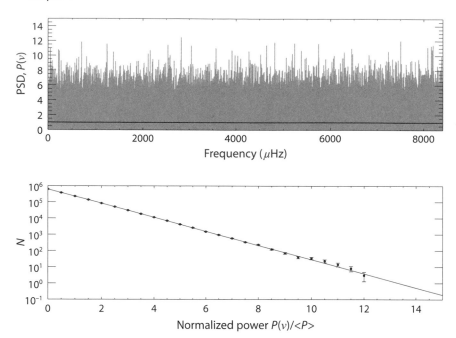

Figure 5.2. Top panel: Frequency-power spectrum of simulated random Gaussian noise, $P(\nu)$ (in gray), with the limit spectrum $\langle P \rangle$ marked in black. Bottom panel: Histogram of the cumulative distribution of the normalized powers $P(\nu)/\langle P \rangle$, showing the number of bins N above a given level.

and both have a zero mean and a variance that reflects that of the noise in the time domain. Each element of the power spectrum comprises the sum of the squares of the real and imaginary parts. When the power spectrum is constructed, the distributions shown by the complex parts are squared and exponentiated. The observed spectrum will therefore be distributed about the limit spectrum with negative exponential statistics, that is, a χ^2 distribution with two degrees of freedom (d.o.f). Because the noise here is white, the limit power spectrum will be flat, and the underlying mean power will reflect the variance of the noise.

The top panel of Figure 5.2 shows an example of such a spectrum (gray), with the limit spectrum marked by the horizontal black line. The visual impact of the negative-exponential noise is quite different from the more commonly encountered case of data scattered with Gaussian noise. Crucially, in any given frequency bin, the variance of the power is equal to the square of the underlying mean power. So, the higher the power, the larger is the scatter about the limit spectrum. The bottom panel of Figure 5.2 shows a histogram of the cumulative distribution of the normalized power in the top panel (i.e., after the observed power has been divided by the limit spectrum). The data follow the expected negative exponential trend (note the logarithmic-linear scale of the plot).

What about the more complicated data content in real observations of a star? Solar-like oscillations are stochastically excited by turbulence associated with the near-surface convection. In the frequency range occupied by the modes, there are also significant contributions from stellar granulation and photon shot noise. We may also regard the granulation as being a stochastic phenomenon, governed by a characteristic amplitude and timescale. The photon shot noise may be described in terms of normally distributed noise. There are also likely to be smaller contributions in the frequency range of interest from other phenomena. Here we invoke the central limit theorem and expect the composite background to have noise approaching that of a normal distribution in the time domain.

So, can we invoke the same statistics as for the simple white-noise spectrum? We still have a potential problem. Take the modes and granulation. Although both are excited by random processes, neither has a flat limit spectrum like the white noise. It would also be reasonable to assume that the same will be true of the other potential contributions in the data. A nonwhite spectrum in frequency implies the presence of nonzero correlations of the signals in time. In spite of this, it may be shown that such processes also give χ^2 with two degrees-of-freedom noise, but the observed power is distributed about a frequency-dependent limit spectrum. The combined spectrum will also have the same noise properties, even if some of the constituent elements are correlated (e.g., the modes and granulation). If we are therefore able to fit a good model of the underlying spectrum, division of the observed spectrum by that model will yield a flat residual spectrum having a constant power spectral density in frequency; any significant departures from a constant power would then point to issues related to the best-fitting model. We shall discuss model fitting in detail in Chapter 6.

Finally, there is the issue of gaps in the data, which will introduce correlations between supposedly independent frequency bins. Provided the duty cycle is not very low, it is usually safe to proceed on the assumption that frequencies at $1/T$ are independent.

We next introduce the frequency-domain characteristics of the different contributing components in the power spectrum, starting with the signatures due to the oscillations and the basic model used to describe each individual mode, a Lorentzian-like profile.

5.3 SOLAR-LIKE OSCILLATIONS IN THE FREQUENCY DOMAIN

5.3.1 Model Based on a Damped, Driven Oscillator

We have already seen that solar-like oscillations are stochastically excited and intrinsically damped by near-surface convection. A useful analogy that allows us to develop a prediction for the expected form of the resonant response in the frequency domain due to the oscillations is then a damped, randomly driven oscillator (e.g., see Section 3.6.2). Here we define our oscillator according to the equation

$$\ddot{x}(t) + 2\eta\dot{x}(t) + (2\pi\nu_0)^2 x(t) = k(t), \qquad (5.35)$$

where $x(t)$ is the displacement, ν_0 the natural frequency of the oscillator, η the linear damping constant, and $k(t)$ is the forcing function. The forcing comprises "kicks" applied at times t_0. The amplitude of each kick is drawn from a random Gaussian distribution of standard deviation a_δ. The forcing may therefore be written as

$$k(t) = a_\delta \delta(t - t_0), \tag{5.36}$$

where $\delta(t - t_0)$ is a delta function of unit amplitude. Provided $|K(\nu)|^2$, the power spectrum of the forcing function, is a slowly varying function of ν, and $\eta \ll 2\pi \nu_0$, the power spectrum in the frequency domain has a Lorentzian profile:

$$P(\nu) \propto \frac{|K(\nu)|^2}{\eta^2} \left(1 + \left(\frac{\nu - \nu_0}{\eta/2\pi}\right)^2\right)^{-1}. \tag{5.37}$$

This holds for both the spectrum of the displacement, $x(t)$, and of the velocity, $\dot{x}(t)$.

We carry the analogy across to the solar-like oscillations: the natural resonances of the star are continually reexcited by small kicks from the near-surface turbulence, which also acts to damp the resonances. We therefore postulate that the solar-like oscillations will have resonant profiles in the frequency domain that are, to good approximation, Lorentzian in shape. This is indeed found to be the case.

The full-width at half-maximum (FWHM) of the Lorentzian profile in cyclic frequency is

$$\Gamma = \eta/\pi = (\pi \tau)^{-1}, \tag{5.38}$$

where τ, which is the e-folding timescale (in amplitude) associated with the damping constant, is usually called the *lifetime* of the mode.

We may define the *height*, H (maximum power spectral density), of the resonant peak profile by

$$P(\nu) = H \left(1 + \left(\frac{\nu - \nu_0}{\Gamma/2}\right)^2\right)^{-1}. \tag{5.39}$$

It then follows that

$$H \propto \frac{|K(\nu)|^2}{\eta^2}. \tag{5.40}$$

The total mean-square power (variance in the time domain) is just the signal amplitude squared and is equal to the area under the Lorentzian profile. Recall that when we use the mean-square calibration of the frequency-power spectrum discussed in Section 5.1.4, we recover $A^2/2$, and not A^2. We therefore have

$$\frac{A^2}{2} \equiv A_{\text{rms}}^2 = \frac{\pi}{2} (H\Gamma). \tag{5.41}$$

It then follows that

$$A^2 \propto \frac{|K(\nu)|^2}{\eta}. \tag{5.42}$$

The energy (kinetic plus potential) of a resonant mode with associated mode inertia I is given by

$$E_{\text{osc}} = IMA^2, \tag{5.43}$$

where M is the mass of the star. With reference to Eqs. 3.170 and 3.171, the rate at which energy is supplied to the mode, $\mathcal{E}$, is

$$\mathcal{E} = 2\eta E_{\text{osc}}, \tag{5.44}$$

from which it follows that

$$\mathcal{E} \propto |K(\nu)|^2 IM. \tag{5.45}$$

5.3.2 Mode Lifetimes and the Duration of Observations

The appearance in the frequency domain of the resonant profile of a mode will depend critically on the relation of the mode lifetime τ to the length T of the dataset.

When $T \gg \tau$, the mode is oversampled in lifetime: the dataset will extend over many independent realizations of the mode, and the Lorentzian profile—as described by Eq. 5.39—will be well resolved (i.e., lie across many bins) in the frequency domain. In the other limit $T \ll \tau$, the phase of the oscillation will not alter significantly over the duration of the observations. In the resulting undersampled lifetime regime we have, in effect, the limiting case of a coherent signal, as described by Eq. 5.20, and the Lorentzian profile is unresolved. The changeover between the two regimes occurs formally at

$$T = 2\tau \equiv 2/(\pi\Gamma). \tag{5.46}$$

A description of H for the two limiting cases, given in units of power per Hertz, is therefore

$$H = \begin{cases} A^2/(\pi\Gamma), & \text{for } T \gg 2\tau \\ TA^2/2, & \text{for } T \ll 2\tau. \end{cases} \tag{5.47}$$

The above descriptions fail in the intermediate regime where τ is neither significantly shorter nor significantly longer than T. A description for H that encompasses both the limiting *and* the intermediate regimes is

$$H \simeq \frac{TA^2}{\pi T\Gamma + 2}. \tag{5.48}$$

This expression, although formally only approximate, does in practice show reasonable accuracy.

Figure 5.3 shows the expected change in the height H (solid line) of a mode of amplitude $A = 3.5\,\text{ppm}$ and lifetime $\tau = 3.7$ days, as a function of the observing time T. Predictions for the two limiting regimes are also plotted: the undersampled

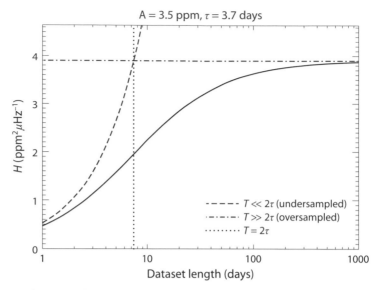

Figure 5.3. Change in the height H (solid line) of a mode of amplitude $A = 3.5$ ppm and lifetime $\tau = 3.7$ days, as a function of the observing time T. Predictions for the two limiting regimes are also plotted: the undersampled ($T \ll 2\tau$, dashed line) and oversampled ($T \gg 2\tau$, dot-dashed line) cases. The vertical dotted line marks the formal transition at $T = 2\tau$.

and oversampled cases. The vertical dotted line marks the formal transition at $T = 2\tau$.

Equations 5.47 and 5.48 suggest an interesting behavior of the height-to-background ratio in the mode peaks (i.e., the height relative to the broadband background power). In the frequency range occupied by detectable p modes, the background power will be dominated by contributions from stellar granulation and shot noise. We have more to say about how the background components are modeled later in the chapter. For now, we assume that both background components are stationary noise sources, meaning the underlying background power spectral density, expressed in units of power per Hertz, will remain unchanged irrespective of the duration T of the observations. Eq. 5.47 then implies that in the undersampled limit $T \ll \tau$, the height-to-background ratio will increase as more data are accumulated and T gets longer. However, once T has increased sufficiently to satisfy $T \gg \tau$, the ratio will no longer change, even if further data are accumulated. Here, further benefit accrues by virtue of the increase in frequency resolution.

5.3.3 Asymmetry of Oscillation Peaks

It is now well established that peaks in the frequency-power spectrum due to the solar p modes are slightly asymmetric and hence are not exactly Lorentzian in shape. We expect p modes observed in other solar-type stars to also show some asymmetry. In this section we consider in detail models that may be used to describe the asymmetry of the mode peaks.

This asymmetry is seen most easily in helioseismic data on modes of high degree, l, since we may average over the many rotationally split m components of each mode

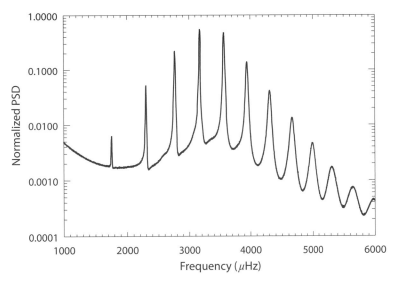

Figure 5.4. Frequency-power spectrum of helioseismic data collected by the HMI instrument on the NASA SDO spacecraft. Plotted is an m-averaged spectrum for solar p modes of $l = 250$, averaged over 28 72-day datasets.

to beat down the noise and reveal the underlying shapes of the peaks. Figure 5.4 shows an example spectrum made from helioseismic data collected by the HMI instrument on the NASA SDO spacecraft. Plotted is an m-averaged spectrum for solar p modes of degree $l = 250$, averaged over 28 72-day datasets. Some of the peaks are clearly asymmetric.

When observing other stars we detect only the low-l modes, and so we do not have large numbers of rotationally split components to average over. It is nevertheless still possible to see the asymmetry visually. The two panels of Figure 5.5 show a zoom of an oscillation spectrum made from BiSON Sun-as-a-star (low-l) data, containing the peak due to the $n = 14$ radial ($l = 0$) p mode. The gray lines show the raw spectrum after light smoothing (a single boxcar of filter width 0.01 μHz). The thin dark lines show the spectrum after application of heavier smoothing (a double boxcar filter, each of width 0.15 μHz); the thick dark lines show the fit of an asymmetric Lorentzian profile to the spectrum (of which more below).

The asymmetry is small, and so we have also plotted the spectrum on a logarithmic scale in power (bottom panel, with the horizontal dashed line included to guide the eye) to accentuate the effect. The coefficient α that is commonly used to measure the asymmetry of the mode peaks (see below) is given by

$$\alpha = \frac{(P_+ - P_-)}{2H}, \tag{5.49}$$

where H is the maximum power spectral density (height) of the peak, and P_+ and P_- are the power spectral densities at frequencies $\Gamma/2$ from the mode frequency ν_0:

$$P_+ \equiv P(\nu_0 + \Gamma/2), \tag{5.50}$$

and

$$P_- \equiv P(\nu_0 - \Gamma/2), \tag{5.51}$$

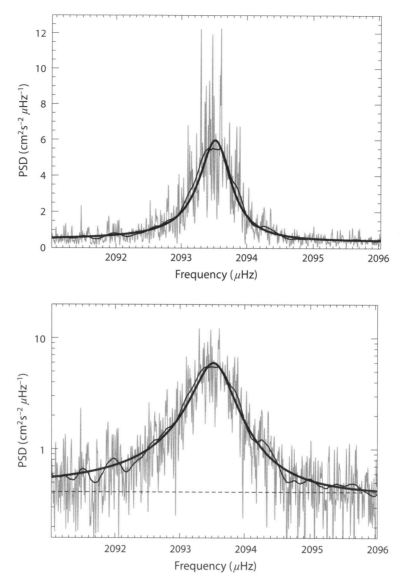

Figure 5.5. Frequency-power spectrum of 20 years of BiSON Sun-as-a-star observations, showing the peak due to the $n = 14$ radial p mode. The gray lines show the raw spectrum after light smoothing (a single boxcar of filter width 0.01 μHz). The thin dark lines are the spectrum after application of heavier smoothing (a double boxcar filter, each of width 0.15 μHz). The thick dark lines show the fit of an asymmetric Lorentzian profile to the spectrum. Top panel: The spectrum on a linear scale in power and frequency. Bottom panel: the spectrum on a logarithmic-linear scale to show more clearly the peak asymmetry (with the horizontal dashed line included to guide the eye).

where $\Gamma = 1/(\pi\tau)$ (Eq. 5.38) is the width of the peak given by the damping time τ. The mode peak in Figure 5.5 has an asymmetry of $\simeq -0.04$. This negative asymmetry is a characteristic of solar oscillation spectra made from Doppler velocity observations. When the solar frequency-power spectrum is made from photometric observations, the sign of the asymmetry is reversed. Later in this section we shall offer a possible explanation for this reversal of asymmetry.

What is the cause of the asymmetry? There are believed to be two contributions. The first contribution is due to the spatially localized nature of the excitation source. The generation of acoustic noise is most efficient in the near-surface layers of the outer convection zone. This is where the Mach number of the turbulent flows is highest. The radial extent of the excitation source is therefore very small compared to the extent of the cavities associated with, for example, the globally coherent modes. As such, we have a situation that is not unlike a Fabry Perot interferometer, in which interference between light rays that have followed different paths in the cavity leads to asymmetry in frequency of the resonant profiles of the cavity. Here, the asymmetry of the p mode profiles arises from the interference of acoustic waves that have been emitted by the same acoustic event but have followed different paths in the interior of the star before subsequently interfering.

Consider, for example, upward- and downward-directed waves. If the excitation source lies in the cavity, then the upward wave undergoes reflection almost immediately. It then travels downward, potentially undergoing multiple reflections in the cavity, like the initially downward-directed wave. When the two waves interfere, they will have accumulated different phase shifts (on reflections at the cavity boundaries), and this modifies the shapes of the resonant profiles in frequency space.

One may model the effect mathematically, for example, by considering wave propagation of different rays in a cavity or by solving for the resonances of an acoustic potential that has a δ-function-like excitation source. Here we adopt a slightly different, illustrative approach that allows us to again make use of the damped oscillator analogy for the modes. Consider a damped oscillator that undergoes two discrete instantaneous excitations from a δ-function source, like the one in Eqs. 5.35 and 5.36. Let the excitations be separated in time by $\Delta t'$. We can think of these excitations as mimicking the upward- and downward-traveling events above; here, we incorporate a phase shift by shifting the two events in time. The expected frequency-power spectrum of the oscillator signal arising from the two kicks is proportional to

$$P(\omega) \propto \left| \frac{1}{\omega - \omega_0 - i\eta} + \frac{\exp(-i\omega\Delta t')}{\omega - \omega_0 - i\eta} \right|^2. \tag{5.52}$$

This expression simplifies, allowing it to be rewritten as

$$P(\nu) \propto \frac{2[1 + \cos(2\pi\nu\Delta t')]}{\eta^2} \left(1 + \left(\frac{\nu - \nu_0}{\eta/2\pi} \right)^2 \right)^{-1}. \tag{5.53}$$

Notice how we have presented the above in a similar form to Eq. 5.37. The rightmost part again describes a Lorentzian, but what about the part that contains the cosinusoidal function? This part describes the impact of the phase shift. When $\omega_0\Delta t' \equiv 2\pi\nu_0\Delta t' = \pi k$, where k is an integer, the excitations are exactly in phase (or exactly in anti-phase), and the resulting peak is a pure Lorentzian. However, when $2\pi\nu_0\Delta t' \neq \pi k$, the peak will show asymmetry.

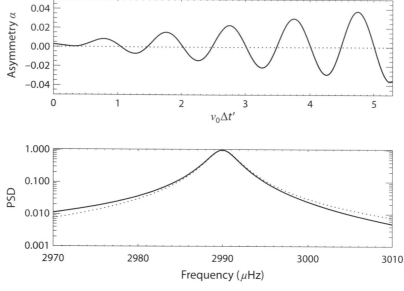

Figure 5.6. Top panel: Fractional peak asymmetry α given at different values of the product $\nu_0 \Delta t'$, for the case of an oscillator of frequency $\nu_0 = 2{,}990\,\mu\text{Hz}$, $\Gamma = 3.5\,\mu\text{Hz}$, and $\Delta t = 4.44\,\text{sec}$. Bottom panel: Resonant profile (solid line) given by the last combination of $\nu_0 \Delta t$ plotted in the top panel. The dotted line shows the simple Lorentzian profile at every combination of $\nu_0 \Delta t'$ where the excitations are in exact phase (or anti-phase).

The top panel of Figure 5.6 shows the fractional peak asymmetry α given at different values of the product $\nu_0 \Delta t'$, for an oscillator of frequency $\nu_0 = 2{,}990\,\mu\text{Hz}$, $\Gamma = 3.5\,\mu\text{Hz}$, and with excitations separated in time by $\Delta t' = 4.44\,\text{sec}$.[1] The bottom panel shows the resonant profile (solid line) given by the last combination of $\nu_0 \Delta t'$ plotted in the top panel. This peak has an asymmetry of $\alpha = -0.035$. The dotted line shows for reference the simple Lorentzian profile given at every combination of $\nu_0 \Delta t'$ where the excitations are in exact phase (or anti-phase).

The second contribution to the asymmetry arises from the correlation between the convective granulation and the signal stemming from the modes themselves. It is of course the convection that excites the modes, and hence some degree of correlation is to be expected. Since the ratio of the granulation signal amplitude to the p-mode amplitudes is significantly higher in photometric as opposed to Doppler velocity observations, this second effect is believed to be much more important in photometric data. Indeed, this offers one possible explanation for the reversal of the observed peak asymmetry referred to early in this section.

To illustrate the impact of this correlation, we consider the various sources in the frequency domain. First, let us again assume that the excitation may be described by a random Gaussian function. Even though the granulation depends on frequency, the Gaussian approximation is a reasonable one, since locally (in frequency) the spectrum of the granulation is nearly white. The complex excitation function in

---

[1] Chosen so that with $N = 1{,}048{,}576$ points in the simulated time series, the period of the oscillator was commensurate with the total length of the dataset.

the frequency domain, $K(\nu)$, may therefore be written as

$$K(\nu) = k_r(\nu) + ik_i(\nu), \tag{5.54}$$

where $k_r(\nu)$ and $k_i(\nu)$ represent the real and imaginary parts, respectively of the random excitation. We again assume that the frequency response or profile of the mode is described by a Lorentzian. The complex profile is then

$$L(\nu) = \frac{\xi(\nu)}{1 + \xi(\nu)^2}\sqrt{H/2} + i\frac{1}{1 + \xi(\nu)^2}\sqrt{H/2}, \tag{5.55}$$

where H is the mode height, and the term $\xi(\nu)$ corresponds to

$$\xi(\nu) = \frac{\nu - \nu_0}{\Gamma/2}, \tag{5.56}$$

with ν_0 and Γ again being the central frequency and linewidth of the mode, respectively. Finally, we include a noise source, which, again for simplicity, is taken to be a random Gaussian function, here multiplied by a frequency-dependent amplitude:

$$B(\nu) = \sqrt{\frac{b(\nu)}{2}}[b_r(\nu) + ib_i(\nu)]. \tag{5.57}$$

The expected response in the complex frequency domain is

$$X(\nu) = L(\nu)K(\nu) + B(\nu). \tag{5.58}$$

If the excitation function and the background noise were uncorrelated, the limit power spectral density would be described by:

$$P(\nu) = \left\langle |X(\nu)|^2 \right\rangle = \frac{H}{1 + \xi(\nu)^2} + b(\nu), \tag{5.59}$$

which is the usual Lorentzian profile added (incoherently) to the background power $b(\nu)$. Here the angle brackets indicate an average over a large number of realizations of the excitation function and of the background noise.

What if instead the excitation function and the background noise are correlated, so that in each frequency bin

$$\langle k_r b_r \rangle = \langle k_i b_i \rangle = \rho_{kb}, \tag{5.60}$$

where ρ_{kb} is the coefficient of correlation between the background noise and the excitation function? The limit power spectral density now takes a more complicated form:

$$P(\nu) = \frac{H}{1 + \xi^2}\left[1 + 2\rho_{kb}\xi(\nu)\sqrt{b/H}\right] + b(\nu). \tag{5.61}$$

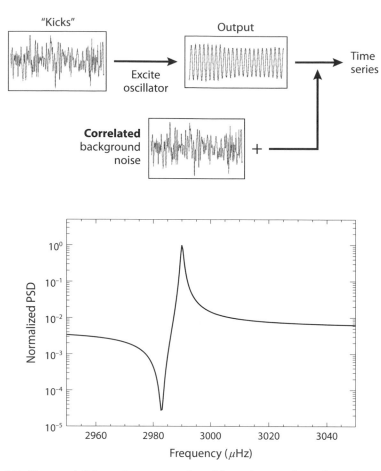

Figure 5.7. Top panel: Schematic representation of the excitation of a single mode. Correlated background noise is added, which is 100% correlated with the excitation function. Bottom panel: The resulting frequency-power spectrum. Reproduced with permission from Chaplin et al., 2008, *A Framework for Describing Correlated Excitation of Solar p-Modes, Astron. Nachr.*, **329**, 440. Copyright 2008 WILEY-VCH Verlag GmbH&Co. KGaA, Weinheim.

For nonzero ρ_{kb}, this gives an asymmetric profile. The asymmetry coefficient α defined according to Eq. 5.49 is then given by

$$\alpha = \rho_{kb}\sqrt{b/H}, \tag{5.62}$$

so that the asymmetric resonant profile may also be written in the more convenient form

$$P(\nu) = \frac{H}{1 + \xi(\nu)^2}\left(1 + 2\alpha\xi(\nu)\right) + b(\nu). \tag{5.63}$$

The above correlated noise scenario is illustrated schematically in Figure 5.7. The excitation function is represented by a time series of kicks of random Gaussian noise. The kicks excite an oscillator, which in the example shown has a natural frequency of $\nu_0 = 2990\,\mu\text{Hz}$ and a linewidth of $\Gamma = 1\,\mu\text{Hz}$. When the time series of kicks is added to the output of the oscillator, it acts as *correlated* background noise and the frequency-power spectrum shows a peak with asymmetry. In the example, shown we scaled the time series of kicks to give a background-to-height ratio of $b/H = 0.5\%$. We did not add any uncorrelated background to the time series, so that the background noise was 100% correlated with the excitation (i.e., $\rho_{kb} = 1$), giving the very extreme shape shown. Because this correlation was positive, the sign of the peak asymmetry was also positive (i.e., more power on the high-frequency side of the resonance). Negative correlation will give a peak showing negative asymmetry.

What happens to the observed power spectral density when the excitation of different modes is correlated? To illustrate what happens, we take the simple case of a frequency-power spectrum comprising two modes whose excitation functions are correlated in time, but there is no background noise. Let the frequency response and excitation function of the first mode be $L_1(\nu)$ and $K_1(\nu)$, respectively. The corresponding functions for the second mode are $L_2(\nu)$ and $K_2(\nu)$. The observed complex frequency spectrum is therefore

$$X(\nu) = L_1(\nu)K_1(\nu) + L_2(\nu)K_2(\nu), \tag{5.64}$$

and the observed power spectral density is

$$P(\nu) = \left\langle |X(\nu)|^2 \right\rangle = [L_1(\nu)K_1(\nu) + L_2(\nu)K_2(\nu)] \left[L_1^*(\nu)K_1^*(\nu) + L_2^*(\nu)K_2^*(\nu) \right]. \tag{5.65}$$

In what follows we shall drop from the notation the explicit dependence of the functions on ν. Multiplying out the terms in Eq. 5.65, we have

$$P = L_1 L_1^* K_1 K_1^* + L_1 L_2^* K_1 K_2^* + L_1^* L_2 K_1^* K_2 + L_2 L_2^* K_2 K_2^*. \tag{5.66}$$

Next, we recall that the excitation of a mode is formulated in terms of the real and imaginary parts of a complex random function. The excitation functions of the two modes may be written as

$$K_1 = k_{1r} + ik_{1i}, \tag{5.67}$$

and

$$K_2 = k_{2r} + ik_{2i}. \tag{5.68}$$

We then define a coefficient ρ_{kk} to describe the correlation of the excitation of the two modes:

$$\langle k_{1r} k_{2r} \rangle = \langle k_{1i} k_{2i} \rangle = \rho_{kk}. \tag{5.69}$$

The coefficient ρ_{kk} is analogous to the coefficient ρ_{kb} above, which described the correlation of the excitation of a mode with the background noise. Here, ρ_{kk} instead fixes the correlation of the excitation of one mode with another.

The cross-term $K_1 K_2^*$ in Eq. 5.66 can be written in terms of the components of Eqs. 5.67 and 5.68:

$$
\begin{aligned}
K_1 K_2^* &= [k_{1r} + i k_{1i}][k_{2r} - i k_{2i}] \\
&= k_{1r} k_{2r} + k_{1i} k_{2i} + i[k_{1i} k_{2r} - k_{2i} k_{1r}] \\
&= 2\rho_{kk} + 0 = 2\rho_{kk}.
\end{aligned}
$$

The other cross-term, $K_1^* K_2$, is also equal to $2\rho_{kk}$. We can also simplify the terms $K_1 K_1^*$ and $K_2 K_2^*$. Let us take the first of these:

$$
\begin{aligned}
K_1 K_1^* &= [k_{1r} + i k_{1i}][k_{1r} - i k_{1i}] \\
&= k_{1r}^2 + k_{1i}^2 = 2.
\end{aligned}
$$

We then also have that $K_2 K_2^* = 2$. The expression for $P(\nu)$ therefore simplifies to

$$
P(\nu) = 2L_1 L_1^* + 2L_2 L_2^* + 2\rho_{kk}[L_1 L_2^* + L_1^* L_2]. \tag{5.70}
$$

We then expand the terms:

$$
P(\nu) = \frac{H_1}{1 + \xi_1(\nu)^2} + \frac{H_2}{1 + \xi_2(\nu)^2} + 2\rho_{kk} \sqrt{H_1 H_2} \frac{1 + \xi_1(\nu)\xi_2(\nu)}{(1 + \xi_1(\nu)^2)(1 + \xi_2(\nu)^2)}. \tag{5.71}
$$

When $\rho_{kk} = 0$, and the excitation of the two modes is independent, we see that Eq. 5.71 reduces to the sum of two Lorentzians. This case is shown in Figure 5.8. When $\rho_{kk} \neq 0$ and the excitation is correlated, the observed power spectral density is modified by the third term in Eq. 5.71, which contains cross-terms between the two modes. The mode peaks are then asymmetric.

Figure 5.9 shows the case $\rho_{kk} = 1$ (i.e., the same time series of random Gaussian noise is used to excite both oscillators). It is worth remembering that the asymmetry here does not come from the influence of correlated background noise, as was the case for the single-mode example illustrated in Figure 5.7, since no background noise was included. Instead, it comes from the interaction of the two modes.

We draw an important conclusion from this discussion: correlated mode excitation can contribute to the observed asymmetry of modes. Is the notion of correlated excitation relevant to solar-like oscillators? We might reasonably hypothesize that the excitation of overtones of a given angular degree and azimuthal order will be correlated in time. This follows from the assumption that the excitation function of a given mode may be described in terms of the component of the granulation that has the same surface spherical harmonic projection (over the corresponding range of temporal frequencies). The cumulative effect of the interactions among all the overtones might then be expected to give rise to significant modifications to the power spectral density. Peak asymmetry arising from these correlations would add to the contribution from the correlated background noise (and from the finite radial extent of the excitation source).

Finally, to illustrate the composite case of correlated excitation and correlated background, we consider two modes whose excitation functions are correlated in time and the presence of correlated background noise. The observed complex

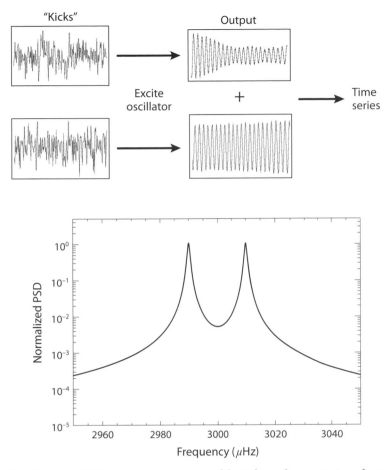

Figure 5.8. Top panel: Schematic representation of the independent excitation of two modes, with no background noise added. Bottom panel: The resulting frequency-power spectrum. Reproduced with permission from Chaplin et al., 2008, *A Framework for Describing Correlated Excitation of Solar p-Modes, Astron. Nachr.*, **329**, 440. Copyright 2008 WILEY-VCH Verlag GmbH&Co. KGaA, Weinheim.

frequency spectrum is now

$$X(\nu) = L_1(\nu)K_1(\nu) + L_2(\nu)K_2(\nu) + B(\nu). \tag{5.72}$$

The observed power spectral density is therefore

$$P(\nu) = 2L_1L_1^* + 2L_2L_2^* + 2\rho_{kk}[L_1L_2^* + L_1^*L_2]$$

$$+ BB^* + B^*[K_1L_1 + K_2L_2] + B[L_1^*K_1^* + L_2^*K_2^*]. \tag{5.73}$$

The term BB^* is just the expectation of the noise background described by Eq. 5.57 (i.e., $BB^* = b$). The final two terms depend on the correlation between the excitation

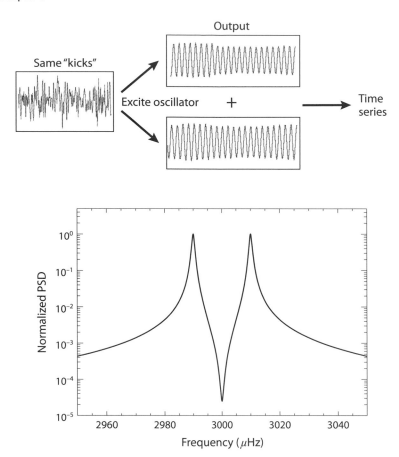

Figure 5.9. Top panel: Schematic representation of the excitation of two modes, where the excitation is 100% correlated but no background noise is added. Bottom panel: The resulting frequency-power spectrum. Reproduced with permission from Chaplin et al., 2008, *A Framework for Describing Correlated Excitation of Solar p-Modes, Astron. Nachr.*, **329**, 440. Copyright 2008 WILEY-VCH Verlag GmbH&Co. KGaA, Weinheim.

and background noise. Let us take the term B^*K_1. It may be written in terms of the real and imaginary components of B^* (cf. Eq. 5.57) and K_1 (Eq. 5.67) respectively:

$$N^*K_1 = \sqrt{\frac{b}{2}}\,[b_r - ib_i][k_{1r} + ik_{1i}]$$

$$= \sqrt{\frac{b}{2}}\,[k_{1r}b_r + k_{1i}b_i + i(k_{1i}b_r - k_{1r}b_i)]$$

$$= 2\rho_{kb}\sqrt{\frac{b}{2}} + 0 = 2\rho_{kb}\sqrt{\frac{b}{2}}. \tag{5.74}$$

The terms B^*K_2, BK_1^*, and BK_2^* simplify to give the same expression. Putting this all together, the expression for P can be written as:

$$P(\nu) = 2L_1L_1^* + 2L_2L_2^* + 2\rho_{kk}[L_1L_2^* + L_1^*L_2]$$
$$+ b + 2\rho_{kb}\sqrt{\frac{b}{2}}[L_1 + L_2 + L_1^* + L_2^*]. \tag{5.75}$$

Equation 5.75 comprises two parts. The first part, which includes the first three terms on the right-hand side of the equation, is just the observed power spectral density for two correlated modes (i.e., Eqs. 5.70 and 5.71). The second part has a contribution, b, which is just the power spectral density of the background noise, and a further term that describes the correlations between the modes and the background noise. These correlations modify the observed power spectral density in a nontrivial manner and contribute to the observed asymmetry of the mode peaks.

For completeness, we can take one more step and expand the terms in Eq. 5.75 to give a final expression for the power spectral density:

$$P(\nu) = \frac{H_1}{1 + \xi_1(\nu)^2}\left(1 + 2\rho_{kb}\xi_1(\nu)\sqrt{\frac{b}{H_1}}\right) + \frac{H_2}{1 + \xi_2(\nu)^2}\left(1 + 2\rho_{kb}\xi_2(\nu)\sqrt{\frac{b}{H_2}}\right)$$
$$+ b + 2\rho_{kk}\sqrt{H_1H_2}\frac{1 + \xi_1(\nu)\xi_2(\nu)}{(1 + \xi_1(\nu)^2)(1 + \xi_2(\nu)^2)}. \tag{5.76}$$

When $\rho_{kb} = 0$ and $\rho_{kk} = 0$, we have uncorrelated excitation and uncorrelated background noise. The resulting frequency-power spectrum, like the one shown in Figure 5.10, is given by the incoherent addition of two Lorentzians with a flat background.

Figure 5.11 shows what happens for the more complicated case when $\rho_{kb} = 1$ and $\rho_{kk} = 1$. We now have two factors contributing to the observed asymmetry of the peaks. The first contribution is the interaction between the two modes, whose excitation is 100% correlated. The second contribution comes from the correlated background, which further distorts the mode peaks.

In Section 6.4 (particularly in Sections 6.4.4 and 6.4.7) we shall use the ideas introduced here when we fit models to the mode peaks.

5.4 GLOBAL PROPERTIES OF THE OSCILLATION SPECTRUM

Having considered the appearance of peaks due to the individual modes, we now consider the overall or global appearance of the oscillation spectrum. We have noted previously how the spectra shown by different solar-like oscillators are at first glance strikingly similar, with the observed powers modulated in frequency by an envelope that in most cases has a Gaussian-like shape. We shall use the frequency-power spectrum of *Kepler*-410 to illustrate the key features and global parameters (Figure 5.12)

The frequency at which the observed power is strongest is ν_{max}, which we have seen appears to scale to very good approximation with the atmospheric acoustic

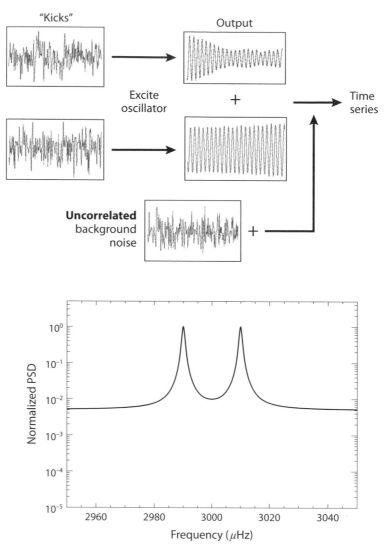

Figure 5.10. Top panel: Schematic representation of the independent excitation of two modes, where the time series also includes uncorrelated background noise. Bottom panel: The resulting frequency-power spectrum. Reproduced with permission from Chaplin et al., 2008, *A Framework for Describing Correlated Excitation of Solar p-Modes, Astron. Nachr.*, **329**, 440. Copyright 2008 WILEY-VCH Verlag GmbH&Co. KGaA, Weinheim.

cut-off frequency ν_{ac} (including a simple scaling derived for an isothermal atmosphere). As stars evolve and the surface gravity and temperature decrease, ν_{ac} and ν_{max} shift toward lower frequencies. The power envelope of the observed oscillation spectrum also gets narrower. Let us use this information to define some useful global parameters associated with the oscillation spectrum.

We know that the dominant observed frequency spacing is the large separation $\Delta\nu$. In main sequence stars, each $\Delta\nu$-wide segment of the spectrum will contain significant power due to the visible $l = 0$, 1, and 2 modes, and, in the highest signal

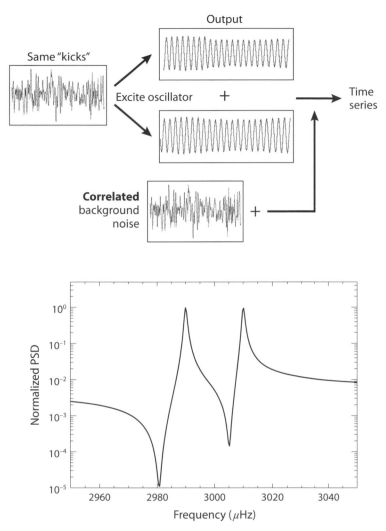

Figure 5.11. Top panel: Schematic representation of the excitation of two modes, where the excitation is 100% correlated, as in Figure 5.10; however, here the time series also includes background noise that is completely correlated with the excitation. Bottom panel: The resulting frequency-power spectrum. Reproduced with permission from Chaplin et al., 2008, *A Framework for Describing Correlated Excitation of Solar p-Modes, Astron. Nachr.*, **329**, 440. Copyright 2008 WILEY-VCH Verlag GmbH&Co. KGaA, Weinheim.

to noise ratio (SNR) observations, small contributions from $l = 3$ (and possibly even higher l) modes. The integrated power in each segment will therefore correspond to the power due to the radial mode multiplied by the sum of the visibilities (in power) over l.

Let us define A_{max} to be the equivalent radial mode amplitude at the center of the p mode envelope (i.e., at ν_{max}). We also define the factor ζ to be the sum of the

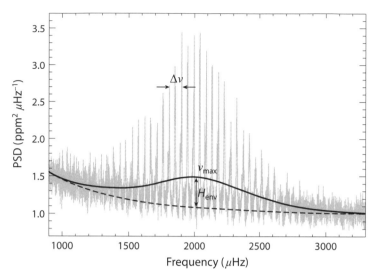

Figure 5.12. Frequency-power spectrum of *Kepler*-410 (gray), showing the Gaussian-like envelope of power due to the p mode oscillations. Key global parameters are also marked on the plot.

normalized mode visibilities (in power; see Section 4.1):

$$\zeta = \sum_l (S_l/S_0)^2. \tag{5.77}$$

If we re-bin the power spectrum into $\Delta\nu$-wide segments, the maximum power spectral density in the segment at the center of the spectrum will be

$$H_{\mathrm{env}} = A_{\max}^2 \zeta / (2\Delta\nu) . \tag{5.78}$$

The magnitude of H_{env} is marked on Figure 5.12. In the figure, the thick dark line corresponds to the heavily smoothed frequency-power spectrum; the dashed line follows the estimated background power (see Section 5.5). It is important to stress that H_{env} is not the same as the height shown by the strongest radial mode. Recall that we derived expressions for mode height in Sections 5.3.1 and 5.3.2, specifically Eqs. 5.41, 5.47, and 5.48. Assuming that the mode peaks are well resolved in frequency, we can write the maximum mode height in terms of $A_{\max}$:

$$H_{\max} = A_{\max}^2 / (\pi\Gamma), \tag{5.79}$$

where Γ is the width of the mode peak. It then follows that we can relate H_{env} to $H_{\max}$:

$$H_{\mathrm{env}} = \left(\frac{\pi\zeta\Gamma}{2\Delta\nu}\right) H_{\max}. \tag{5.80}$$

If we adopt a Gaussian function to describe the $\Delta\nu$-binned spectrum, the height of the Gaussian will be H_{env}. We also introduce the term Γ_{env} to describe the FWHM of the Gaussian. The total integrated power due to the modes, P_{tot}, is then just the integral underneath the Gaussian, which may be written in terms of its height and FWHM as

$$P_{\text{tot}} = \left(\frac{\pi}{4 \ln 2} \right)^{1/2} H_{\text{env}} \Gamma_{\text{env}}. \tag{5.81}$$

To predict H_{env} and hence P_{tot} for different stars, we need to know $A_{\max}$ and $\Delta \nu$. We may estimate the latter from the mean stellar density. To estimate $A_{\max}$, we refer back to Section 4.2.3.

First, we must consider the type of observations that have been, or are to be, made; here, we shall consider the photometric case. Next, as shown in Section 4.2.1, the instrumental response affects the amplitude of the measured oscillation signal. Let $A_{\max,\text{bol}}$ and $A_{\max,\lambda}$ be the maximum amplitudes for bolometric and narrow-band photometric observations, respectively, with the already introduced $A_{\max}$ defined to be the maximum amplitude for a given extended-wavelength response (e.g., that of *Kepler*). Eqs. 4.48, 4.49 and 4.50, taken together with Eq. 4.39, then suggest scaling relations of the form

$$A_{\max,\text{bol}} \propto (L/M)^s \, T_{\text{eff}}^{-(r-1)}, \tag{5.82}$$

$$A_{\max,\lambda} \propto (L/M)^s \, \lambda^{-1} T_{\text{eff}}^{-r}, \tag{5.83}$$

and

$$A_{\max} \propto (L/M)^s \, \mathcal{C}_{\text{bol}}^{-1} T_{\text{eff}}^{-(r-1)}, \tag{5.84}$$

where $\mathcal{C}_{\text{bol}}$ describes the bolometric correction for the chosen instrumental response (see Eq. 4.31), and $r = 1.5$ for adiabatic oscillations. If the bolometric correction can be described as a power law in T_{eff}, that is,

$$\mathcal{C}_{\text{bol}} = \left(\frac{T_{\text{eff}}}{T_{\text{bol}}} \right)^{\beta_{\text{bol}}}, \tag{5.85}$$

where the power-law index is β_{bol}, and T_{bol} is the effective temperature at which the observed amplitude is the same as the bolometric amplitude, then Eq. 5.84 becomes

$$A_{\max} \propto (L/M)^s \, T_{\text{eff}}^{-(r-1+\beta_{\text{bol}})}. \tag{5.86}$$

Using the bolometric scale allows for a proper comparison of data collected by different instruments. For example, we could correct different observations to the bolometric scale, using the appropriate $\mathcal{C}_{\text{bol}}$, and then test the simple bolometric model described by Eq. 5.82 using the corrected amplitudes. However, this does have the disadvantage of using estimated temperatures, which are uncertain, to compute the bolometric corrections. What is arguably a better approach is to instead correct the bolometric model to the observed scale, which is in effect what Eq. 5.86 achieves.

Using solar values to calibrate the amplitude relation then gives:

$$A_{\max} \simeq A_{\max,\odot}\, C_{\text{bol},\odot}^{-1}\, \left(\frac{L/L_\odot}{M/M_\odot}\right)^s \left(\frac{T_{\text{eff}}}{T_{\text{eff},\odot}}\right)^{-(r-1+\beta_{\text{bol}})}, \tag{5.87}$$

where $A_{\max,\odot}$ is the solar amplitude in the adopted bandpass, and we have used the fact that

$$C_{\text{bol}} \equiv \left(\frac{T_{\text{eff}}}{T_{\text{eff},\odot}}\right)^{\beta_{\text{bol}}} C_{\text{bol},\odot}, \tag{5.88}$$

where $C_{\text{bol},\odot}$ is the bolometric correction at the solar effective temperature $T_{\text{eff},\odot} = 5{,}777$ K. Eq. 5.87 may then be written equivalently as

$$A_{\max} \simeq A_{\max,\odot}\, C_{\text{bol},\odot}^{-1}\, \left(\frac{\nu_{\max}}{\nu_{\max,\odot}}\right)^{-s} \left(\frac{T_{\text{eff}}}{T_{\text{eff},\odot}}\right)^{3.5s-r+1-\beta_{\text{bol}}}. \tag{5.89}$$

In the case of *Kepler*, $A_{\max,\odot} \simeq 3.5$ ppm, and the bolometric correction may be described to good approximation by $T_{\text{bol}} = 5{,}934$ K and $\beta_{\text{bol}} = 0.8$.

There is some spread in the best-fitting coefficients given by fits to real data, depending for example on the evolutionary state of the stars. Approximate values from *Kepler* data of the coefficients for cool main sequence and subgiant stars are $s \simeq 0.7$ and $r \simeq 3.5$; there is evidence to suggest that the best-fitting r for red giants is larger. Other, more complicated models have been tested against *Kepler* data. Those which include some additional dependence on mass tend to fare better than the above model. All relations tend to overestimate the amplitudes of hotter solar-like oscillators. A suitable multiplicative correction may then be added, for example, Eq. 5.87 then becomes:

$$A_{\max} \simeq A_{\max,\odot}\, C_{\text{bol},\odot}^{-1}\, \mathcal{F}(L, T_{\text{eff}}) \left(\frac{L/L_\odot}{M/M_\odot}\right)^s \left(\frac{T_{\text{eff}}}{T_{\text{eff},\odot}}\right)^{-(r-1+\beta_{\text{bol}})}, \tag{5.90}$$

where the additional corrective term has the form

$$\mathcal{F}(L, T_{\text{eff}}) = 1 - \exp\left(-\frac{T_{\text{red}} - T_{\text{eff}}}{1{,}550\ \text{K}}\right), \tag{5.91}$$

with

$$T_{\text{red}} = T_{\text{red},\odot}\, (L/L_\odot)^{-0.093}. \tag{5.92}$$

Here, T_{red} is the temperature of the red edge of the instability strip (for δ-Scuti pulsators) at the luminosity of the star; and $T_{\text{red},\odot} \simeq 8{,}900$ K. The red edge provides an approximate guide to the locations at which discernible near-surface convection zones are established. The closer the star lies in the HR diagram to the red edge (on the cool side), the stronger is the attenuation of the predicted maximum mode amplitude.

The final step needed to predict P_{tot} for other stars is a prediction for Γ_{env}, the width of the oscillation power envelope. Results from *Kepler* data have suggested

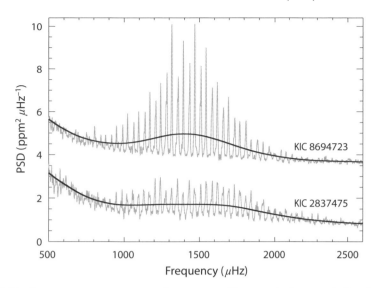

Figure 5.13. Frequency-power spectra of two stars with very similar ν_{max} (offset in power spectral density for clarity). Black lines are the heavily smoothed spectra. The top star (KIC 8694723) is an early G-type star, which has a classic Gaussian-like oscillation power envelope. The bottom star (KIC 2837475) is an early F-type star—about 600 K hotter than KIC 8694723—which has a much flatter envelope.

a relation of the form

$$\Gamma_{\mathrm{env}} \simeq \begin{cases} c\nu_{\mathrm{max}}^{k}, & \text{red giants} \\ \nu_{\mathrm{max}}/2, & \text{main sequence stars and subgiants,} \end{cases} \tag{5.93}$$

where the coefficients take values $c \simeq 0.66$ and $k \simeq 0.88$. The transition between these two regimes is yet to be fully explored. Moreover, analysis of the *Kepler* data suggests that there is probably an additional temperature dependence in Γ_{env} not captured by Eq. 5.93. For example, hotter F-type stars show flatter envelopes (see Figure 5.13) to the extent that a Gaussian is no longer a good description of the shape.

Using the amplitude scaling described by Eq. 5.89 with the coefficients s and r, but neglecting the high-temperature correction here for simplicity and using $\Gamma_{\mathrm{env}} \propto \nu_{\mathrm{max}}^{k}$ and $\Delta\nu \propto \nu_{\mathrm{max}}^{\beta}$, we can write an approximate scaling relation for P_{tot}:

$$P_{\mathrm{tot}} \simeq P_{\mathrm{tot},\odot}\, \mathcal{C}_{\mathrm{bol},\odot}^{-2} \left(\frac{\nu_{\mathrm{max}}}{\nu_{\mathrm{max},\odot}}\right)^{t} \left(\frac{T_{\mathrm{eff}}}{T_{\mathrm{eff},\odot}}\right)^{u}, \tag{5.94}$$

where $t \equiv -2s - \beta + k$, and $u \equiv 7s - 2r + 2 - 2\beta_{\mathrm{bol}}$. For cool main sequence and subgiant stars we have $t \approx -1.2$ and $u \approx -1.7$.

Finally, we make a few remarks about the variation of peak line width, Γ, across the oscillation spectrum. All solar-like oscillators show an increase in damping, and hence Γ, with increasing frequency. This increase is not always monotonic and often shows a small dip around ν_{max}. Line widths of detectable modes can vary by an order of magnitude or more (depending on the SNR of the data). It is therefore

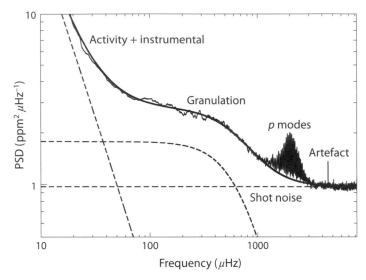

Figure 5.14. Frequency-power spectrum of *Kepler*-410 (gray), with annotations indicating the various contributions to the observed power spectral density. The heavy dark line shows a best-fitting model of the composite background, which comprises of shot noise, granulation, and activity and very-low frequency instrumental noise (dashed lines follow the contributions of the individual components).

easier to resolve closely spaced peaks—such as the rotationally split components of a nonradial mode—at lower frequencies than at higher frequencies. The peak line width at ν_{max} also appears to be a strong function of T_{eff}, with hotter stars presenting larger line widths. For example, the most prominent radial solar p modes have $\Gamma \simeq 1\,\mu$Hz; the hottest (F-type) stars showing solar-like oscillations have corresponding line widths that can be 5 μHz or more.

An approximate scaling relation for the expected line width at ν_{max} is given by an exponential relation of the form:

$$\Gamma = \Gamma_0 \, \exp\left(\frac{T_{eff} - 5{,}777\,\mathrm{K}}{T_0}\right), \qquad (5.95)$$

where $\Gamma_0 \approx 1.1\,\mu$Hz and $T_0 \approx 500$ K.

5.5 FREQUENCY SPECTRUM OF GRANULATION AND ACTIVE-REGION SIGNAL

We now look at the frequency-domain properties of intrinsic stellar signals due to granulation and activity, whose basic characteristics were introduced in Sections 4.3.1 and 4.3.2.

Figure 5.14 again shows the frequency-power spectrum of *Kepler*-410, but this time both axes are plotted on a logarithmic scale to emphasize the background components. The plot annotations indicate the various contributions to the observed power spectral density. The heavy dark line shows a best-fitting model of the composite background, which comprises components that model granulation, activity, and shot noise. Any very-low frequency instrumental noise is assumed to be captured by the activity component of the model.

Let us begin by considering a suitable form for the granulation component. We made predictions in Section 4.3.1 for the amplitude and timescale of the granulation signal in both photometry and Doppler velocity. We start with a simple phenomenological model in which the stochastic time-domain signal is modeled as a random superposition of exponentially decaying signals of the form

$$
x(t) \propto
\begin{cases}
0, & t < t_0 \\
\exp\left[-(t - t_0)/\tau_c\right], & t \geq t_0,
\end{cases}
\tag{5.96}
$$

where τ_c is the characteristic timescale, and individual "pulses" are centered on random times t_0. The composite signal may be modeled as a random walk of the form

$$
x(t + \delta t) = x(t) \exp\left(-\delta t/\tau_c\right) + e(t + \delta t),
\tag{5.97}
$$

where samples are separated in time by δt, and the $e(t)$ are normally distributed random variables with zero mean and standard deviation $\sigma_c \sqrt{\delta t/\tau_c}$, with σ_c being the characteristic amplitude of the granulation signal. The frequency-power spectrum of the time-domain signal then takes the form

$$
P_c(\nu) = \frac{4\sigma_c^2 \tau_c}{1 + (2\pi \nu \tau_c)^2}.
\tag{5.98}
$$

With a power-law exponent of two, this is a so-called *red-noise* process. This simple model can provide reasonable fits to the observed spectra, although at higher frequencies—where the expression will tend to a power law of -2—the observed fall-off in power tends to be steeper.

Another simple description for the granulation follows from the slightly modified time-domain model:

$$
x(t) \propto
\begin{cases}
\exp\left[(t - t_0)/\tau_c\right], & t < t_0 \\
\exp\left[-(t - t_0)/\tau_c\right], & t \geq t_0,
\end{cases}
\tag{5.99}
$$

the power spectrum of which is, to good approximation

$$
P_c(\nu) \simeq \frac{4\sqrt{2}\sigma_c^2 \tau}{1 + (2\pi \nu \tau_c)^4}.
\tag{5.100}
$$

This model can also provide good fits to data. Indeed, it is hard to systematically discriminate against any one model. The common practice is to use a more general expression of the form

$$
P_c(\nu) = \frac{b_c}{1 + (2\pi \nu \tau_c)^{\beta_c}},
\tag{5.101}
$$

with the timescale τ_c, power b_c, and exponent β_c parameters to be fitted. The granulation component that was fitted in Figure 5.14 has this more general form and yielded a best-fit exponent of $\beta_c \simeq 2.4$. It is important to note that once the value of the exponent changes, so too does the meaning of the timescale τ_c, in the context of the time-domain signal (this should be obvious from the two simple models for the time-domain signal that we considered above).

Given a best-fitting timescale and exponent, can we relate the fall-off of the power to a characteristic scale in frequency? There are several options, one being the frequency

$$\nu_\tau = (2\pi\tau_c)^{-1}, \tag{5.102}$$

at which the power falls to half its maximum value. Another is a less obvious, but equally useful, characteristic frequency that we call $\nu_{\tau,0}$. It marks the location of the "knee" of the granulation signal in the frequency domain, where a point of inflection occurs in the modeled form of the power spectral density. We can find this frequency by differentiating Eq. 5.101 twice with respect to frequency and then setting the resulting expression equal to zero. From this we derive

$$\nu_{\tau,0} = \frac{1}{2\pi\tau_c} \left(\frac{\beta_c - 1}{\beta_c + 1} \right)^{1/\beta_c}. \tag{5.103}$$

When, for example, $\beta_c = 2$, we have

$$\nu_{\tau,0} = \frac{1}{2\sqrt{3}} (\pi\tau_c)^{-1}. \tag{5.104}$$

What of the signal due to stellar activity? We know of course to expect a narrow-band signal from periodic signatures in the lightcurve due to the rotational modulation of spots and active regions. The main panel of Figure 5.15 shows part of the *Kepler* long-cadence lightcurve of the F-type star KIC 2837475, whose oscillation spectrum was shown in Figure 5.13. Periodic signatures due to the rotational modulation of spots are readily apparent in the lightcurve. The measured period is just under 3.7 days. The inset shows the very-low-frequency part of the spectrum of the lightcurve, which reveals strong peaks due to the surface rotation.

Since the spots and active regions that contribute a detectable signal evolve and decay on certain characteristic timescales, we would also expect a broad-band, frequency-dependent contribution to the background, like that arising from granulation. The timescales governing the evolution of active features tend to be longer than those governing the main component of granulation, and the amplitudes are typically higher. This means that the frequency-domain signatures of activity tend to be concentrated at very low frequencies in the frequency-power spectrum, where we will also see peaks from rotational modulation. It is worth stressing that this is also where we might expect contributions from long-term instrumental drifts. The model used in Figure 5.14 to fit the low-frequency rise in power due to activity had a simple power-law form with a fixed exponent of -2. This fit likely also captured contributions from low-frequency instrumental drifts, in addition to those from

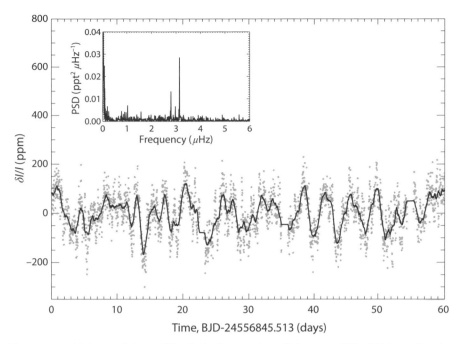

Figure 5.15. Main panel: Part of the *Kepler* long-cadence lightcurve of KIC 2837475, showing periodic signatures due to the rotational modulation of starspots. The measured period is just under 3.7 days. Inset: Frequency spectrum of the lightcurve. The prominent peaks around 3 μHz are the signatures of the surface rotation.

stellar activity. Needless to say, disentangling contributions due to instrumental variations (including how those variations are handled when the data are prepared for analysis) is very challenging.

5.6 HARMONIC STRUCTURE FROM TRANSITS, ECLIPSES, AND STELLAR ROTATION

There may also be periodic signatures in the data due to transiting exoplanets or eclipsing binaries. Like the signatures arising from rotational modulation of starspots and active regions, these signatures are not simple sinusoids. This means that the signatures will manifest in the frequency-power spectrum not as a single peak at the main period, but as a power-modulated pattern of peaks lying at the harmonics of the underlying period of the phenomenon (i.e., some combination of the rotation period of the star and the orbital period of the binary or planet). These peaks of course carry rich information on the underlying phenomena. But as we shall see in this section, they can also present complications for the analysis of the asteroseismic data.

Detailed asteroseismic analysis of the frequency spectrum requires careful fitting of models not only of the overtone structure due to the oscillations but also of the background power, with granulation usually being the key component in the latter. If harmonic structure from transits, eclipses, or rotation encroaches on the frequencies containing not only the oscillations but also the granulation background,

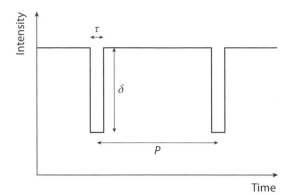

Figure 5.16. Schematic representation of a transit signal in the idealized form of equally spaced boxcar functions of depth δ and width (duration in time) τ, having a repeat period P.

it can complicate the asteroseismic analysis. Here we make some predictions of the expected structure in the frequency-power spectrum, subject to some basic simplifying assumptions regarding the shapes of time-domain signatures. We do so for photometric observations; long, continuous data of this type are already available that allow us to illustrate the effects of these signatures.

Let us begin by predicting the structure arising from periodic transits or eclipses of a bound planet or companion star that is significantly smaller than the target star. This allows us to approximate each transit as a box-shaped signature. It neglects the curvature shown in real transits, due to the basic geometry of one disc eclipsing another and the contribution of limb darkening. For now we ignore the possibility of a grazing transit or eclipse, which will tend to produce a triangular-shaped signature (a situation we consider to later in this section).

Figure 5.16 shows the resulting time-domain signatures, where δ is the fractional depth of each transit, τ is the duration of each transit, and P is the period at which the transits recur. Note that this simple model also ignores the presence of secondary transits—when the planet is obscured by the host star—although they of course have much more modest depths than the main transit signatures. We can describe each transit by the normalized boxcar function:

$$\Pi_\tau(t) \overset{\text{def}}{=} \begin{cases} 1, & -\tau/2 \leq t \leq \tau/2 \\ 0, & \text{Otherwise.} \end{cases} \tag{5.105}$$

A full description of the time-domain signal in Figure 5.16 is given by convolving this boxcar function (after multiplication by minus the depth, $-\delta$) with a Dirac comb, $\text{III}_P(t)$, of repetition period P and, finally, multiplying by another, wider boxcar $\Pi_T(t)$ corresponding to the full duration T of the observations. This may be written as

$$x(t) = -\left[\delta \Pi_\tau(t) * \text{III}_P(t) \right] \times \Pi_T(t). \tag{5.106}$$

The Fourier transform is then

$$X(v) = \left[\delta \mathrm{sinc}(\pi v \tau) \times \frac{1}{P} \mathrm{III}_{1/P}(v) \right] * \mathrm{sinc}(\pi v T). \tag{5.107}$$

We therefore expect to see a pattern of narrow peaks at harmonics of the period P, whose amplitudes are modulated by the $\mathrm{sinc}(\pi v \tau)$ function. The longer the transit duration τ, the more rapid will be the fall-off of the peak amplitudes with increasing frequency. The first minimum of the sinc function lies at the frequency τ^{-1}.

What we are most interested in here is how prominent these signatures will be in the final, calibrated power spectrum. To answer this question we must determine the expected power of the sinc-squared profile in the calibrated spectrum (i.e., the power level that will satisfy Parseval's theorem). In what follows we use the terminology introduced in Section 5.1.4.

To begin, we consider a sinc-squared profile in frequency having unit power at zero frequency. If there are N points in the time domain, we have $N/2 + 1$ independent bins in the frequency domain separated by $\Delta_T = T^{-1} \equiv (N\Delta t)^{-1}$ (Eq. 5.5), where Δt is the cadence. Summing the profile over the $N/2 + 1$ frequencies v_k running from zero up to the Nyquist frequency gives

$$\sum_{k=0}^{N/2} P'_k \equiv \sum_{k=0}^{N/2} \mathrm{sinc}^2(\pi \tau v_k \tau) = 1/(2\tau \Delta_T). \tag{5.108}$$

The multiplicative factor of $1/2$ arises because we are only interested in the positive-frequency side of the spectrum. Implicit in Eq. 5.108 is also the assumption that the Nyquist frequency $(2\Delta t)^{-1}$ is much higher than τ^{-1}, so that the spectrum in effect captures all but an insignificant fraction of the power due to the sinc-squared profile. This is equivalent to the condition $\tau \gg 2\Delta t$.

However, the actual spectrum given by the transit-like signatures does not have power at all frequencies, only those frequencies which lie at the harmonics of the period P. This power will occupy P/τ bins up to the first minimum of the sinc function at τ^{-1}. The total number of bins up to this frequency is $(\tau \Delta_T)^{-1}$, hence the fraction of bins occupied by power due to the transits is $P\Delta_T \equiv P/T$. Correcting by this factor gives

$$\sum_{k=0}^{N/2} P'_k = P/(2\tau). \tag{5.109}$$

The above tells us the expected ratio of the power at zero frequency to the total power is unity divided by $P/(2\tau)$, that is, the reciprocal of Eq. 5.109.

The final ingredient we need to estimate the calibrated power of the transit signatures is the mean-square power in the time-domain lightcurve. For our adopted model (Eq. 5.106) it is

$$\frac{1}{N} \sum_{j=1}^{N} x_j^2 = \delta^2 \tau / P. \tag{5.110}$$

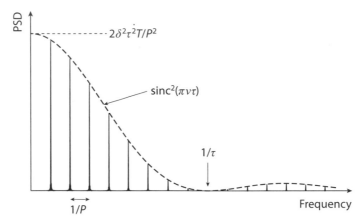

Figure 5.17. Schematic representation of the frequency-domain signatures arising from the idealized time-domain transit model.

The required calibration factor that will give the predicted power at zero frequency is therefore

$$c_P = \left(\frac{1}{N} \sum_{j=1}^{N} x_j^2 \right) \Big/ \left(\sum_{k=0}^{N/2} P_k' \right) = 2\delta^2 \tau^2 / P^2 \,. \qquad (5.111)$$

A full description of the calibrated power H_{tran} of the sinc-squared envelope, which fixes the prominence of the observed harmonics, is therefore

$$H_{\text{tran}}(\nu_k) = \begin{cases} \left[2\delta^2 \tau^2 / P^2 \right] \, \text{sinc}^2(\pi \nu_k \tau), & \text{power per bin} \\[2mm] \left[2\delta^2 \tau^2 T / P^2 \right] \, \text{sinc}^2(\pi \nu_k \tau), & \text{power per Hz.} \end{cases} \qquad (5.112)$$

Figure 5.17 shows a schematic representation of the expected frequency-domain signatures arising from the idealized time-domain transit model.

If we relax the condition that the transiting or eclipsing object is significantly smaller than the target star or allow the possibility of a grazing transit, the resulting profiles tend to a "V" or triangular shape. Let us therefore consider first the situation illustrated in Figure 5.18, where the transit profiles are now triangles having a baseline equal to τ:

$$\wedge_\tau(t) \overset{\text{def}}{=} \begin{cases} 1 - |2t/\tau|, & |2t/\tau| \le 1 \\[2mm] 0, & \text{Otherwise.} \end{cases} \qquad (5.113)$$

The lightcurve is now

$$x(t) = -\left[\delta \wedge_\tau (t) * \text{III}_P(t) \right] \times \Pi_T(t). \qquad (5.114)$$

The triangular function $\wedge_\tau(t)$ can be constructed from the convolution of two identical boxcar functions, each half the width (i.e., $\tau/2$) of the baseline of

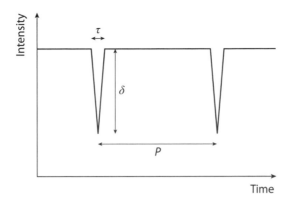

Figure 5.18. Schematic representation of a transit due to an eclipsing binary or grazing planetary transit, in the idealized form of equally spaced triangular functions of depth δ, width (duration in time) τ, and repeat period P.

the triangle. We therefore have

$$x(t) = -\left[\delta\{\Pi_{\tau/2}(t) * \Pi_{\tau/2}(t)\} * \mathrm{III}_P(t)\right] \times \Pi_T(t), \tag{5.115}$$

the Fourier transform of which is

$$X(\nu) = \left[\delta\mathrm{sinc}^2(\pi\nu\tau/2) \times \frac{1}{P}\mathrm{III}_{1/P}(\nu)\right] * \mathrm{sinc}(\pi\nu T). \tag{5.116}$$

The calibration differs slightly from the box-shaped transit case considered above because of the changes to the first step

$$\sum_{k=0}^{N/2} P'_k \equiv \sum_{k=0}^{N/2} \mathrm{sinc}^4(\pi\nu_k\tau/2) = 2/(3\tau\Delta_T), \tag{5.117}$$

and also

$$\frac{1}{N}\sum_{j=1}^{N} x_j^2 = \delta^2\tau/(3P). \tag{5.118}$$

Following the steps to calibrate the spectrum correctly gives

$$H_{\mathrm{eclp}}(\nu_k) = \begin{cases} \left[\delta^2\tau^2/(2P^2)\right]\mathrm{sinc}^4(\pi\nu_k\tau/2), & \text{power per bin} \\ \left[\delta^2\tau^2 T/(2P^2)\right]\mathrm{sinc}^4(\pi\nu_k\tau/2), & \text{power per Hz.} \end{cases} \tag{5.119}$$

The triangular-shaped profiles are what we expect for an eclipsing binary system. However, we then also need to remember that there will be significant secondary eclipses—when the more luminous star eclipses its companion—which are not shown in the simple schematic in Figure 5.18. The predicted frequency spectrum due to the eclipses is then given by the addition of the predicted spectrum due to the

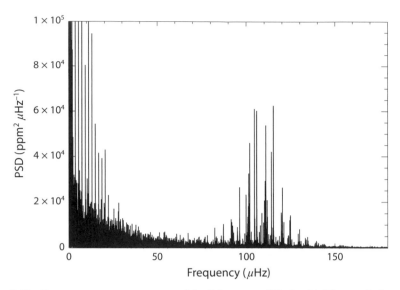

Figure 5.19. Frequency-power spectrum of the lightcurve of *Kepler*-91. The regularly spaced peaks at the lowest frequencies are the harmonic signatures of a transiting exoplanet. The oscillation spectrum of the star appears at higher frequencies (centered on $\approx 110\,\mu$Hz).

primary eclipses (a function of P, τ_1, and δ_1) and the predicted spectrum due to the secondary eclipses (a function of P, τ_2 and δ_2).

Figures 5.19 and 5.20 show two examples from the *Kepler* database where strong transit signals have the potential to bias fits to the granulation background and the oscillation spectrum. Figure 5.19 shows the frequency-power spectrum of *Kepler*-91, a low-luminosity red giant. It has a single transiting Jupiter-sized planet, which orbits the parent star at a period of $P \simeq 6.25$ days. The transits have durations of just over 11 hours, implying that the first minimum of the sinc-squared modulation in power (see Figure 5.17) will lie at around $\tau^{-1} \simeq 25\,\mu$Hz. The peaks in Figure 5.19 lying at the orbital harmonics are indeed seen to vanish into the granulation power at around this frequency. The prominence of the peaks at frequencies below this would nevertheless have a significant impact on any fits to the granulation background were they not first removed (e.g., by fitting the transit signatures in the time domain), or included in any fitting model.

Figure 5.20 shows the spectrum of another low-luminosity red giant, *Kepler* Object of Interest (KOI) 5220. Its lightcurve shows transits, which, if associated with a body orbiting the red giant, would indicate an object more than twice the size (radius) of Jupiter. There is a nearby stellar companion, with which the transit signal might be associated. Irrespective of the origin of the transit, this case nevertheless presents an excellent example of transit signatures ringing into the frequency range occupied by the detected oscillations.

Finally, let us consider the case of rotational modulation of starspots. Here we use a very simplistic model comprising a cosinusoidal-shaped signal arising from a single spot traversing close to the chord through the center of the visible disc of the star, as shown in Figure 5.21. This model neglects two things. First, limb darkening will affect the shape, particularly toward the extremes of the transit when the spot

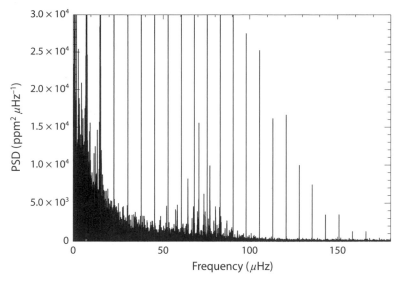

Figure 5.20. Frequency-power spectrum of the lightcurve of *Kepler* Object of Interest (KOI) 5220. Here the strong pattern of regularly spaced peaks due to a transiting body cuts across the oscillation spectrum of the star (the latter centered around 75 μHz).

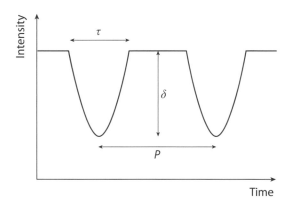

Figure 5.21. Schematic representation of the signal due to a spot crossing the visible stellar disc at the stellar equator—with the star viewed edge-on ($i_s = 90$ degrees)—in the idealized form of equally spaced cosinusoidal transits of depth δ, width (duration in time) τ, and repeat period P.

is close to the limb of the stellar disc. Second, the signal will be affected by the path taken by the spot as it traverses the disc, which in turn depends on its latitude on the stellar surface and the angle of inclination i_s offered by the star (i.e., the orientation of the rotation axis with respect to the line of sight).

In the example shown in Figure 5.21, the basic component of the lightcurve is

$$\sqcap_\tau(t) \stackrel{\text{def}}{=} \begin{cases} \cos(\pi t/\tau), & -\tau/2 \le t \le \tau/2 \\ 0, & \text{Otherwise.} \end{cases} \tag{5.120}$$

so that

$$x(t) = - [\delta \cap_\tau (t) * \text{III}_P(t)] \times \Pi_T(t). \qquad (5.121)$$

Now we can write the basic lightcurve component as

$$\delta \cap_\tau (t) = \delta \cos[2\pi t/P] \times \Pi_\tau(t). \qquad (5.122)$$

The Fourier transform is therefore

$$X(\nu) = \left[\delta \text{sinc}[\pi(\nu - 1/P)\tau] \times \frac{1}{P} \text{III}_{1/P}(\nu) \right] * \text{sinc}(\pi \nu T). \qquad (5.123)$$

Notice how the convolution in the frequency domain implied by the multiplication in Eq. 5.122 adds a shift of $1/P$ to the sinc function.

The first step of the calibration gives

$$\sum_{k=0}^{N/2} P'_k \equiv \sum_{k=0}^{N/2} \text{sinc}^2[\pi(\nu_k - 1/P)\tau] \approx 1/(2\tau\Delta_T), \qquad (5.124)$$

where we have assumed that $P^{-1} \ll \tau^{-1}$. The mean-square power in the time domain is now

$$\frac{1}{N} \sum_{j=1}^{N} x_j^2 = \delta^2 \tau/(2P). \qquad (5.125)$$

The calibrated spectrum is therefore given approximately by

$$H_{\cos}(\nu_k) \approx \begin{cases} [\delta^2 \tau^2/P^2] \ \text{sinc}^2[\pi(\nu_k - 1/P)\tau], & \text{power per bin} \\ [\delta^2 \tau^2 T/P^2] \ \text{sinc}^2[\pi(\nu_k - 1/P)\tau], & \text{power per Hz.} \end{cases} \qquad (5.126)$$

We note that the assumption $P^{-1} \ll \tau^{-1}$ leads to larger errors in the prediction the closer the spot transit chord is to the disc center (when by definition we have $P = 2\tau$).

The rotational modulation signal exhibited by KIC 2837475 (see Figure 5.15), although clearly detectable, is nevertheless not strong enough to ring significantly into the region where the granulation-dominated background and oscillations are fitted. Growing evidence suggests that strong activity suppresses the amplitudes of the oscillations, making them harder to detect; hence, very active stars with extremely strong signals due to rotational modulation—which have the potential to show prominent harmonics at higher frequencies—are much less likely to have detectable oscillations. That said, borderline active cases, where detections are possible, are then likely to be difficult to analyze.

Kepler-63 is a young and very active Sun-like star. The top panel of Figure 5.22 shows its *Kepler* long-cadence lightcurve, which is dominated by rotational modulation. Notice how the variation looks much smoother than our truncated cosinusoidal model, reflecting its simplified nature (recall the discussion above of the various aspects that were neglected). Also visible in the lightcurve are the transits of *Kepler*-63b, a Neptune-sized planet in a 9.4-day orbit around the star. Notice the scale on the ordinate: the spot modulation has an amplitude above 10^4 ppm, between one and two orders of magnitude higher than the spot modulation signal shown by KIC 2837475.

The bottom panel of Figure 5.22 shows the frequency-power spectrum of the short-cadence lightcurve of *Kepler*-63. It is dominated by the harmonics of the rotation period, which ring through the spectrum all the way to the short-cadence Nyquist frequency. Attempts were made to detect oscillations in these data, but none were detected. Needless to say, the analysis had to confront the complications caused by the strong spot modulation signal.

5.7 SUPER-NYQUIST ASTEROSEISMOLOGY

Finally in this chapter, we consider some of the practicalities of working in the super-Nyquist regime. As noted in Section 5.1.3, we are not necessarily limited to studying periodic signals below the notional Nyquist frequency. There is plenty of sensitivity in the super-Nyquist regime (i.e., at frequencies $\nu > \nu_{\rm Nyq}$). However, to make use of this sensitivity we must be able to discriminate the true frequencies from the aliased frequencies. Let us consider how this might be done, using *Kepler* long-cadence observations to illustrate the challenges involved.

The *Kepler* long-cadence (hereafter LC) data are composed of $\Delta t =$ 29.4-min cadences, which are exactly regular in the spacecraft frame of reference. This establishes a notional LC Nyquist frequency of $\nu_{\rm Nyq} \simeq 283\,\mu$Hz. Each 29.4 min cadence is in turn a summation of 270 individual integrations of $\simeq 6$ sec each. Most of each cadence is therefore given over to the collection of photons. The very high fractional duty cycle for the *Kepler* integrations means the signal attenuation follows closely that described by Eq. 5.9.

The left panel of Figure 5.23 shows the predicted frequency-power spectrum of several undersampled sinusoids for idealized observations made on a regular *Kepler*-like LC cadence. The sinudoids all have unit amplitude and frequencies that lie in the super-Nyquist regime, between $310\,\mu$Hz and $460\,\mu$Hz (marked with black lines on the figure). The vertical dashed lines lie at multiples of $\nu_{\rm Nyq}$. Peaks in gray are therefore aliases of the true frequencies. The spectrum is repeated every $2\nu_{\rm Nyq}$ because of the discrete nature of the calculation. This spectrum was calibrated so that a sinusoid of unit amplitude with an infinite sample rate would show a maximum power per bin of unity. The observed powers at the true frequencies are much lower than unity, the true power. This is due to the sinc-function attenuation described by Eq. 5.9. The dotted line marks the sinc-squared suppression envelope in power (i.e., η^2). In spite of this suppression, the true signals show reasonable sensitivity in the super-Nyquist regime.

The above is not quite the whole story as far as timing issues for the *Kepler* observations are concerned. *Kepler* was inserted into a 372.5-day heliocentric, Earth-trailing orbit. Observations of the pulsations of a *Kepler* target—be it one in the original field, or in the K2 fields in the ecliptic plane—are phase modulated in the spacecraft frame of reference. By this we mean that there is a component of the orbital motion of *Kepler* along the line-of-sight (target) direction that delays or advances the arrival time of light from the star. The annual size of the effect is approximately

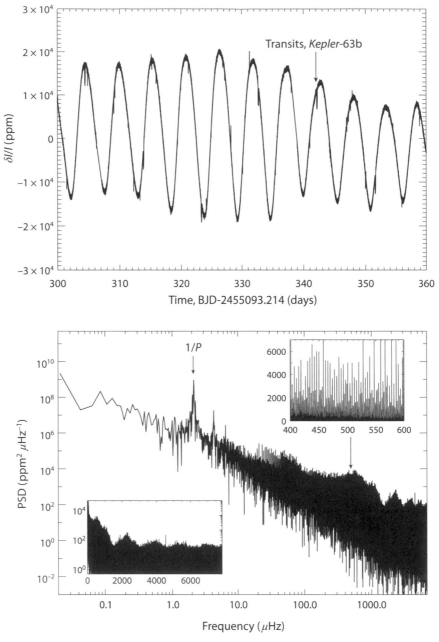

Figure 5.22. Top panel: Long-cadence *Kepler* lightcurve of *Kepler*-63. Bottom panel: Frequency-power spectrum of the short-cadence lightcurve of *Kepler*-63. The insets show the envelope of the power due to the harmonics (left-hand inset) and a snippet of one part of the spectrum (right-hand inset) showing the individual harmonics.

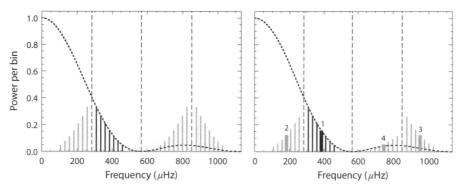

Figure 5.23. Left panel: Frequency-power spectrum for a series of sinusoids of unit amplitude having frequencies between 310 μHz and 460 μHz, sampled on a regular cadence. Peaks due to the true frequencies are rendered as black lines. The vertical dashed lines mark multiples of ν_{Nyq}. Peaks in gray are aliases of the true frequencies. The dotted line marks the sinc-squared attenuation envelope (i.e., η^2). Right panel: Spectrum now given by irregular sampling of the undersampled sinusoids. The true peak and the aliases of one of the frequencies are rendered in thick black and gray lines, respectively. Reproduced from Chaplin et al., 2014, *Super-Nyquist Asteroseismology of Solar-Like Oscillators with Kepler and K2—Expanding the Asteroseismic Cohort at the Base of the Red Giant Branch, Mon. Not. R. Astro. Soc.,* **445**, 946.

± 190 sec for targets in the original field, and approximately ± 500 sec for targets in the ecliptic.

To compensate for this effect[2] the *Kepler* time stamps are corrected to Barycentric arrival times (actually Barycentric Dynamical Time), that is, times pegged to the solar-system barycenter. Intervals between the corrected time stamps are no longer regular but are modulated periodically on a ∼1 yr timescale. This periodic modulation splits the aliased peaks in the frequency spectrum into many components.

The right panel of Figure 5.23 shows the result of simulating LC observations of an idealized star in the original *Kepler* field. The artificial data span 4 simulated years and include the aforementioned timing effects. The simulated pulsations are the same, undersampled sinusoids from the left panel. But notice how the aliases are no longer exact, reflected copies about multiples of ν_{Nyq}. Some copies have different maximum power spectral densities (i.e., peak heights in the spectrum) than others. This is because the power has been split into several components, thereby reducing the maximum heights in Figure 5.23 relative to those expected for simple reflected aliases (the total integrated power in the aliases is conserved).

The top-left panel of Figure 5.24 shows a zoom of the peak due to the true frequency marked "1" in the right panel of Figure 5.23. It is not split but retains its form as a single peak. Since we have oversampled this spectrum, the peak shows the sinc-squared profile due to the window function of the observations (see Section 5.1.5 and Eq. 5.28). The other panels show zooms of the aliases marked "2", "3" and "4" on Figure 5.23; they are all split into several components.

---

[2] Aside from the varying component due to the orbital motion of the spacecraft around the Sun, we assume that there are no other line-of-sight components showing significant variation on the timescale of the observations (e.g., variations due to the target being in a short-period binary).

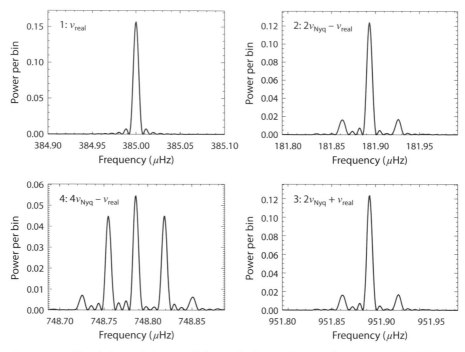

Figure 5.24. Top-left panel: Zoom of the peak due to the true frequency marked "1" on the right panel Figure 5.23. Other panels: Zooms of the aliases marked "2", "3" and "4" in Figure 5.23. Reproduced from Chaplin et al., 2014, *Super-Nyquist Asteroseismology of Solar-Like Oscillators with Kepler and K2—Expanding the Asteroseismic Cohort at the Base of the Red Giant Branch, Mon. Not. R. Astro. Soc.*, **445**, 946.

The exact manner in which the alias peaks are affected by the timing modulation depends on the relation of their frequencies to the sampling frequency:

$$\nu_s \overset{\text{def}}{=} (\Delta t)^{-1} \equiv \nu_{\text{Nyq}}/2. \tag{5.127}$$

If the true frequency is ν, then aliases at $n'\nu_s \pm \nu$, or equivalently, $2n'\nu_{\text{Nyq}} \pm \nu$ (where n' is an integer) will share the same structure (e.g., what is predominantly a triplet structure when $n' = 1$, or a quintuplet structure when $n' = 2$, with the frequency splitting between adjacent components being $\sim 1 \text{ yr}^{-1}$).

Thus for coherent pulsations that have high SNRs, the introduction of sideband structure provides the means to discriminate the real (no splitting) and aliased (split) peaks. But what about solar-like oscillations, which we know are not coherent and typically have much lower amplitudes (and hence lower SNRs in the frequency spectrum) than coherent pulsations? The rich spectra of overtones shown by solar-like oscillators provides the necessary, a priori information needed to select the true spectrum from the aliased spectra. To illustrate how, we use data on the low-luminosity red giant KIC 4351319, which was observed in both long and short cadence during the nominal *Kepler* mission.

The top panel of Figure 5.25 shows the frequency power spectra of simultaneous LC (upper plot) and SC (lower plot) data. The nominal LC Nyquist frequency, ν_{Nyq}, is

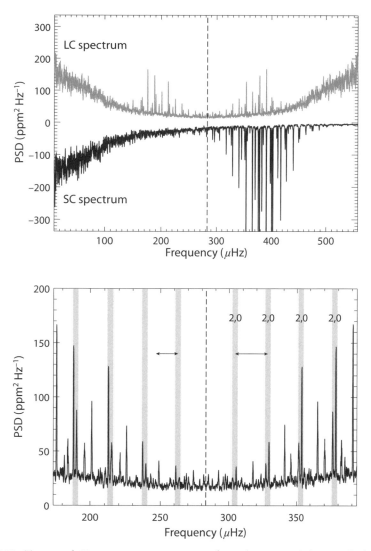

Figure 5.25. Top panel: Frequency power spectra of simultaneous LC (upper plot) and SC (lower plot) data collected by *Kepler* on the low-luminosity red giant KIC 4351319. Bottom panel· Zoom of the LC spectrum above and below ν_{Nyq}. The lengths of the horizontal arrows in the bottom panel correspond to expected average separations, assuming a ν_{max} of $\simeq 190\,\mu$Hz (peaks below ν_{Nyq}) and $\simeq 380\,\mu$Hz (peaks above ν_{Nyq}). The shaded regions mark the locations of pairs of adjacent $l = 2$ and $l = 0$ modes. Reproduced from Chaplin et al., 2014, *Super-Nyquist Asteroseismology of Solar-Like Oscillators with Kepler and K2—Expanding the Asteroseismic Cohort at the Base of the Red Giant Branch, Mon. Not. R. Astro. Soc.,* **445**, 946.

marked by the vertical dashed line. Because SC observations are also available for this star, we are able to tag the peaks in the LC spectrum above ν_{Nyq} as the true oscillation frequencies (see the figure). However, two checks based on use of the LC spectrum alone allow us to make the same call.

The bottom panel of Figure 5.25 shows a zoom of the LC spectrum above and below ν_{Nyq}. Our first check uses the known correlation between the large frequency separation, $\Delta\nu$, and the frequency of maximum oscillation power, ν_{max}. We have seen how this relation can be expressed in the form $\Delta\nu \propto \nu_{max}^{\beta}$. While there is evidently spread in the relation due to, for example, dependences on mass and metallicity, the strong correlation nevertheless allows us to discriminate the true from the aliased spectrum.

The lengths of the horizontal arrows in the bottom panel correspond to expected average separations, assuming a ν_{max} of $\simeq 190\,\mu$Hz (peaks below ν_{Nyq}) and $\simeq 380\,\mu$Hz (peaks above ν_{Nyq}). The shaded regions mark the locations of pairs of adjacent $l = 2$ and $l = 0$ modes. The frequency intervals between these regions provide a visual estimate of the observed large separation. Evidently, it is the spectrum above ν_{Nyq} that conforms to the expected average separation.

The second check relates to the relative power shown by adjacent $l = 2$ and $l = 0$ modes. In each pair, the $l = 0$ mode lies at a higher frequency. We expect to see more power in the $l = 0$ mode than in its $l = 2$ counterpart (recall our discussion in Section 4.1). Although predictions of the exact power ratios are rendered uncertain by the complexities of nonadiabatic calculations, observed ratios are usually not too far from the nearly 2-to-1 ratio expected from the assumption of energy equipartition and geometric cancellation of the perturbations on the visible stellar disc. Unless the star is observed with the rotation axis along the line of sight, power shown by the $l = 2$ modes will be spread across several components (recall Eq. 4.14). The ratio in total power then tends to be exaggerated in the frequency-power spectrum, because it is the *maximum* power spectral densities (i.e., the heights of the mode peaks) that are immediately apparent from a visual check. Inspection of the mode pairs in the bottom panel of Figure 5.25 again implies that the spectrum above ν_{Nyq} is the true one: the $l = 0$ modes, the higher-frequency modes in each marked pair, have the higher observed powers.

5.8 FURTHER READING

Several good textbooks cover in detail the fundamentals of Fourier analysis. This is a good example:

- James, J. F., 2011, *A Student's Guide to Fourier Transforms: With Applications in Physics and Engineering*, Cambridge University Press.

The next paper describes in detail how the analysis of asteroseismic data collected by a ground-based network would be affected by gaps, with results given for different assumed network configurations:

- Arentoft, T., et al., 2014, *Benefits of Multiple Sites for Asteroseismic Detections*, Mon. Not. R. Astron. Soc., **437**, 1318.

The following papers provide excellent reviews of the excitation and damping of solar-like oscillations:

- Houdek, G., 2010, *Stellar Turbulence and Mode Physics*, Astrophys. Sp. Sci., **328**, 237.
- R. Samadi, K. Belkacem, and T. Sonoi, 2015, *Stellar Oscillations II. The Nonadiabatic Case*, in *Ecole Evry Schatzman 2014: Asteroseismology and Next*

Generation Stellar Models, EAS Publications Series **73**, 111, eds. Michel, E., Charbonnel, C., and Dintrans, B., EDP Sciences.

Various papers in the literature, both theoretical and observational, discuss the asymmetry of the oscillation peaks in the frequency-power spectrum. Here we select a few examples. The first and third papers below discuss the origins of the asymmetry in the context of the interference of waves in the mode cavity; the second paper does so for correlations with the background noise. Our discussion of the impact of correlations is based on the presentation in the fifth paper (and other papers by the same authors); whilst our discussion of the influence of the excitation source is based on the treatment presented in the final paper. That paper, and the fourth one, also present alternative strategies for fitting asymmetric peaks that go beyond fitting the modified Lorentzian formalism.

- Gabriel, M., 1995, *On the Profile of the Solar p-Mode Lines, Astron. Astrophys.*, **299**, 245.
- Roxburgh, I. R., and Vorontsov, S. V., 1997, *On the Formation of Spectral Line Profiles of Solar p Modes, Mon. Not. R. Astron. Soc.*, **292**, L33.
- Kumar, P., and Basu, S., 1999, *Line Asymmetry of Solar p-Modes: Properties of Acoustic Sources, Astrophys. J.*, **519**, 396.
- Chaplin, W. J., and Appourchaux, T., 1999, *Depth of Excitation and Reversal of Asymmetry of Low-l Solar p Modes: A complementary Analysis of BiSON and VIRGO/SPM Data, Mon. Not. R. Astron. Soc.*, **309**, 761.
- Toutain, T., Elsworth, Y., and Chaplin, W. J., 2006, *Peak Asymmetry of Solar p Modes: A Framework to Explain the Effects of the Correlated Noise from the Acoustic Source, Mon. Not. R. Astron. Soc.*, **371**, 1731.
- Vorontsov, S. V., and Jefferies, S. M., 2013, *Modeling Solar Oscillation Power Spectra. II. Parametric Model of Spectral Lines Observed in Doppler-Velocity Measurements, Astrophys. J.*, **778**, 75.

The next two papers report on state-of-the-art analyses of the *Kepler* data and look in detail at scaling relations for the solar-like oscillations and granulation:

- Huber, D. et al., 2011, *Testing Scaling Relations for Solar-Like Oscillations from the Main Sequence to Red Giants Using Kepler Data, Astrophys. J.*, **743**, 143.
- Kallinger, T. et al., 2014, *The Connection between Stellar Granulation and Oscillation as Seen by the* Kepler *Mission, Astron. Astrophys.*, **570**, 41.

Finally, these papers discuss in detail the practical application of super-Nyquist asteroseismology:

- Murphy, S. J., Shibahashi, H., and Kurtz, D. W., 2013, *Super-Nyquist Asteroseismology with the* Kepler *Space Telescope, Mon. Not. R. Astron. Soc.*, **430**, 2986.
- Chaplin, W. J., Elsworth, Y., and Davies, G. R., 2014, *Super-Nyquist Asteroseismology of Solar-Like Oscillators with* Kepler *and K2—Expanding the Asteroseismic Cohort at the Base of the Red Giant Branch, Mon. Not. R. Astron. Soc.*, **445**, 946.

5.9 EXERCISES

1. Starting from the relevant scaling relations, derive the expression for the total observed oscillation power, P_{tot} (Eq. 5.94), in the form applicable to Sun-like stars:

$$P_{tot} \simeq P_{tot,\odot} \left(\frac{\nu_{max}}{\nu_{max,\odot}} \right)^{-1.2} \left(\frac{T_{eff}}{T_{eff,\odot}} \right)^{-1.7}.$$

2. Show that the total background power underneath the oscillation spectrum of a Sun-like star may be approximated by the expression:

$$B_{tot} \approx 2\nu_{max}\sigma^2 \Delta t + \nu_{max}^{(1-\beta_c)} \left(\frac{b_c}{(2\pi \tau_c)^{\beta_c}} \right).$$

(Hint: Two simplifying assumptions must be made regarding the use of the expression for the granulation contribution.)

3. The file `oneMsolar.txt` contains fundamental properties for a $1M_\odot$ stellar evolutionary track. Use the scaling relations in the chapter to predict the global SNR in the expected oscillation spectrum—that is, the ratio of the total observed oscillation power P_{tot} and the total background power underneath the oscillation spectrum, B_{tot} (see Exercise **2**)—along the evolutionary track, assuming observations made by the *Kepler* spacecraft at *Kepler* apparent magnitudes of

    (a) Kp = 8;
    (b) Kp = 10; and
    (c) Kp = 12.

You may take the maximum RMS solar amplitude in the *Kepler* bandpass to be 2.5 ppm and the granulation amplitude to be 40 ppm. Take the shot noise per 58.85 sec short-cadence integration to be 70 ppm at Kp = 9, and assume that the noise follows Poisson statistics to good approximation over the range of magnitudes under consideration.

4. Consider two evolved stars to be observed by TESS. They have the same mass ($1.1M_\odot$), but different ages. The first star, which lies at the base of the red-giant branch, has a radius of $R = 2.84R_\odot$ and an effective temperature of $T = 4,864$ K. The second star, which is slightly more evolved, has $R = 4.00R_\odot$ and $T = 4,800$ K. Contemporaneous data were collected in long (30 min) *and* short (2 min) cadences on both stars. You may assume that the maximum oscillation amplitudes of the two stars in the TESS bandpass are 13.4 ppm, and 21.8 ppm respectively. You may also assume that the background power spectral density at ν_{max} is dominated for both stars by shot noise and takes the observed value of $\simeq 300$ ppm^2 μHz^{-1}.

    (a) For each star, estimate the expected ν_{max} and the height of the oscillation power envelope H_{env} for observations made in both the long- and short-cadence data. State clearly all assumptions made in your calculations.

    (b) Comment for each star on the implications of your results for analysis using either the long- or short-cadence lightcurves.

6 Observational Data: Analysis of the Frequency Spectrum

We now come to the analysis of the frequency-power spectrum. Asteroseismic modeling of the fundamental properties and internal structures of stars requires data on asteroseismic parameters as input. These parameters are usually estimated from an analysis of the frequency-power spectrum of the photometric or Doppler velocity observations.

As we shall see in this chapter, automated extraction of global or average asteroseismic parameters—such as $\nu_{\max}$ and $\Delta\nu$—is relatively straightforward. These parameters provide collective and easy-to-measure signatures of the oscillation spectrum. However, the quality of the observations that are now available is such that we would be selling the data short by using only these average parameters: significantly richer and more powerful information is contained in the individual oscillation mode frequencies and other individual mode parameters. When used, this individual information leads to significant improvements in the precision (in particular, of stellar ages) and accuracy of estimated fundamental properties. It also allows detailed tests to be made of the evolution of the structure and dynamics of stars, and of the physics of stellar interiors, tests that are not possible using the global parameters alone.

The colloquial label *peak bagging* has now acquired wide use as a description of this analysis, a term that has its origins in the hill-walking and mountaineering communities (where one ticks off, or "bags," those mountains or hill "peaks" that one has climbed). Here we are bagging the parameters of the individual mode peaks in the power spectrum.

The characteristics of the solar-like oscillations place particular demands on the analysis, beyond what would be needed for coherent modes. One does not need to fit a complicated model to estimate parameters of a coherent mode: simple spectrum estimation techniques that involve fitting sinusoids to the data will return suitable estimates of the frequency, amplitude, and phase of the signal. We have already seen that there is one limit where solar-like oscillations can be treated in the same way—namely when the mode lifetime τ significantly exceeds the duration T of the observations (Section 5.3.2). Since phase is then maintained over the duration of the observations, one may make use of the extra information (and constraints) provided by the phase, something that is lost if the analysis uses the power spectrum only.

However, in most cases we do not have that luxury. Detectable modes of solar-type stars have lifetimes that are typically of the order of days. Multi-month observations provide the frequency resolution needed to resolve the oscillation peaks, and the phase in each bin of the spectrum encompassing the mode is effectively randomized or scrambled. A common approach then adopted is to fit the observed peaks in the power spectrum to multiparameter models based on the damped oscillator model in Section 5.3. This is the approach we describe in some detail below. Moreover, we shall assume that long-duration, high duty-cycle observations are available to us, like those available from *Kepler*, K2, and CoRoT (and which will be available in future from TESS and PLATO).

Short-duration, ground-based observations present challenges for such an approach. One then has to judge whether a simpler, coherent approach to the analysis is warranted (dependent on the ratio of the expected values of τ to the known T). Unfortunately, for main sequence and subgiant stars, one is often in the gray area where the mode peaks are neither confined to one frequency bin (effectively coherent over the observation length T) nor fully resolved.

Aside from fitting a model to describe the oscillation peaks, one must also account for contributions arising from other phenomena. The observed power spectrum includes contributions from stellar granulation, perturbations due to stellar activity, photon shot noise, and instrumental noise. Finally, there is also the added subtlety that the oscillations and granulation are correlated and not independent phenomena (see Section 5.3.3), something that is often ignored in analyses.

The layout of this chapter is as follows. We start in Section 6.1 with a discussion of probabilistic methods for detecting signatures of individual modes in the spectrum, including schemes that incorporate a Bayesian framework for the analysis. This first section—which describes tests that may be used to provide first-guess frequencies for the peak-bagging analysis—also lays some of the important statistical groundwork for the sections that follow, beginning in Section 6.2 with the automated detection of the ensemble signatures of the oscillations, where we also discuss estimation of the global or average asteroseismic parameters. In Sections 6.3 through 6.5 we consider the detailed peak-bagging analysis, where a multiparameter model is fitted to the spectrum to provide robust estimates of parameters of the individual modes. Section 6.3 introduces the fundamentals; Section 6.4 discusses the practical implementation and challenges, using main sequence stars as examples; while Section 6.5 considers the additional challenges posed by subgiants and red giants showing mixed modes.

6.1 STATISTICAL TESTS: EXAMPLE OF DETECTION OF SIGNATURES OF INDIVIDUAL MODES

6.1.1 The $H0$ (False-Alarm) Approach

In the false-alarm approach we seek to estimate the probability that any prominent narrow-band power in the observed frequency-power spectrum is merely part of the broad-band background (which we have seen has both stellar and instrumental contributions). We do not therefore obtain an estimate indicating that the power is likely to be a mode; rather, we hope for a result that tells us it is very unlikely to be part of the background. This is a test of the *null* or *H*0 hypothesis.

We begin by recalling from Section 5.2 that the observed frequency-power spectrum will be distributed about the underlying limit spectrum with negative exponential statistics (i.e., a χ^2 distribution with two degrees of freedom). We shall assume that the spectrum is sampled at its natural resolution $1/T$ (where T is the length of the dataset in the time domain), so that frequencies in the discrete transform are uncorrelated modulo the binwidth $\Delta_T = T^{-1} \equiv (N\Delta t)^{-1}$ (Eq. 5.5). Here we assume that there are no significant interruptions the data. We may then write the probability p that the observed power in a given frequency bin ν will be equal to or greater than[1] $P(\nu)$ as

$$p\,[P(\nu)] = \frac{1}{P_{\text{lim}}(\nu)} \exp\left(-\frac{P(\nu)}{P_{\text{lim}}(\nu)}\right),\tag{6.1}$$

where $P_{\text{lim}}(\nu)$ is the limit spectrum. Underlying the approach to peak finding is the assumption that, from the observed power spectrum, we can obtain a good estimate of the background limit spectrum. This is crucial if we are to provide robust estimates from our probabilistic tests.

Let $\langle P(\nu) \rangle$ be our estimate of the background power, which we assume is close to the true, underlying background. We also introduce the terminology "spike" to describe power lying prominently above the background level in any one frequency bin. The probability p of observing a spike having a power spectral density greater than or equal to $P(\nu)$ is then simply

$$p\,[P(\nu)] = \frac{1}{\langle P(\nu) \rangle} \exp\left(-\frac{P(\nu)}{\langle P(\nu) \rangle}\right).\tag{6.2}$$

In what follows we shall formulate the approach in terms of the relative power,

$$s_\nu = P(\nu)/\langle P(\nu) \rangle \tag{6.3}$$

(i.e. we consider observed power relative to the mean "background" level). The equation for p may then be presented in the more concise form

$$p(s_\nu) = \exp\left(-s_\nu\right).\tag{6.4}$$

In practice we are interested in searching or testing a range of frequencies in the frequency spectrum. Let us test the frequency range Δ_p, which contains

$$N = \Delta_p T \tag{6.5}$$

independent frequency bins. The probability that we fail to find a spike at or above the level s_ν in any one bin is simply $1 - p(s_\nu)$; hence, the probability that we fail to

---

[1] These probabilities are the same, because the integral of the negative exponential distribution—which gives the cumulative probability—is also a negative exponential.

find at least one such event over the N bins is $[1 - p(s_\nu)]^N$. The final step is to turn this probability around, to give the sought-for estimate of observing at least one such spike in the N bins:

$$p_N = 1 - [1 - p(s_\nu)]^N. \tag{6.6}$$

A low value for p_N indicates that the spike is unlikely to be a statistical fluctuation in background power and may therefore be considered as a candidate mode.

We can also formulate the probability in terms of the cumulative binomial (Bernoulli) distribution. Generalizing to the probability of observing at least r spikes in N bins at or above the relative power s_ν, we have

$$p[r; p(s_\nu), N] = \sum_{r=r}^{N} p(s_\nu)^r [1 - p(s_\nu)]^{N-r} \frac{N!}{r!(N-r)!}. \tag{6.7}$$

The probability p_N given by Eq. 6.6 therefore corresponds to $p[1; p(s_\nu), N]$.

The utilization of the $H0$ statistic as a test for candidate modes requires us to make choices about the range of bins N to be searched and the threshold probability required to flag any detected excess power as a potential candidate mode. Although objective criteria may be used to fix the values, the choices made are to some extent arbitrary. Let us consider the impact of those choices.

In the limit when $p(s_\nu)$ is small, the probability p_N may be written as:

$$p_N \equiv p[1; p(s_\nu), N] \simeq Np(s_\nu) \equiv N \exp(-s_\nu). \tag{6.8}$$

Let p_{thr} be the chosen threshold probability. The relative height threshold required to satisfy $p(s_\nu) \leq p_{thr}$ is then given by:

$$s_\nu(p_{thr}) \simeq \ln N - \ln p_{thr}. \tag{6.9}$$

With p_{thr} fixed, the smaller N—and, therefore, the corresponding range in frequency, Δ_p—the lower, and less demanding, will be the required threshold power $s_\nu(p_{thr})$.

Now, we might have available to us a priori information that would allow us to say with some confidence that a particular range of frequencies should be expected to contain a mode (or modes). This allows us to narrow the range Δ_p. However, we must be confident about our assumptions. Suppose we had estimates of the expected frequency of maximum power, ν_{max}, and the average large separation, $\Delta\nu$, available to us. We might, quite reasonably, elect to fix $N = \Delta\nu T$ and search N-bin ranges centered on the estimated ν_{max} (since we have the expectation that each range will contain several, hopefully detectable, low-degree modes).

Let us take an example, with $\nu_{max} = 3{,}090\,\mu$Hz, $\Delta\nu = 135\,\mu$Hz, and $T = 1$ year. This fixes $N \simeq 4{,}260$. Suppose we set $p_{thr} = 10\%$, we would then have a required threshold of $s_\nu(p_{thr}) \simeq 10.7$. If we searched across a ν_{max}-wide window, there would be $\nu_{max}/\Delta\nu \simeq 23$ independent N-bin ranges to test, but this would mean that we would expect to find (on average) approximately $23 \times 0.1 \simeq 2$ spikes by chance on account of statistical fluctuations in the background alone. These are unwanted false-positive detections. (There would in fact be a 91% chance of finding one false-alarm spike, a 68% chance of finding two such spikes, and even a 40% chance of finding three such spikes.)

Narrowing Δ_p would further exacerbate the problem. For example, taking $\Delta_p = 10\,\mu$Hz would decrease $s_\nu(p_{\text{thr}})$ to $\simeq 8.1$, and the expected number of false-positive detections would rise to around 30, which is clearly unacceptable. To reduce the expected false-positive rate, we might choose to keep $\Delta_p = 135\,\mu$Hz but decrease p_{thr} to, say, 1%. There would now only be a 20% chance of returning a single false-positive detection, with only a 2% chance of returning two such rogue detections.

The bottom line is that if one adopts a basic test like the one described here, one must be careful to assess the consequences of the choices made for p_{thr} and N. The availability of prior information clearly has an important role to play, which suggests that we should consider a Bayesian approach to the problem. We shall introduce Bayesian statistics in Section 6.1.2. Before doing that, we complete this section by introducing some simple extensions of the basic $H0$ test.

We have seen that the basic test searches for prominent power in individual bins. This is a sensible approach to find lightly damped modes, which, depending on the length of the dataset being used, might not be resolved in frequency (see Section 5.3.2). However, if the modes one is searching for are expected to be resolved, one can take advantage of the fact that power will be spread over a range of bins commensurate with the corresponding mode linewidth Γ. While prominent spikes need not necessarily lie in consecutive bins—under low SNR conditions, statistical fluctuations may rule this out—they must nevertheless all fall in some sensible range in frequency to pass for mode-like structure. Here, prior information about the target star may be used to estimate a suitable range that will cover the expected peak linewidths of the sought-for modes.

Let us suppose that we choose a range N_Γ. The probability that r spikes are found at or above the level s_ν in a range of N_Γ bins is

$$p_\Gamma(s_\nu) = p[r; \exp(-s_\nu), N_\Gamma]. \tag{6.10}$$

One can then calculate the probability that at least one such "concentration" will appear by chance over the wider search range of N bins. This final probability depends not only on p_Γ but also the number of possible arrangements of the N_Γ bins across the N bins of Δ_p. There are $N - N_\Gamma + 1$ possible arrangements, and so the required probability is

$$p_{\Gamma,N}(s_\nu) = p[1; p_\Gamma(s_\nu), N - N_\Gamma + 1]$$
$$\simeq p[1; p_\Gamma(s_\nu), N], \tag{6.11}$$

where we have assumed that $N \gg N_\Gamma$. The threshold required to satisfy $p_{\Gamma,N}(s_\nu) \le p_{\text{thr}}$, hence flagging the possible presence of a mode, will vary according to the number of spikes r found above it over a given width range. The threshold needed when $r = 2$ is higher than for $r = 3$: the greater the number of strong spikes over a narrow range of bins, the lower the likelihood will be that these are the result of the broad-band noise.

Another approach is to search for power being present above a particular threshold in two or more consecutive bins. Consider the probability of u consecutive bins all having spikes at or above s_v. Since there are $N - u + 1$ possible arrangements across the full Δ_p, the probability of this occurring at least once somewhere in the range is

$$p_{u,N}(s_v) = p[1; \exp(-us_v), N - u + 1]$$
$$\simeq p[1; \exp(-us_v), N], \tag{6.12}$$

where we assume that $N \gg u$. The multiplication in the exponent follows, because the probability of finding a single spike must be raised to the power u to give the estimate for u spikes.

Finally, we might choose to average the spectrum into u-binwide averages. This changes the statistics of the spectrum. For any one u-bin average, the probability is now given by the integral

$$p(s_v' \geq s_v, u) = \int_{s_v}^{\infty} \frac{\exp(-s_v')}{\gamma(u)} s_v'^{(u-1)} \, ds_v', \tag{6.13}$$

where γ is the gamma function. Across the full N-bin range, there are now N/u averages (average bins), and so the probability of finding at least one average power exceeding the chosen threshold somewhere in the range is

$$p_{u,N}(s_v) = p[1; p(s_v' \geq s_v, u), N/u]. \tag{6.14}$$

6.1.2 The $H1$ (Odds Ratio) Approach

The procedures outlined in Section 6.1.1 tested the null or $H0$ hypothesis (i.e., the probability that any prominent power corresponds simply to a statistical fluctuation in the background noise). A more comprehensive approach is one for which we also test the competing $H1$ hypothesis: that the tested range of frequency contains a mode. Such a test requires that we make assumptions about the nature of the underlying mode signal we are searching for. Incorporating this prior information, and testing the odds of $H1$ being true versus $H0$ being true, may naturally be cast in terms of Bayes' Theorem.

We wish to use all the information that is available to us to estimate the probability that a particular hypothesis H is true: in our case, that we have detected a mode. This depends on the information D we extract from the data, in addition to any prior information I that is available or assumed. The prior information could, for example, be a set of assumed, reasonable limits on certain parameters describing the underlying mode signal in the frequency domain. Consideration of this information gives the prior probability distribution $p(H|I)$, that is, the probability of the hypothesis *given* the set of prior information. Next, there is the information we obtain from the data D, which was of course central to our discussion in Section 6.1.1. The distribution $p(D|H, I)$ gives the likelihood of the data *given* the assumed hypothesis.

Using Bayes' Theorem, we may write the posterior probability $p(H|D, I)$ of the tested hypothesis H in terms of the other probabilities:

$$p(H|D, I) = \frac{p(H|I)p(D|H, I)}{p(D|I)}. \tag{6.15}$$

Notice that there is an additional normalization term, $p(D|I)$, which is the so-called *global likelihood* for all competing hypotheses. The sum of the posterior probabilities of the competing hypotheses is unity, and so the normalization may be estimated from

$$p(D|I) = \sum_i p(H_i|I)p(D|H_i, I). \tag{6.16}$$

In some cases estimation of $p(D|I)$ may be nontrivial, given the potential number of competing hypotheses. Here, matters are much more straightforward, because we have just two competing hypotheses: in the range of the spectrum we are searching we either have ($H1$ hypothesis) or have not ($H0$ hypothesis) detected oscillation power.

So, we can write the posterior probability for the $H1$ hypothesis in any one bin of the searched range as

$$p(H1|s_\nu) = \frac{p(H1)p(s_\nu|H1)}{p(H0)p(s_\nu|H0) + p(H1)p(s_\nu|H1)}, \tag{6.17}$$

where $p(H0)$ and $p(H1)$ express our prior belief in each of the competing hypotheses, with $p(H0) + p(H1) = 1$.

Let us begin by considering the equivalent of the simple bin-by-bin test introduced in Section 6.1.1. There we met the likelihood for $H0$, which is just

$$p(s_\nu|H0) = \exp(-s_\nu). \tag{6.18}$$

To estimate the likelihood $p(s_\nu|H1)$ for the competing $H1$ hypothesis, we first write our model for the spectrum:

$$P(\nu) = H\delta(\nu - \nu_0) + B(\nu). \tag{6.19}$$

We have a spike (hence the δ-function) of height H in a single bin at frequency ν_0, and a background power $B(\nu)$. Dividing through by the background—where again we assume that in practice we can extract a reasonable estimate of the background—the normalized power in any given bin is $s_\nu = h + 1$. An obvious problem we must now confront is that we do not know a priori the true normalized height h of the searched-for mode. We must therefore integrate over the plausible range of heights we expect for any mode we hope to uncover. This is the process of *marginalization*.

If h sets the maximum on the prior, and $w(h')$ is a weighting function on the heights h' in the range $0 \leq h' \leq h$, we have

$$p(s_\nu|H1) = \int_0^h \frac{w(h')}{h' + 1} \exp\left[-s_\nu/(h' + 1)\right]\mathrm{d}h' \Big/ \int_0^h w(h')\mathrm{d}h' \tag{6.20}$$

The simplest choice for $w(h')$ is a flat prior, where we assume that all heights up to h are equally likely. Eq. 6.20 then simplifies to

$$p(s_\nu|H1) = \frac{1}{h} \int_0^h \frac{1}{h'+1} \exp\left[-s_\nu/(h'+1)\right]dh'. \tag{6.21}$$

Rather than having a hard cut on the prior, we might instead choose a function that falls off more gradually. For example, if we have an estimate for an uncertainty σ_h on the maximum height, we might choose a Gaussian model for the prior of the form

$$w(h') = \begin{cases} 1, & \text{for } h' \leq h \\ \exp\left[-(h'-h)^2/(2\sigma_h^2)\right], & \text{for } h' > h. \end{cases} \tag{6.22}$$

Substitution into the expression for the $H1$ likelihood then requires us to solve

$$p(s_\nu|H1) = \left(h + \sigma_h\sqrt{\pi/2}\right)^{-1} \int_0^\infty \frac{w(h')}{h'+1} \exp\left[-s_\nu/(h'+1)\right]dh'. \tag{6.23}$$

Other choices are also possible.

Generalizing the above to the case of testing for the probability of finding a prominent spike across N bins, we can write

$$p(s_\nu|H00) = 1 - [1 - p(s_\nu|H0)]^N, \tag{6.24}$$

and

$$p(s_\nu|H11) \equiv p(s_\nu|H1). \tag{6.25}$$

Comparison of $p(s_\nu|H11)$ and $p(s_\nu|H00)$ will therefore give the odds of one hypothesis versus the other.

One aspect of this test that now should be obvious is that the outcome can clearly be biased—potentially very severely—by our choice of $p(H0)$, $p(H1)$, and h, and our assumption that all heights up to h are equally probable. The simplest approach for the prior probabilities is to assume that $p(H0)$ and $p(H1)$ are equally probable. We then have

$$p(H1|s_\nu) = \left[1 + \frac{p(s_\nu|H0)}{p(s_\nu|H1)}\right]^{-1}, \tag{6.26}$$

and

$$p(H11|s_\nu) = \left[1 + \frac{p(s_\nu|H00)}{p(s_\nu|H11)}\right]^{-1}. \tag{6.27}$$

Finally in this section, let us expand on the above to account for power from the damped modes being spread over several bins. This can be approached in a variety of ways. Here we present two options, the first based on the false-alarm test for prominent power in u consecutive bins outlined in Section 6.1.1. From that test, we know already that

$$p(s_\nu|\text{H0}) = \exp(-us_\nu). \tag{6.28}$$

What about $p(s_\nu|\text{H1})$? This is now a little trickier. Once we allow for power being spread over several bins, we need to make assumptions about how that power is distributed. We know that a good model for the underlying mode power is a Lorentzian; we will allow fully for this in the second test below. Here we make the assumption that, provided the range in frequency covered by the u bins is significantly narrower than the peak linewidth Γ, the underlying mode power across the u bins is to first approximation very similar. The expression for the $H1$ likelihood then follows straightforwardly. Taking the simpler case of a uniform prior on the height h, we have

$$p(s_\nu|H1) = \frac{1}{h}\int_0^h \frac{1}{h'+1}\exp[-us_\nu/(h'+1)]\mathrm{d}h'. \tag{6.29}$$

We may then use the above expressions for the H0 and H1 likelihoods to estimate the posterior probability for H1.

In the second test, we allow explicitly for the Lorentzian mode profile. The model we test is

$$P(\nu) = \frac{H}{1+\xi(\nu)^2} + B(\nu), \tag{6.30}$$

where $\xi(\nu) = (\nu - \nu_0)/(\Gamma/2)$ (Eq. 5.56), and $B(\nu)$ is again the background power. The model for the normalized power is therefore

$$s_\nu' = \frac{h}{1+\xi(\nu)^2} + 1. \tag{6.31}$$

We now seek the joint probability for the presence of a mode across the frequency range we are testing, implying we must take the product of the individual likelihoods of each bin. For H0, this gives the joint likelihood

$$p(s_\nu|H0) = \prod_N \exp(-s_\nu), \tag{6.32}$$

over the N bins being tested. For $H1$, not only do we not know the normalized height h, we also do not know the underlying linewidth Γ. We must therefore marginalize over two parameters (height and width). Taking the simple case of uniform priors on h and Γ leads to

$$p(s_\nu|H1) = \prod_N \left(\frac{1}{h\Gamma}\int_0^h \int_0^\Gamma \frac{1}{s_\nu'}\exp[-s_\nu/s_\nu']\mathrm{d}h'\mathrm{d}\Gamma' \right). \tag{6.33}$$

Given Eqs. 6.32 and 6.33, we are in a position to estimate the posterior probability for $H1$ (e.g., using Eq. 6.26).

6.2 AUTOMATED DETECTION OF OSCILLATION SPECTRA

Starting from an analysis-ready lightcurve or time series, the first stage in any analysis is to establish whether there are detectable signatures of oscillations in the frequency-power spectrum. This question may be answered by application of the mode-detection codes that we introduce in this section, which test for ensemble or joint signatures from the entire oscillation spectrum.

If a detection is made, we are then in a position to extract estimates of the average or global asteroseismic parameters $\Delta\nu$ and ν_{max} (of which more in a few paragraphs). Individual-mode detection tests—like those discussed in Sections 6.1.1 and 6.1.2—can then be used to provide initial estimates of the individual oscillation frequencies for input to the peak bagging analysis, which we shall come to in Sections 6.3, 6.4 and 6.5.

We consider two classes of approach to establish the presence of detectable modes. The first approach (and the most common) seeks to capitalize on the almost equal spacing in frequency of overtones of the same angular degree. We consider two methods under this heading: use of the power spectrum of the power spectrum[2] (PSPS), which extracts signatures that are periodic at harmonics of $2/\Delta\nu$; and a method that combines (or folds) the spectrum modulo $\Delta\nu/2$, and which therefore reinforces the signal due to the modes against the incoherent noise background. In the second class of approach, one searches the spectrum for evidence of excess power, over and above the broad-band background, whose properties—the total power excess and width in frequency—are commensurate with expectations based on prior knowledge of the appearance of solar-like oscillation spectra.

Both classes of approach capitalize on the use of a priori information. We know that the power envelope of a solar-like oscillation spectrum has a FWHM Γ_{env} that depends on ν_{max}; the higher ν_{max}, the wider will be the power envelope (see Section 5.4). If we are searching a part of the spectrum at low frequencies, we therefore know that any detected oscillations will have a narrow envelope; similarly, successful searches at higher frequencies go hand-in-hand with a much wider power envelope. It is therefore pointless to search too wide a frequency range at low frequencies, since a large fraction of the search range will then contain no mode power. In contrast, if one searches too narrow a range at high frequencies, one risks omiting a significant fraction of detectable mode signal. A sensible approach is therefore to fix the range in frequency being searched as some multiple of the center of the search window. For example, a range fixed at twice the Γ_{env} expected for that central frequency will cover all but an insignificant fraction of the potentially detectable mode spectrum.

Methods that search for signatures of $\Delta\nu$ can also make use of the fact that solar-like oscillations follow the relation $\Delta\nu \propto \nu_{max}^{\beta}$ to very good approximation. This means that if the frequency range being searched is centered on the actual ν_{max} of the star, we have a strong constraint on the expected $\Delta\nu$. This information can be used as a belt-and-braces check on any claimed detection and also to narrow the values of $\Delta\nu$ one tests for. Here one must be careful not to be too restrictive, because that risks missing genuine and interesting outlying stars.

---

[2]This is equivalent to computation of the envelope of the autocorrelation of the time-domain data (i.e., averaged over the signal frequency), a result known as the Wiener-Khinchin theorem.

Methods that search for signatures of excess power can use scaling relations for the total oscillations power and the width of the oscillation envelope (again, see Section 5.4), which may be expressed as functions of $\nu_{\max}$. This helps us judge the robustness of any claimed detection. Before we discuss the various tests in detail, we add a few remarks on the preparation of the power spectrum for the detection codes and on the estimation of $\Delta \nu$ and $\nu_{\max}$.

Starting with the preparation of the spectrum, in some cases it may be advantageous to remove significant gradients due to the background (i.e., the increase of power toward low frequencies). A seemingly obvious approach would be to actually fit a model of the background power spectral density—including expressions for the granulation and activity components like those discussed in Section 5.5—to the spectrum. However, the presence of significant mode power may bias the fit, if no allowance is made for the modes themselves. That of course presents a catch: the presence of detectable modes must first be established.

There are various ways around this problem. The more evolved the star, the larger will be the amplitudes due to its oscillations and granulation. If these signals dominate over the shot noise, an estimate of the variance of the lightcurve can in principle be used to provide an approximate estimate of $\nu_{\max}$. This logic follows from the various scaling relations we have met in Chapters 4 and 5: the maximum oscillation amplitude and the granulation amplitude may both be expressed as functions that depend on $\nu_{\max}$. We could then fit a background model to the smoothed spectrum that includes a Gaussian to model the contribution of the mode power, with the first-guess parameters of the Gaussian (including its frequency) fixed by the prior on $\nu_{\max}$. This approach is less viable for main sequence stars, which have weaker oscillation and granulation amplitudes. In all but the very brightest targets, it is then shot noise that usually dominates the variance of the lightcurve.

A more sophisticated approach would be to attempt to fit simultaneously the background power and a Gaussian-like function, without any prior knowledge of whether oscillation power is present at detectable levels. This is not as unconstrained a problem as it might at first seem: remember that the properties of the granulation background (amplitude and timescale) and the oscillation power envelope (height and width) can all be expressed in terms of functions that include $\nu_{\max}$.

Another, simpler approach is to obtain an estimate of the background using a method that is relatively insensitive to the presence of the oscillation spectrum. The utilization of a moving-median filter satisfies this requirement. Provided the median filter is set carefully, it will of course be less susceptible than a moving-mean filter to the narrow-band, high-power excursions that are present across frequencies occupied by the modes. If $P_{\mathrm{mm}}(\nu)$ is the power spectral density provided by the moving-median filter, then a proxy for the mean background power spectral density is

$$B_{\mathrm{mm}}(\nu) = P_{\mathrm{mm}}(\nu)/(\ln 2), \qquad (6.34)$$

where the normalization in the denominator corrects for the difference between the mean and median of data distributed with χ^2, two degrees-of-freedom statistics. A filter that is applied in log-log space can work reasonably well—with the width of the filter set to vary from a few tenths of a dex at low frequencies to narrower widths at higher frequencies—and helps mitigate the impact of the inevitable rise in filtered power across the frequency range occupied by the oscillation spectrum. This rise is due to the slowly decaying Lorentzian tails of the modes, which pull up the medians. The effects are more problematic in stars with lower $\Delta\nu$, or higher peak linewidths Γ, and in spectra made from shorter-duration datasets (which give low frequency resolution).

Finally, some comments on estimating $\Delta\nu$ and ν_{max} are in order. The PSPS and matched-filter tests outlined in Section 6.2.1 both provide robust estimates of $\Delta\nu$, together with a crude guide to ν_{max} (from the range of the spectrum over which the detection of the modes is made). In contrast, the global detection test outlined in Section 6.2.2 provides a better proxy for ν_{max} (from the location in frequency of the global SNR in the modes). Further refinement of the parameters (notably ν_{max}) may be achieved as follows.

The range of the spectrum containing the modes may be searched to find signatures of the peaks due to the radial modes, using the tests outlined in Sections 6.1.1 and 6.1.2, along with the additional guidance covered in Section 6.4.3. The best-fitting gradient from a linear regression of the extracted frequencies on the relative overtone number—the absolute order does not matter—then provides a refined estimate of $\Delta\nu$, with the scatter of the data giving a robust estimate of the uncertainty on the parameter. Further refinement may be contemplated if the spectrum is then subjected to a full peak-bagging analysis, using the frequencies extracted from the resulting multiparameter model fits.

Matters are a little trickier where ν_{max} is concerned, not necessarily with respect to the technical aspects of the analysis, but rather in relation to the actual definition of the parameter itself. One must of course allow for the observed realization of the noise. But even having compensated for that (e.g., by averaging the spectrum over frequency), should one then regard ν_{max} as being strictly defined as the exact location in frequency of the maximum of the resulting, smoothed power? Or should one make use of the available data across the entire oscillation spectrum and fit a suitable model to the power envelope to estimate ν_{max}?

As noted in Section 5.4, when the spectrum is averaged over segments of width $\Delta\nu$, the resulting binned spectrum follows, to reasonable approximation, a Gaussian function. This suggests a common approach to the analysis: fitting the binned spectrum to a multiparameter model comprising a Gaussian plus additional terms to represent the broad-band background (see Section 5.5 and Section 6.4.1). Provided the number of bins being averaged across each $\Delta\nu$ is at least 30 or so, the averaged values will (to good approximation) be normally distributed with the uncertainty on each average being the usual error on the mean (i.e., the standard deviation of the contributing powers divided by the square root of the number of bins). A weighted, nonlinear least-squares fit to the data provides the sought-for estimate of ν_{max} and its formal uncertainty. To avoid quantization affecting the accuracy of the solution, a useful trick is to actually fit overlapping $\Delta\nu$-wide averages and then apply a standard correction to the estimated uncertainties to allow for the partial correlation of the averages.

As per the $\Delta\nu$ parameter, subsequent peak-bagging of the individual mode peaks may be used to provide the data needed to refine the estimated ν_{max} (e.g., from a fit of the individual mode powers as a function of frequency).

6.2.1 Detection of Signatures of Near-Regular Frequency Spacings of Modes

Power Spectrum of Power Spectrum Test

We begin our description of specific methods by looking in detail at the power spectrum of the power spectrum (PSPS) test. This is a commonly used approach, and assessing objectively whether or not significant signatures of the large-separation $\Delta\nu$ have been detected is relatively straightforward. Methods based on use of the autocorrelation of the power spectrum are in contrast beset by strong correlations between the elements of the output, which makes handling the statistical significance of the result nontrivial.

Let us consider first the expected form of the second spectrum. To help explain, we start with an idealized case in which each mode is confined to a single bin of the first spectrum (i.e., the power spectrum of the lightcurve). This might approximate the form expected in a u-bin averaged spectrum, where u is sufficiently large to average over the Lorentzian profiles. Were the first spectrum to indeed be an averaged spectrum, we would need to take account of the averaging when considering statistical aspects of the test, as we shall explain later. We also begin by assuming that we have selected a portion of the first spectrum that contains detectable signatures of the modes. We shall come back later in this section to discuss the application of the test in an automated way (i.e., one in which the spectrum is searched systematically to *detect* the modes).

In our entry-level example here, the selected portion of the first spectrum will present a dominant frequency spacing of $\simeq \Delta\nu/2$, that is, the approximate average spacing between the $l = 0$ and $l = 1$ modes (the degrees that provide the most prominent modes in the spectrum). Let Δ_s be the range in frequency that is selected from the first spectrum. Ignoring for the moment the Gaussian-like power envelope of the oscillation spectrum—we incorporate it a bit later—and assuming that any signatures of mixed modes do not significantly disrupt the near-regular overtone pattern, the dataset thereby presented to compute the second spectrum will, to first order, be described by a Dirac comb of spacing $\Delta\nu/2$, multiplied by a boxcar function of width Δ_s. When the first spectrum is sampled at its natural frequency resolution, there will be

$$N_s = \Delta_s/\Delta_T \equiv T\Delta_s \tag{6.35}$$

independent frequency bins in the selected frequency range.

The second Fourier spectrum is in time. It will contain $N_s/2 + 1$ bins, running from zero up to a Nyquist limit of $(2\Delta_T)^{-1} \equiv T/2$ s. It comprises a Dirac comb of peaks located at harmonics of the inverse spacing $2/\Delta\nu$. And each peak will be described by the function $\mathrm{sinc}\,(\pi t\Delta_s)$ (i.e., the transform of the boxcar from the frequency domain).

This is not quite the whole story. There will also be peaks in the second spectrum at $1/\Delta\nu$ and its harmonics. That is because the first spectrum also contains signatures that repeat at multiples of $\Delta\nu$, since each degree l has a different visibility in the first spectrum (there will also be detectable power due to the $l = 2$ modes and possibly higher l too). However, the power in the $1/(\Delta\nu)$ harmonics is relatively weak, and the signatures in the $2/(\Delta\nu)$ harmonics dominate. Clearly, matters will be even less ideal if a star shows strong signatures of mixed modes in its oscillation spectrum.

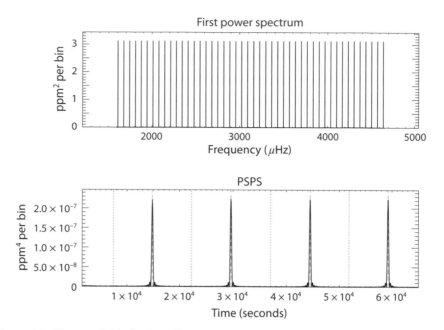

Figure 6.1. Top panel: Idealized oscillation spectrum, comprising peaks separated by $\Delta\nu/2$. Each peak occupies a single bin, and all have the same power. Bottom panel: Power spectrum of the first spectrum (i.e., the PSPS). The first peak lies at $2/(\Delta\nu)$, and all subsequent peaks lie at its harmonics. The vertical dotted lines mark $1/(\Delta\nu)$, and its harmonics.

Figures 6.1, 6.2, and 6.3 illustrate the simple entry-level case. The top panel of Figure 6.1 shows an idealized spectrum comprising peaks separated by $\Delta\nu/2$ that mimick the dominant spacing of the $l = 0$ and $l = 1$ modes. All peaks have the same power. The bottom panel shows the PSPS of the first spectrum. The first peak lies at $2/(\Delta\nu)$, and all subsequent peaks lie at its harmonics. Note the absence of power at $1/(\Delta\nu)$ and its harmonics (locations marked by the vertical dotted lines).

Next, in Figure 6.2 the peaks in the first spectrum have alternating maximum powers (in the ratio 1.0:1.5) to mimic the different visibilities of $l = 0$ and $l = 1$ modes. The most prominent peaks in the PSPS still lie at $2/(\Delta\nu)$ and its harmonics. However, some power is now present at $1/(\Delta\nu)$ and its harmonics. Finally, Figure 6.3 introduces weaker peaks (relative visibility of 0.5) to mimic $l = 2$ modes. Power in $1/(\Delta\nu)$ and its harmonics remains weak. But notice that there is now also a slow modulation of power across the PSPS. The period of this modulation corresponds to the small frequency separation between the $l = 0$ and 2 modes.

With the basics covered, let us now impose the frequency-dependent oscillation power envelope on the first spectrum. We have seen that a Gaussian function is a reasonable choice to describe the variation of mode power with frequency. If we now choose our window in the first spectrum so that it spans the Gaussian profile, sufficiently far into the wings that only negligible mode power remains, the Gaussian in effect acts as an apodizing window, like the simple frequency (boxcar) cut in our illustrative example of Figure 6.1. That initial example led in the second spectrum to a convolution with the sinc function transform of the boxcar. With the Gaussian power envelope now included, we get a convolution with a Gaussian

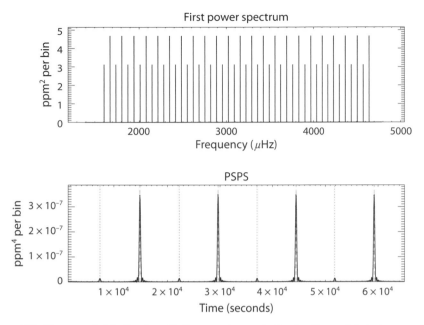

Figure 6.2. Top panel: Idealized oscillation spectrum, similar to Figure 6.1, but now with alternating maximum powers (in the ratio 1.0:1.5) to mimic the different visibilities of $l = 0$ and $l = 1$ modes. Bottom panel: Power spectrum of the first spectrum (i.e., the PSPS). The most prominent peaks lie at $2/(\Delta\nu)$ and its harmonics. Note how, in contrast to Figure 6.1, some power is now present at the locations marked by the vertical dotted lines, that is, $1/(\Delta\nu)$ and its harmonics.

function: remember that the Fourier transform of a Gaussian is another Gaussian. Formally, if the Gaussian describing the power envelope in the first spectrum has a width parameter equal to $c = \Gamma_{\text{env}}/(2\sqrt{2\ln 2})$, where Γ_{env} is the FWHM of the envelope (see Section 5.4), then its Fourier transform is

$$\exp\left[-\nu^2/(2c^2)\right] \xleftrightarrow{\mathcal{F}} c\sqrt{2\pi}\exp\left(-2\pi^2 c^2 t^2\right). \tag{6.36}$$

The Fourier transform here is a Gaussian in time, characterized by a time-domain FWHM of $(4\ln 2)/(\pi\Gamma_{\text{env}})$. Since the PSPS is in *power*, it will therefore comprise a set of peaks that are each described by a Gaussian function having a FWHM that is a factor of $\sqrt{2}$ narrower, that is,

$$\Gamma_2 = \left(\frac{4\ln 2}{\sqrt{2\pi}}\right)(\Gamma_{\text{env}})^{-1}. \tag{6.37}$$

The number of independent bins spanning each Gaussian peak depends on the binwidth, Δ_{s}^{-1}, in the second spectrum. A suitable choice to sample in the first spectrum would be $\Delta_{\text{s}} = 2\Gamma_{\text{env}}$ (i.e., twice the FWHM of the oscillation power envelope). It then follows that the number of bins lying across one FWHM of each

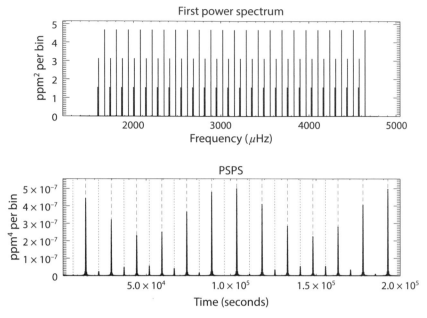

Figure 6.3. Top panel: Idealized oscillation spectrum, similar to Figure 6.1, but now with an extra set of weaker peaks to simulate the presence of $l = 2$ modes. Bottom panel: Power spectrum of the first spectrum (i.e., the PSPS). As per Figure 6.2, we see power at $2/(\Delta \nu)$ and its harmonics, and $1/(\Delta \nu)$, and its harmonics. Note also the slower modulation of power across the spectrum, which is fixed by the small frequency separation between the $l = 0$ and $l = 2$ modes.

Gaussian peak in the second spectrum will always be

$$n_{2,\Gamma} = \Delta_s \, \Gamma_2 \equiv \left(\frac{8 \ln 2}{\sqrt{2\pi}} \right). \tag{6.38}$$

If we choose to oversample the second spectrum by a factor m, that number will increase to $mn_{2,\Gamma}$ bins.

Figure 6.4 illustrates the impact of including the Gaussian modulation of power in the first spectrum, and the resulting Gaussian peaks in the PSPS. How do we relate the expected maximum power spectral density of each of these Gaussians to the properties of the mode peaks in the first spectrum? We must again apply Parseval's Theorem. Let A be the equivalent radial-mode amplitude. Since for now we assume that the modes in the first spectrum are unresolved, the $n_s = \Delta_s / \Delta \nu$ bins that contain a radial mode will have a power per bin of $A^2/2$. Recall that calibration of the first spectrum means that we recover the mean-square power $A^2/2$ and not A^2. Note that we work here with a power-per-bin scaling in the first spectrum, which makes some of the material below a little easier to develop and follow.

If the mode powers in the first spectrum were not modulated by a Gaussian-like envelope, all radial modes would have the same power, and the mean-square power

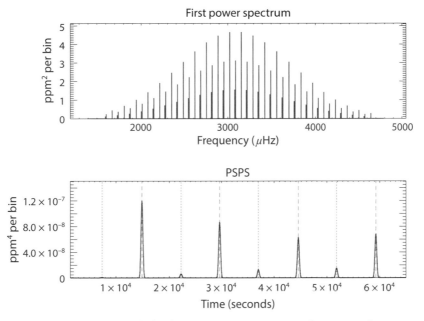

Figure 6.4. Top panel: Same idealized spectrum as in Figure 6.3, but now with a Gaussian envelope applied to the mode powers. Bottom panel: PSPS, with each peak now taking the form of a Gaussian.

presented to the second spectrum on account of the radial modes alone would be

$$\frac{1}{N_s} \sum_{k=1}^{N_s} P_k^2 = \left(\frac{A^2}{2}\right)^2 \left(\frac{n_s}{N_s}\right) \equiv \left(\frac{A^4}{4T\Delta\nu}\right), \qquad (6.39)$$

where we have summed over the N_s selected bins in the first spectrum, which have powers P_k. Notice how the units of the above will be $(\text{ppm})^4$ for photometric data, or $(\text{ms}^{-1})^4$ for Doppler velocity data, since we sum the squares of data from the first *power* spectrum. We must now correct for the Gaussian modulation of the mode powers in that first spectrum and also include the power from the nonradial modes. The corrected mean-square power becomes

$$\frac{1}{N_s} \sum_{k=1}^{N_s} P_k^2 = \left(\frac{\pi}{8\ln 2}\right)^{1/2} \left(\frac{\Gamma_{\text{env}}}{\Delta_s}\right) \left(\frac{\zeta_2 A_{\text{max}}^4}{4T\Delta\nu}\right), \qquad (6.40)$$

with A_{max} now the maximum mode amplitude at the center of the oscillation spectrum, and ζ_2 the sum of the squares of the radial-mode normalized mode visibilities (Eq. 5.77) in power, that is,

$$\zeta_2 = \sum_l (S_l/S_0)^4. \qquad (6.41)$$

Most of this power goes into the $(T/2)/(2/\Delta\nu) = T\Delta\nu/4$ Gaussian peaks in the second spectrum, which lie at $2/\Delta\nu$ and its harmonics. To get an approximate estimate of the expected maximum power $P_{2,\max}$ of the $2/\Delta\nu$ harmonics in the second spectrum, we must divide the total power in Eq. 6.40 by the number of peaks; we must also correct for the fact that in each peak of the second spectrum, power is spread across a Gaussian profile. This gives:

$$P_{2,\max} \lesssim \left(\frac{\pi}{4\ln 2}\right)\left(\frac{\Gamma_{\text{env}}}{\Delta_{\text{s}}}\right)^2\left(\frac{\zeta_2 A_{\max}^4}{T^2\Delta\nu^2}\right). \tag{6.42}$$

Notice the inequality: the reason for it is not only because some power goes into the $1/\Delta\nu$ harmonics, and the power is modulated by the small frequency separation, but also because the frequency spacings in the first spectrum will be close to, but not exactly, regular. Moreover, as stars evolve into the subgiant phase, the appearance of avoided crossings in the nonradial modes, and later additional signatures of mixed modes, will remove power from the main harmonics in the second spectrum.

Next, let us drop the constraint that the mode peaks in the first spectrum are unresolved and include explicitly the Lorentzian profiles. We shall make one approximation: we assume that the most prominent modes in the first spectrum—which provide the largest contribution to the signal in the second spectrum—all have approximately the same associated damping time τ. We therefore elaborate our model of the first spectrum to include mode peaks with Lorentzian profiles, each with a FWHM of $\Gamma = (\pi\tau)^{-1}$ (Eq. 5.38); in effect, each peak is now convolved with a Lorentzian function.

We must first modify the expression for the total mean-square power (Eq. 6.40). Power in each mode peak in the first spectrum will now be spread over several (potentially many) bins, which can alter significantly the sum of the squares of the powers (even though the total power over all bins is the same). It turns out that the predicted mean-square power will be diminished, relative to the prediction in Eq. 6.40, by the factor $1/(\pi\Gamma T) \equiv (\tau/T)$:

$$\frac{1}{N_{\text{s}}}\sum_{k=1}^{N_{\text{s}}} P_k^2 = \left(\frac{\pi}{8\ln 2}\right)^{1/2}\left(\frac{\Gamma_{\text{env}}}{\Delta_{\text{s}}}\right)\left(\frac{\zeta_2\tau A_{\max}^4}{4T^2\Delta\nu}\right). \tag{6.43}$$

The complex second transform is now modulated by an exponentially decaying function of the form $\exp(-t/\tau)$. The maximum *powers* of the peaks in the second power spectrum are therefore attenuated by the multiplicative factor $\exp(-2t/\tau)$. Peaks lying at increasingly longer times in the second spectrum will now have diminishing powers relative to the first peak. However, each still has a Gaussian profile.

To predict the power of the first peak, we can no longer simply divide Eq. 6.43 by the number of overtones in the second spectrum (i.e., $T\Delta\nu/4$), as we did above for the unresolved case; the peaks now all have different powers. As τ approaches to T, so we need our prediction to tend to the unresolved case. Multiplying $T\Delta\nu/4$ by the extra factor τ/T and including the additional exponential attenuation,

$$\exp\left[-4/(\Delta\nu\tau)\right] \equiv \exp\left(-2t/\tau\right),$$

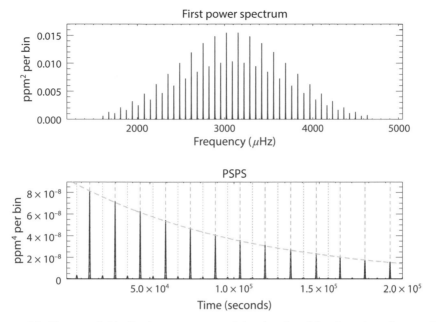

Figure 6.5. Top panel: Idealized spectrum, containing $l = 0$ and $l = 1$ modes, where each peak is now a Lorentzian profile. Bottom panel: PSPS, showing the resulting attenuation due to the Lorentzian profiles in the first spectrum (dashed gray line).

determined at the location $t = 2/(\Delta \nu)$ of the first peak, gives us a reasonable prediction for the power of the first peak. We obtain:

$$P_{2,\mathrm{max}} \lesssim \left(\frac{\pi}{4\ln 2}\right) \left(\frac{\Gamma_{\mathrm{env}}}{\Delta_{\mathrm{s}}}\right)^2 \left(\frac{\zeta_2 A_{\mathrm{max}}^4}{T^2 \Delta \nu^2}\right) \exp\left[-4/(\Delta \nu \tau)\right]. \tag{6.44}$$

Notice that, aside from the exponential attenuation factor, this corresponds to the expression for the unresolved case (Eq. 6.42).

Figure 6.5 shows what happens when we introduce into the first spectrum Lorentzian profiles, each having a width of $\Gamma = 0.7\,\mu\mathrm{Hz}$ (equivalent to a damping time of $\tau = 5.3$ days). The dashed gray line in the PSPS follows the resulting attenuation. (Note that we used a simulated spectrum having $l = 0$ and 1 modes, but no $l = 2$ modes, since this case illustrates the damping attenuation more clearly than the case with $l = 2$ modes included.)

What about the expected background level in the second spectrum? This depends on the variance (scatter) about the underlying limit spectrum in the selected range of the first spectrum. Typically, the variance of the background in the first spectrum is most important here, so that we have, to reasonable approximation,

$$P_{2,\mathrm{noise}} \approx \frac{2}{N_{\mathrm{s}}} \left(\frac{1}{N_{\mathrm{s}}} \sum_{k=1}^{N_{\mathrm{s}}} B_k^2\right), \tag{6.45}$$

that is, the total mean-square power due to the background components in the first spectrum will be spread over the $N_s/2$ bins of the second spectrum. (We assume the mean power of the first spectrum is removed, prior to computation of the second spectrum, so that the first component of the second spectrum has zero power.)

The background level in the second spectrum will be the same irrespective of whether we present the data in the form of a pristine, unaveraged first spectrum or an averaged (or co-added) version. If we present a u-bin averaged spectrum, or a spectrum made from co-adding u independent short spectra, the variance in the first spectrum will be reduced by a factor u compared to the pristine, unaveraged case. However, that power will then be divided over a factor of u fewer bins than in the pristine case, giving the same mean background level.

This result indicates that it is the impact of the presentation of the first spectrum on the signal, not the background noise, that has the potential to influence the SNR of the peaks in the second spectrum. Earlier in the section we saw that to maximize the signatures of $\Delta \nu$ in the second spectrum, it is advantageous either to average the first spectrum over several bins or to present a spectrum that has been made by co-adding several shorter spectra. Both approaches will give spectra that tend to the unresolved case (subject of course to a suitable choice for the averaging frequency or co-adding duration), which, all other things being equal, gives the most prominent peaks.

How might we run the PSPS procedure to detect oscillations in a blind, automated fashion, given only the power spectrum of the target star but no other information? We can search the spectrum systematically, beginning at low frequency and moving progressively through the spectrum. At each location we must select a range of frequencies to search around the central frequency ν_s (i.e., we must choose a suitable value for $\Delta_s \equiv N_s T^{-1}$). The range is chosen based on the expected width Γ_{env} of the oscillation power envelope, assuming a ν_{max} corresponding to the center of the search window (i.e., $\nu_{\mathrm{max}} \equiv \nu_s$). Recall from Section 5.4 that Γ_{env} may be expressed as a power law in ν_{max}, meaning that here we can compute expected values $\Gamma_{\mathrm{env}}(\nu_s)$ based simply on the center of the search window, ν_s. The tested range at each ν_s is then fixed to the value

$$\Delta_s(\nu_s) = 2\Gamma_{\mathrm{env}}(\nu_s). \tag{6.46}$$

Assuming that a spectrum of modes is present with a ν_{max} close to the center of the search window, this width will capture all but an insignificant fraction of the detectable mode power. In sum, the higher the frequency of the search window, the wider will be the range Δ_s.

We can also use the known power-law dependence of $\Delta \nu$ on ν_{max} (i.e., $\Delta \nu \simeq \alpha \nu_{\mathrm{max}}^{\beta}$) to help judge whether prominent peaks in the second spectrum have periods Π that tally with the expected value of $2/(\Delta \nu)$ and its harmonics. That expected period will be given approximately by

$$\Pi_s \equiv 2/\Delta \nu(\nu_s) \simeq \left(\frac{2}{\alpha}\right) \nu_s^{-\beta}. \tag{6.47}$$

The statistics of the PSPS need to be handled with a little care. If the first spectrum is flat (i.e., white) and unaveraged, then the PSPS will also be flat and have χ^2 two-degrees-of-freedom statistics, just like the first spectrum. Provided we have

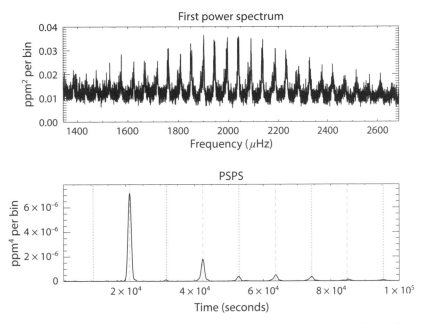

Figure 6.6. Frequency-power spectrum (top panel) and PSPS (bottom panel) for *Kepler*-410.

done a reasonably good job of dividing out the nonwhite background profile, we may therefore test the PSPS against the $H0$ or null hypothesis (i.e., that the first spectrum is consistent with being composed of white noise only), employing tests like those presented in Section 6.1.1. In contrast, if we were to present a u-bin averaged white-noise spectrum, or a co-added white-noise spectrum made from u independent spectra, the PSPS would instead be described by χ^2, $2u$ degrees-of-freedom statistics. Again, we saw earlier in the chapter how to handle the $H0$ hypothesis test for this case (e.g., see Eq. 6.13).

If the first spectrum is not white, things are a bit more complicated. Let us begin by presenting an unaveraged first spectrum to compute the PSPS. Regions of the PSPS that contain features arising from nonwhite components in the first spectrum will then tend toward showing Gaussian, and not χ^2 two-degrees-of-freedom, statistics. The more prominent the feature against the background, the tighter will be the scatter in the observed power about the feature. This can be seen clearly in Figure 6.6, which shows the PSPS of *Kepler*-410. As in the previous plots, $1/(\Delta\nu)$ and its harmonics are marked by vertical dotted lines, whereas $2/(\Delta\nu)$ and its harmonics are marked by vertical dashed lines. Notice how the noise on the prominent peaks that dominate the PSPS is notably Gaussian-like—there is very little scatter about the underlying Gaussian peak profiles—and the overall appearance differs strongly from the ragged, negative-exponential statistics shown by the first spectrum. (We also note in passing that power in the PSPS peaks of *Kepler*-410 falls off much more rapidly with increasing timescale than does the power in Figure 6.5. This is because the typical mode lifetime is shorter than in our idealized example.) If the first spectrum is averaged, then depending on the value of u, we can end up having a PSPS that presents Gaussian-like statistics irrespective of the balance between white and nonwhite components in the first spectrum.

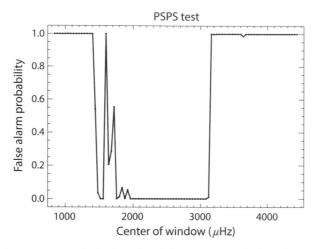

Figure 6.7. False-alarm probability for a PSPS $H0$ test, as applied to the frequency-power spectrum of *Kepler*-410.

Figure 6.7 shows the result of applying a systematic search of the spectrum of *Kepler*-410 for the presence of significant signatures in the PSPS. The estimated false-alarm probability, given by applying an $H0$ test (see above), is plotted as a function of the frequency at the center of the range being tested; as expected, the probability falls to very low values across the range where we know significant mode power is present.

More evolved stars present added complexities associated with the frequency pattern of the nonradial modes and even suppressed $l = 1$ amplitudes, which can redistribute power in the PSPS. Figures 6.8 and 6.9 show two such examples. KIC 8561221 (Figure 6.8) is one of a number of more-evolved stars—this is a star at the base of the red-giant branch—that have been found by *Kepler* to show suppressed $l = 1$ mode amplitudes. We therefore see much stronger signatures in the PSPS at $1/(\Delta\nu)$ and its harmonics, compared to the classic main sequence case of *Kepler*-410. The suppressed amplitudes are believed to result from the scattering of $l = 1$ waves into higher spatial wavenumbers in the core of the star. KIC 9574283 (Figure 6.9) is a star of very similar mass and evolutionary state, which does not show the same $l = 1$ suppression. Here the mixed $l = 1$ modes manifest as an irregular pattern in frequency, and in spite of the very high SNR, this is seen to have an impact on the appearance of the PSPS, notably in relation to which of the harmonics appear most prominently.

Finally in this section, we may also use the formulas above to make a prediction of the SNR in the PSPS given a priori knowledge of the target and the expected duration of the observations. The fundamental stellar properties may be used as input to the asteroseismic scaling relations to compute the oscillation parameters, allowing us to predict $P_{2,\mathrm{max}}$. Computations of the expected A_{max} should take into account the instrumental response function (see Chapter 4). The contribution to the background from granulation can be predicted using the scaling relations and equations in Sections 4.3.1 and 5.5. The shot noise contribution can be estimated using the apparent magnitude of the target star and the known instrumental performance

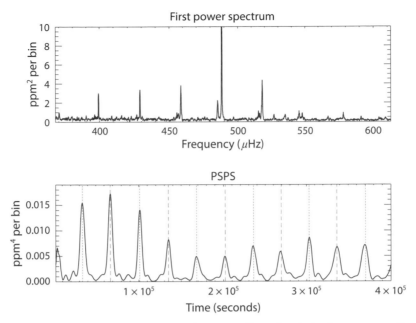

Figure 6.8. Frequency-power spectrum (top panel) and PSPS (bottom panel) for KIC 8561221.

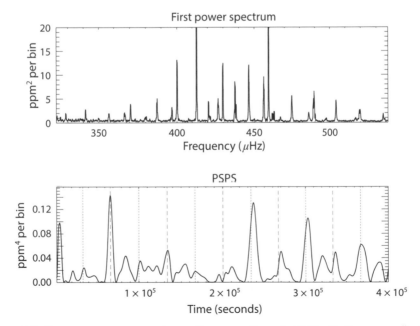

Figure 6.9. Frequency-power spectrum (top panel) and PSPS (bottom panel) for KIC 9574283.

properties. Even if we have a good estimate of the shot noise per cadence, σ, at just one apparent magnitude, we can make approximate estimates for other magnitudes. If V_{ref} is the reference magnitude, with its known noise, then

$$\sigma(V_{\mathrm{obs}}) = \sigma(V_{\mathrm{ref}}) \, 10^{\,[(V_{\mathrm{obs}}-V_{\mathrm{ref}})/5]}. \tag{6.48}$$

Here the magnitudes should be those in the bandpass of the instrument. The above expression neglects a variety of subtle factors, such as the impact of the temperature of the star and that of the crowding of stars in the target field (for which the instrumental point spread function is relevant). Nevertheless, Eg. 6.48 does provide some guidance in the absence of other information.

We may therefore predict $P_{2,\mathrm{noise}}$ and hence the underlying SNR of the $2/(\Delta\nu)$ peak. At this point it is worth remembering that this prediction will be uncertain, in the sense of being an upper limit only (for the reasons outlined above). In Section 6.2.2, we outline a test that is much less dependent on any fine detail of the structure of the oscillation spectrum—as is the case for the PSPS test—and therefore provides a somewhat more robust way of making predictions of the asteroseismic detectability of the modes.

Matched Filter Test

The *matched filter* test takes its inspiration from signal processing. We aim to build a suitable template in frequency of the oscillation spectrum and then correlate that template with the observed spectrum. The optimal filter is the one that maximizes the SNR of the resulting correlated response.

The following formulation of the test again seeks to infer the presence of a significant signature of $\Delta\nu$, like the spectrum of the power spectrum test discussed above. We shall also assume that application to the observed spectrum of a suitable moving-median filter has already provided a reasonable estimate of the broadband background, so that we may work with relative (background-normalized) powers, s_ν, as per the tests in Section 6.1. Here we write the relative powers as $s(\nu_k)$, with ν_k denoting the finite frequencies of the observed spectrum.

Let us begin by considering the solar-like oscillation spectra presented by cool main sequence stars. A reasonable starting point for the template is then the asymptotic relation for high-overtone p modes, which we met in Chapter 3:

$$\nu_{nl} = \Delta\nu \, (n + l/2 + \epsilon) - \delta\nu_{0l}. \tag{6.49}$$

It follows that the simplest filter we can construct is one composed of a regular grid of frequencies modulo the tested value of $\Delta\nu/2$, with the exact placement of the frequencies depending on the tested value of ϵ. The template then has nonzero values at the putative locations of successive $l = 0$ and $l = 1$ modes. We might also choose to incorporate some width at each frequency to account for the finite widths of the mode peaks and rotational splitting of the $l = 1$ modes.

The matched-filter response is given by summing the observed powers at the frequencies specified by the template. Working in relative powers, we can write the response to the filter described above—a regular grid, with some width incorporated at each frequency—as a double summation,

$$s_{\mathrm{mf}} = \frac{1}{N} \left(\sum_{k=n_{\mathrm{ref}}-N_{\mathrm{ord}}/2}^{n_{\mathrm{ref}}+N_{\mathrm{ord}}/2} \sum_{j=k-u/2}^{j=k+u/2} s(\nu_j) \right), \qquad (6.50)$$

where the basic grid of the template is set by

$$\nu_k = \Delta\nu \left(k/2 + \epsilon \right), \qquad (6.51)$$

and

$$N = uN_{\mathrm{ord}} \qquad (6.52)$$

corresponds to the total number of independent frequency bins in the summation. The first summation in Eq. 6.50 runs over N_{ord} radial orders centered on a reference order n_{ref}; the second summation runs over u bins at the putative test location of each $l = 0$ or 1 mode.

Assuming that the frequencies in the observed spectrum are independent, the summed metric s_{mf} follows χ^2, $2N$-degrees-of-freedom statistics. The probability of obtaining a value of the response greater than or equal to the observed s_{mf} is then given by

$$p(s'_{\mathrm{mf}} \geq s_{\mathrm{mf}}, N) = \int_{s_{\mathrm{mf}}}^{\infty} \frac{\exp\left(-s'_{\mathrm{mf}}\right)}{\gamma(N)} s'^{(N-1)}_{\mathrm{mf}} \, ds'_{\mathrm{mf}}. \qquad (6.53)$$

We therefore seek the combination of estimates of $\Delta\nu$ and ϵ that minimizes the false-alarm probability; or, put more simply, that maximizes s_{mf} (for a given value of N).

The blind application of the test to the frequency-power spectrum again implies a systematic search, with the center of the frequency range being tested and the known prior on the width of the oscillation power envelope being used to select suitable values of $\Delta\nu$ (and numbers of overtones N_{ord}) to test. We refer to the discussion in Section 6.2.1: the frequency at the center of the range being tested fixes typical values expected for $\Delta\nu$, from which a reasonable choice for the number of orders to include in the matched-filter response would be

$$N_{\mathrm{ord}} = 2\Gamma_{\mathrm{env}}/\Delta\nu. \qquad (6.54)$$

Figure 6.10 shows the matched-filter response s_{mf} given by the frequency-power spectrum of *Kepler*-410. The sharp peak in the response marks the $\Delta\nu$ of the star's oscillation spectrum; the line with the highest response provides an indication of where in frequency the oscillation spectrum is centered.

There is nothing to stop us from adding further complexity to the filter. We might, for example, include the second-order terms $\delta\nu_{0l}$, implying additional optimization over small-spacing parameters $\delta\nu_{01}$, $\delta\nu_{02}$, and (if appropriate) $\delta\nu_{13}$. We could also incorporate a linear variation of the large separation with radial order n,

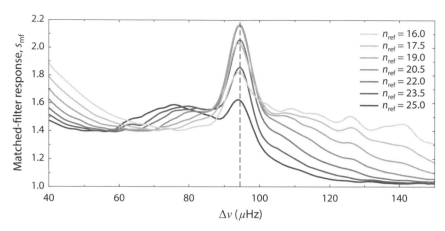

Figure 6.10. The matched-filter response s_{mf} given by the frequency-power spectrum of *Kepler*-410. Each line follows the maximum response for a given n_{ref} (i.e., the maximum over a range of possible values of ϵ at the tested $\Delta\nu$).

that is,

$$\Delta\nu_n = \Delta\nu \left[1 + \alpha_{\Delta\nu}(n - n_{ref})\right],\tag{6.55}$$

controlled by the coefficient $\alpha_{\Delta\nu}$, which implies an asymptotic relation of the form

$$\nu_{nl} = \Delta\nu \left[n + \alpha_{\Delta\nu}/2 \left(n - n_{ref}\right)^2 + l/2 + \epsilon\right] - \delta\nu_{0l}.\tag{6.56}$$

Finally, note that the above filters may not be optimal in more evolved stars; there, we might choose to limit the filter to follow the pattern of radial modes only.

6.2.2 Detection of Total Power Due to the Modes

In this section we present an alternative, and simpler, approach to detection of the modes, which is based on the total observed power due to the oscillations, P_{tot}, introduced in Section 5.4.

The approach relies on having a reasonably robust estimate of the broad-band background. Any region of the spectrum over which the total power significantly exceeds the background level may then be a good candidate to contain the spectrum of modes. However, that power excess must be consistent with prior expectations for solar-like oscillations. This so-called *global detection* approach is applicable both as a robust means of predicting whether modes might be detectable in a given target—given some prior knowledge of the fundamental properties, and the apparent magnitude and instrumental response—or as a means of actually detecting modes in the spectrum given no other information. We begin by describing the method as a predictive tool, in a form that has been successfully applied to inform asteroseismic target selection for *Kepler* and K2, and is being used to plan for TESS. Then we describe its implementation as a detection tool, in which we use a Bayesian formulation to assess the significance of potential excess power due to modes.

As per the previous tests, we select N_s-bin wide segments of the frequency-power spectrum, again assuming that each segment is suitably chosen so that it would contain all but an insignificant fraction of any oscillation power. We recall that P_{tot} is the integrated mode power: it may be predicted using Eq. 5.81 and the simple scaling relations in Section 5.4, which are based on fundamental stellar properties.

We define the integrated background power by

$$B_{\text{tot}} = \Delta_T \sum_{k=1}^{N_s} B(\nu_k) = T^{-1} \sum_{k=1}^{N_s} B(\nu_k), \tag{6.57}$$

where $B(\nu_k)$ is the total power spectral density due to the background terms (here, assumed to be calibrated in power per Hertz). As noted in Section 6.2.1, use of scaling relations and information on the apparent magnitude of the target and the instrumental response allows us to predict the background.

From the above we may then construct a global measure of the SNR in the oscillations signal:

$$\text{SNR}_{\text{tot}} = P_{\text{tot}}/B_{\text{tot}}. \tag{6.58}$$

When the observed power in the oscillations relative to the background power is high, the power excess due to the oscillations will be clearly visible, and SNR_{tot} will be high. However, at much lower SNR_{tot} the statistical fluctuations in the background power may swamp the oscillation signal, so that the excess is much harder to distinguish from the expected variability in the background.

A total of N_s independent frequency bins contributes to estimation of the underlying P_{tot} and B_{tot}, and hence to SNR_{tot}. We may therefore test SNR_{tot} against χ^2 $2N_s$ degrees-of-freedom statistics to determine whether the predicted SNR would be hard to explain by chance alone. In general, the probability of obtaining any value SNR' above a given level SNR is given by

$$p(\text{SNR}' \geq \text{SNR}, N_s) = \int_x^\infty \frac{\exp(-x')}{\gamma(N_s)} x'^{(N_s-1)} \, dx', \tag{6.59}$$

where $x = 1 + \text{SNR}$, and γ is the Gamma function. Note the normalization of x: as SNR tends to zero (i.e., no oscillation signal present), we require that x tends to unity. To flag a possible detection, we demand that the prediction of the *observed* SNR of the star exceed some SNR threshold, SNR_{thr}, corresponding to a fractional false-alarm probability, p_{SNR} (e.g., $p_{\text{SNR}} = 0.01$, or 1%). The required false-alarm threshold follows by solving

$$p(\text{SNR})' \geq \text{SNR}_{\text{thr}}, N_s) = p_{\text{SNR}}. \tag{6.60}$$

The probability that the observed SNR_{tot} would exceed SNR_{thr} is then given by

$$p_{\text{final}} = \int_y^\infty \frac{\exp(-y')}{\gamma(N_s)} y'^{(N_s-1)} \, dy', \tag{6.61}$$

with

$$y = (1 + \text{SNR}_{\text{thr}})/(1 + \text{SNR}_{\text{tot}}). \qquad (6.62)$$

We may regard p_{final} as providing an estimate of the probability of detecting solar-like oscillations in the star. The sensitivity of the detection probability to the length of the observations, T, is captured by the presence of N_s in the formalism. As T (and hence N_s) increases, so the relative statistical fluctuations in B_{tot} will decrease in magnitude. Hence, for a given underlying SNR_{tot}, the excess power due to the oscillations will be more clearly visible against the background (as reflected by a reduction in the size of the required SNR_{thr}).

We may also cast the global detection method as a test to be applied in a blind, automated fashion to the actual observations to search for significant signatures of the oscillations. Again, we begin at low frequencies and work systematically through the spectrum. With ν_s the frequency at the center of each tested frequency range $\Delta_s \equiv N_s T^{-1}$, the observed, global SNR over that range is computed from

$$\text{SNR}(\nu_s) = T^{-1} \sum_{k=1}^{N_s} P(\nu_k)/B(\nu_k). \qquad (6.63)$$

where the $P(\nu_k)$ are the observed power spectral densities, and the $B(\nu_k)$, are the estimated background power spectral densities (e.g., as determined from application of the moving-median filter; see Eq. 6.34). We may then compute the corresponding false-alarm ($H0$) probability for $\text{SNR}(\nu_s)$ from

$$p(\text{SNR}(\nu_s), N_s | H0) = \int_x^\infty \frac{\exp(-x')}{\gamma(N_s)} x'^{(N_s - 1)} \, dx', \qquad (6.64)$$

with

$$x = 1 + \text{SNR}(\nu_s). \qquad (6.65)$$

To compute the alternative $H1$ hypothesis, we need a predicted SNR against which to compare the observed $\text{SNR}(\nu_s)$. This we know how to do, courtesy of the description above on estimating the predicted ratio SNR_{pred}. The $H1$ probability may therefore be written as

$$p(\text{SNR}(\nu_s), \text{SNR}_{\text{pred}}, N_s | H1) = \int_x^\infty \frac{\exp(-x')}{\gamma(N_s)} x'^{(N_s - 1)} \, dx', \qquad (6.66)$$

with

$$x = (1 + \text{SNR}(\nu_s))/(1 + \text{SNR}_{\text{pred}}). \qquad (6.67)$$

The posterior probability for the $H1$ hypothesis is then given by

$$p(H1 | \text{SNR}(\nu_s), \text{SNR}_{\text{pred}}, N_s) = \left[1 + \frac{p(\text{SNR}(\nu_s), N_s | H0)}{p(\text{SNR}(\nu_s), \text{SNR}_{\text{pred}}, N_s | H1)} \right]^{-1}, \qquad (6.68)$$

from which we may judge whether a detection of the oscillations has been made.

Figure 6.11 shows the results of applying a systematic search of the spectrum of *Kepler*-410 for the presence of significant signatures in SNR_{tot}. The top panel shows the frequency-power spectrum of the star after dividing by the median-filter estimated background. Note the two prominent artifact peaks at high frequencies. The middle-left panel shows the measured SNR_{tot} as a function of the frequency at the center of the range being tested (dark line, joined by points at which the test was applied). The dashed line marks the predicted SNR_{pred}. The middle-right and bottom-left panels show, respectively, the resulting estimates of $p(SNR(\nu_s), N_s | H0)$ (Eq. 6.64) and $p(SNR(\nu_s), SNR_{pred}, N_s | H1)$ (Eq. 6.66). Finally, the bottom-right panel plots the posterior probability for H1, $p(H1 | SNR(\nu_s), SNR_{pred}, N_s)$ (Eq. 6.68). The posterior shows the hoped-for high values across the frequency range occupied by the modes.

6.3 FUNDAMENTALS OF PEAK-BAGGING: EXTRACTION OF INDIVIDUAL MODE PARAMETERS

We now come to the extraction of individual mode parameters, using the peak-bagging approach of fitting a multiparameter model to the observed oscillation spectrum. In this section we cover the fundamentals of the method. We shall then discuss practical implementations of the peak-bagging in Sections 6.4 and 6.5.

6.3.1 Fundamentals: Maximum Likelihood Estimation

Peak-bagging demands not only that we choose an appropriate model to describe the observed oscillation spectrum but also, crucially, that the fitting procedure incorporate the correct noise statistics for the spectrum. We know from previous sections that the observed spectrum will be distributed about the underlying limit spectrum with χ^2 two-degrees-of-freedom statistics. In what follows we shall write the modeled power as $M(\mathbf{m}, \nu)$, where $\mathbf{m}$ is the array of model parameters needed to describe the combination of the modes (Lorentzian-like functions) and the other contributions (e.g., those from the broad-band background). Assuming a good model for the limit spectrum, we then have

$$ p\left[P(\nu)\right] = \frac{1}{M(\mathbf{m}, \nu)} \exp\left(-\frac{P(\nu)}{M(\mathbf{m}, \nu)}\right). \tag{6.69} $$

Since in reality we deal with discrete quantities, we rewrite the above as

$$ p\left(P_k\right) = \frac{1}{M(\mathbf{m}, \nu_k)} \exp\left(-\frac{P_k}{M(\mathbf{m}, \nu_k)}\right), \tag{6.70} $$

where again we use k to tag each of the N independent frequency bins of interest.

To fit the model, we need to find the combination of model parameters that makes the data most likely (given the assumed model). Under the assumption that the frequency bins are independent and uncorrelated, we need to maximize the

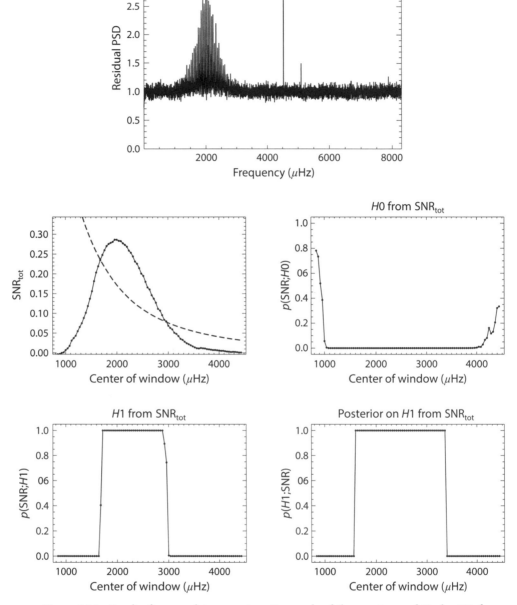

Figure 6.11. Results from applying a systematic search of the spectrum of *Kepler*-410 for the presence of significant signatures in SNR_{tot}. Top panel: Power spectrum of the star after dividing by the median-filter estimated background. Middle-left panel: Measured SNR_{tot} as a function of the frequency at the center of the range being tested (dark line, joined by points at which the test was applied). The dashed line marks the predicted SNR_{pred}. Middle-right panel: Resulting estimates of $p(SNR(\nu_s), N_s | H0)$ (Eq. 6.64). Bottom-left panel: $p(SNR(\nu_s), SNR_{pred}, N_s | H1)$ (Eq. 6.66). Bottom-right panel: posterior probability for H1, on $p(H1 | SNR(\nu_s), SNR_{pred}, N_s)$ (Eq. 6.68).

following joint probability:

$$\mathcal{L}(\mathbf{m}) = \prod_{k=1}^{N} p(P_k; \mathbf{m}). \tag{6.71}$$

This is called *maximum likelihood estimation*, or MLE. Substitution of Eq. 6.70 into Eq. 6.71 gives

$$-\ln \mathcal{L}(\mathbf{m}) = \sum_{k=1}^{N} \left[\ln M(\mathbf{m}, \nu_k) + \left(\frac{P_k}{M(\mathbf{m}, \nu_k)} \right) \right] \equiv S(\mathbf{m}). \tag{6.72}$$

Maximization of the natural logarithm of $\mathcal{L}(\mathbf{m})$—it is in practice much easier to work in logarithmic space, to avoid the potential instability of numerical algorithms—is equivalent to minimization of the quantity $S(\mathbf{m})$. Since a host of minimization algorithms are available "off the shelf," MLE is usually coded as a minimization problem. We therefore seek the set of parameters $\tilde{\mathbf{m}}$ that gives

$$\tilde{\mathbf{m}} = \operatorname*{argmin}_{\mathbf{m}} [S(\mathbf{m})], \tag{6.73}$$

which is equivalent to satisfying

$$\tilde{\mathbf{m}} = \operatorname*{argmax}_{\mathbf{m}} [\ln \mathcal{L}(\mathbf{m})]. \tag{6.74}$$

The covariance of the best-fitting parameters $\tilde{\mathbf{m}}$ is given by

$$\operatorname{cov}(\tilde{\mathbf{m}}) = \mathrm{H}^{-1}(\tilde{\mathbf{m}}), \tag{6.75}$$

where the Hessian matrix $\mathrm{H}(\tilde{\mathbf{m}})$ has elements

$$h_{ij} = \frac{\partial^2 S(\tilde{\mathbf{m}})}{\partial m_i \partial m_j}, \tag{6.76}$$

which may be computed numerically. The uncertainties on the best-fitting parameters are then given by the diagonal elements of the inverse Hessian:

$$\sigma_{m_i}^2 = \mathrm{H}^{-1}[\tilde{m}_i]. \tag{6.77}$$

In the limit of N tending to infinity, the asymptotic properties of MLE mean that the estimated parameters $\tilde{\mathbf{m}}$ are unbiased, normally distributed, and give the minimum variance. By unbiased, we mean that they tend to the true, underlying parameters (assuming of course an accurate model); and by minimum variance, we mean that no other combination of parameters will give smaller uncertainties. This is the *Cramér-Rao lower bound*, and it represents a fundamental limitation on the precision to which the parameters may be estimated.

Whether in practice we approach these asymptotic limits depends on the characteristics of the dataset in hand. An observed spectrum that has a very high frequency resolution (hence large N) in which the mode peaks appear prominently against the slowly varying (in frequency) power due to the other contributions clearly presents favorable conditions for MLE. However, such conditions are not always satisfied, yet the spectrum may still warrant a peak-bagging analysis to extract parameters for its constituent modes. We then run the risk of not finding or coming close to the minimum in $S(\mathbf{m})$ where the true parameters are located: our solution may be biased. Moreover, the estimated uncertainties on the parameters may be unreliable, and in practice they may be characterized by distributions that are far from normally distributed. The implicit assumption that the best-fitting parameters have normally distributed uncertainties is clearly a notable limitation and a potential cause for concern when the fitting problem is nontrivial.

There is also the added complication of gaps in the data. This may introduce significant correlations between the frequency bins, leading to potentially significant deviations from the assumption of strict statistical independence. In practice, the extent of the correlations is generally not a major cause for concern. However, the regular gap structure given by ground-based observations demands that additional complexity be included in the fitting model $M(\mathbf{m}, \nu_k)$, to represent the resulting aliased power.

6.3.2 Fundamentals: Bayesian Description of Parameter Estimation

Next, let us consider something that did not explicitly enter our previous discussions, and that is any prior information I that we have on the fitting model. This could be, for example, a set of reasonable limits on the frequency of a mode. Consideration of this information gives us a *prior probability distribution* $p(\mathbf{m}|I)$, the probability of the model given the set of prior information. To show how to use this prior information, let us recast the previous presentation of peak-bagging parameter estimation in the context of Bayes' theorem.

Our goal is to assess how well some model $M(\mathbf{m}, \nu_k)$, calculated from an array of model parameters $\mathbf{m}$, fits the observed data. There may be a set of competing (mutually exclusive) models that we wish to assess. We can think of the assessment in terms of utilizing the available information. In addition to the prior information, we of course have the information we obtain from the data D. The distribution $p(D|\mathbf{m}, I)$ gives the probability of the data given the assumed model and set of model parameters. This is just the likelihood function $\mathcal{L}(\mathbf{m})$ that we met in Section 6.3.1.

We want to use the above information—the prior information and the information obtained from the data—to estimate $p(\mathbf{m}|D, I)$, the posterior probability for the model. This is a probabilistic statement of the appropriateness of the model given all the available information. Bayes' theorem tells us how to write this in terms of the other probabilities, that is,

$$p(\mathbf{m}|D, I) = \frac{p(\mathbf{m}|I)p(D|\mathbf{m}, I)}{p(D|I)} \equiv \frac{p(\mathbf{m}|I)\mathcal{L}(\mathbf{m})}{p(D|I)}, \tag{6.78}$$

where the global likelihood $p(D|I)$, which provides the normalization, may be estimated from

$$p(D|I) = \sum_j p(\mathbf{m}_j|I)p(D|\mathbf{m}_j, I). \tag{6.79}$$

For some problems, computation of the normalization is straightforward (as for the mode detection problem we met in Section 6.1.2). The global likelihood for an individual peak-bagging model can be assessed numerically; however, the process will usually be computationally expensive. There is also the issue of whether the range of models considered is in any way complete. In practice these issues are not major, since, just as in Section 6.3.1, we choose to work in logarithmic space, meaning $\ln p(D|I)$ is merely a constant offset:

$$\ln p(\mathbf{m}|D, I) = \ln p(\mathbf{m}|I) + \ln \mathcal{L}(\mathbf{m}) + \text{constant}. \qquad (6.80)$$

Complications associated with computing the global likelihood do not affect our ability to properly compare and hence assess one model against another. Consider, for example, two competing models $\mathbf{m}_i$ and $\mathbf{m}_j$. The ratio of the probabilities of the models is

$$O_{ij} = \frac{p(\mathbf{m}_i|D, I)}{p(\mathbf{m}_j|D, I)} = \frac{p(\mathbf{m}_i|I)p(D|\mathbf{m}_i, I)}{p(\mathbf{m}_j|I)p(D|\mathbf{m}_j, I)} = \frac{p(\mathbf{m}_i|I)}{p(\mathbf{m}_j|I)} B_{ij}, \qquad (6.81)$$

where in the equation the ratios of the likelihoods have also been written as the *Bayes' factor*, B_{ij}. Having chosen a favored model, the posterior parameter distributions may be assessed without worrying about the global likelihood normalization. It is often the case that one has no prior information favoring one model over another, in which case the ratio of the prior probabilities is unity. However, great care is needed if the models are nested (i.e., similar models with different numbers of parameters).

One can use Eq. 6.80, implying explicit use of the prior information, to formulate a *maximum a posteriori* (MAP) approach, in which one seeks to find the parameters that satisfy

$$\tilde{\mathbf{m}}_{\text{MAP}} = \underset{\mathbf{m}}{\text{argmax}} \left[p(\mathbf{m}|I)\mathcal{L}(\mathbf{m}) \right]. \qquad (6.82)$$

The MAP approach is tantamount to finding the *mode* of the posterior distribution. However, it suffers from the same limitations as MLE in giving only a "point estimate" of the outputs, with no information on the underlying distributions.

6.3.3 Fundamentals: Use of Sampling Algorithms

We could potentially gain some insights on the trustworthiness of our best-fitting parameters, and the distributions of the model parameters, with Monte Carlo simulations. We would construct artificial data, whose limit and statistical properties mimic as closely as possible those of the real observations, and from fits to many independent realizations of those data we would assess the statistical distributions of the best-fitting parameters. This approach clearly has some merit, a noteworthy point being that one knows the true input parameters and the true limit spectrum. However, translating lessons learned to the real data clearly rests on the fidelity of the Monte Carlo simulations. Although quite sophisticated simulations are now possible, uncertainty is inevitable, and some caution is always indicated.

This is where a different approach, involving the use of *sampling algorithms*, can help. These techniques are becoming very popular across a range of scientific fields, including asteroseismology. They provide a way to estimate the statistical distributions of the model parameters—the *posterior distributions*—from just a single dataset. This means the methods may be applied to real data, where only one realization of the observational dataset is available. Here, we introduce one class of such methods, the *Markov Chain Monte Carlo* (MCMC) method.

Ideally, we wish to map out the probability distribution $p(\mathbf{m}|D, I)$. This can be done in practice by performing a suitable pseudo-random walk in parameter space, with the density of the random-walk samples reflecting the probability. The construction of a *Markov chain* satisfies these requirements. Various numerical algorithms have been designed to achieve the desired sampling, the best-known being the Metropolis-Hastings algorithm, which we summarize briefly here. We direct the reader to other texts for more detailed discussions of MCMC methods.

To construct a Markov chain, we need to draw a set of random samples of $\mathbf{m}$ that are representative of the posterior distribution $p(\mathbf{m}|D, I)$. Let us suppose that we begin our chain at "time" t by sampling the parameters at some suitable first-guess locations, giving the starting set $\mathbf{m}_t$. We must then calculate the next sample in the chain, $\mathbf{m}_{t+1}$, which depends on $\mathbf{m}_t$. We first propose a new sample, $\mathbf{m}'$, which we draw from the *proposal distribution* $Q(\mathbf{m}'|\mathbf{m}_t)$. A common choice for the proposal distribution is a normal (Gaussian) distribution:

$$Q(\mathbf{m}'|\mathbf{m}_t) = \frac{1}{\sigma\sqrt{2\pi}} \exp\left[-\frac{(\mathbf{m}' - \mathbf{m}_t)^2}{2\sigma^2}\right]. \tag{6.83}$$

The proposal distribution is then symmetric [i.e., $Q(\mathbf{m}'|\mathbf{m}_t) = Q(\mathbf{m}_t|\mathbf{m}')$]. The proposed sample $\mathbf{m}'$ depends on the previous sample $\mathbf{m}_t$, because the proposal distribution is centered on $\mathbf{m}_t$. This makes it more likely that samples having similar values to $\mathbf{m}_t$ will be drawn from the distribution.

Whether we accept the proposed sample as the next step in the Markov chain depends on the Metropolis ratio r, which is defined as

$$r = \frac{p(\mathbf{m}'|D, I)}{p(\mathbf{m}_t|D, I)} \frac{Q(\mathbf{m}_t|\mathbf{m}')}{Q(\mathbf{m}'|\mathbf{m}_t)}. \tag{6.84}$$

The proposed sample is accepted with a probability that is given by

$$p(\mathbf{m}_t, \mathbf{m}') = \min(1, r). \tag{6.85}$$

If $r \geq 1$, we always accept the proposed sample. If the proposal distribution is symmetric, this corresponds to $p(\mathbf{m}'|D, I) \geq p(\mathbf{m}_t|D, I)$. If $r < 1$, the proposed sample will be accepted with a probability r. Acceptance implies that the new sample in the chain is

$$\mathbf{m}_{t+1} = \mathbf{m}'. \tag{6.86}$$

If instead the new sample is rejected, we will have

$$\mathbf{m}_{t+1} = \mathbf{m}_t. \tag{6.87}$$

The above steps are repeated until we build a chain large enough to robustly sample the posterior probability distribution. Posterior distributions for each parameter are then trivially obtained by marginalizing over the other parameters:

$$p(m_j|D, I) = \int_{i \neq j} p(m_i|D, I)\mathrm{d}m_i. \qquad (6.88)$$

In practice this means constructing suitable histograms of the samples from the final Markov chain.

While the chain will eventually converge to the sought-for distribution, care must be taken when using early samples from the chain (which may provide a rather poor distribution if they map only the very low parts of the probability hyperspace). Additions to the above recipe also include the implementation of *parallel tempering*, where one seeks to prevent the chain becoming stuck in subsidiary maxima in probability space, which lie away from the true maximum. This is achieved by running several chains in parallel, with different chains governed by progressively more distorted (flattened) versions of the posterior distribution $p(\mathbf{m}|D, I)$. Again, we direct the reader to other texts for more details on implementing MCMC in practice.

6.4 PEAK-BAGGING: PRACTICALITIES AND CHALLENGES

6.4.1 Basic Peak-Bagging Model

Now that we have covered the technical aspects of peak-bagging, let us discuss in more detail the practical steps associated with the full analysis of the power spectrum, including the challenges associated with construction of the fitting model.

Assuming that oscillations have been detected in the frequency-power spectrum—using tools like those discussed in Section 6.2—we must judge whether the detections have been made at a level that will allow extraction of robust parameters on individual modes. We must then formulate an appropriate fitting model, construct a suitable list of first-guess parameters for the modes and the background contributions, and decide on the optimization strategy (i.e., MLE, MAP, or a sampling approach). The spectrum is then fitted using the chosen model and strategy.

A first-guess list of frequencies may be constructed using results from, for example, the tests outlined in Section 6.1, which identify the locations of potential candidate modes. Candidate mode peaks that have been judged as significant must then be associated with a particular angular degree l, so that we can build an accurate fitting model. In many cases, construction of an échelle diagram using the candidate frequencies provides an easy way to make the correct identification. The $l = 0$ and $l = 2$ overtones lie in close proximity to one another, as do the $l = 1$ and $l = 3$ overtones. This means that, for example, in the case of photometric observations—where the $l = 3$ modes are very weak—the dominant pattern in the oscillation spectrum is one of $l = 1$ modes alternating with closely separated pairs of $l = 0$ and $l = 2$ modes. Three clear ridges will then be apparent in the échelle diagram, with the two closely separated ridges corresponding to the $l = 0$ and $l = 2$ modes.

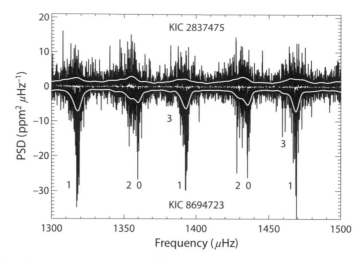

Figure 6.12. Frequency-power spectra of the two the stars from Figure 5.13. Shown in black are the raw spectra—which have the same native frequency resolution—and in white spectra smoothed with a 3.5 μHz double-boxcar filter. The bottom star (KIC 8694723) is an early G-type star: the odd-and even-degree modes are straightforward to discriminate visually (see annotations in the figure). The top star (KIC 2837475) is a hotter, early F-type star, which shows much wider oscillation peaks.

If the oscillations are heavily damped, modes that are adjacent in frequency cannot be so easily resolved. Rapid rotation may also confuse the pattern of nonradial mode components (of which more in Section 6.4.4). Figure 6.12 shows zooms of parts of the oscillation spectra of the stars shown in Figure 5.13: an early G-type star (lower part) and an F-type star (upper part), some 600 K hotter than its counterpart. Although visual identification of the different degrees is relatively easy for the G-type star, the much wider peaks shown by the F-type star means that discriminating in its case is much less straightforward. Two options to resolve the identification then present themselves.

The first is a statistical approach: one peak-bags the spectrum twice, with the mode identification swapped, and then compares metrics that reflect the quality of the best-fitting solutions (e.g., using the Bayesian odds ratio). In the second approach, we rely on prior knowledge of the effective temperature T_{eff}. Here, one uses the fact that if the detectable radial-mode frequencies are fitted to the asymptotic relation for high-overtone p modes, the inferred ϵ follows a fairly well behaved trend in T_{eff}. The trend is curved, with a maximum in ϵ at around $T_{\text{eff}} \simeq 5{,}500$ K. If ν_j are the candidate mode frequencies, mode-by-mode estimates of ϵ may be computed using

$$\epsilon_j = \begin{cases} \epsilon'_j + 1, & \text{if } \epsilon'_j < 0.5 \text{ and } \Delta\nu > 3\,\mu\text{Hz} \\ \epsilon'_j, & \text{otherwise,} \end{cases} \tag{6.89}$$

where

$$\epsilon'_j = \left(\frac{\nu_j}{\Delta\nu} \bmod 1 \right). \tag{6.90}$$

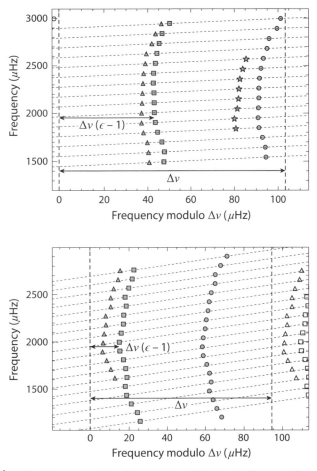

Figure 6.13. Échelle diagrams of the frequencies of 16 Cyg A (top panel) and *Kepler*-410 (bottom panel). The annotations shows how the offset of the $l = 0$ ridge in the diagram relates to ϵ. Dotted lines follow the sloping orders, while the vertical dashed lines appear at multiples of $\Delta\nu$.

The above deals appropriately with the wrap-around effect inherent in the modulus operation, providing the correct radial order match for the solar p modes and a smooth transition in the inferred ϵ values across different evolutionary states.

It is then a matter of assessing which set of frequencies in the échelle diagram has ϵ_j values that match the expectation for the radial modes of the star. Figure 6.13 shows the échelle diagrams of two other stars we have met, 16 Cyg A (top panel) and *Kepler*-410 (bottom panel). We see that the estimate of ϵ given by Eq. 6.89 is equivalent to measuring the offset of the $l = 0$ ridge in the échelle diagram. 16 Cyg A is the cooler of the two stars—having $T_{\mathrm{eff}} \simeq 5{,}800\,\mathrm{K}$, versus $T_{\mathrm{eff}} \simeq 6{,}300\,\mathrm{K}$ for *Kepler*-410—and has a larger ϵ.

Once the candidate modes have been l-tagged, we can then set about constructing a fitting model that contains the requisite number of modes. The basic building block of the fitting model is the profile used to describe the power spectral density in frequency of each individual mode component. We have seen that a good time-domain model for the modes is a damped, randomly driven oscillator, which suggests the approach most commonly adopted to describe the resonant response in frequency: the use of a Lorentzian profile (see Eq. 5.39).

The generalized formalism for the power spectral density of a mode of a given radial order n and angular degree l must take account of rotational splitting that lifts the frequency degeneracy in l—thereby resulting in up to $2l + 1$ potentially detectable m components (magnetic fields may also contribute to the splitting)—as well as the visibilities of components of differing l and azimuthal order m. Incorporating the Lorentzian building block, the power spectral density of the multiplet may be written as

$$\mathcal{M}_{nl}(\nu) = \sum_{m=-l}^{l} \mathcal{E}_{lm}(i_{\mathrm{s}}) H_{nl} \left(\frac{1}{1 + \xi_{nlm}^2(\nu)} \right), \qquad (6.91)$$

with

$$\xi_{nlm}(\nu) = \frac{\nu - \nu_{nl} - m\delta\nu_{nlm}}{\Gamma_{nl}/2}. \qquad (6.92)$$

Here, ν_{nl} is the frequency of the mode, and $\delta\nu_{nlm}$ is the rotational frequency splitting of each m component. The parameter H_{nl} defines a single height (maximum power spectral density) for the multiplet, with the relative heights of the m components fixed by $\mathcal{E}_{lm}(i_{\mathrm{s}})$ (Eq. 4.14), which is defined as being a function of the stellar angle of inclination i_{s}. We remind the reader that Eq. 4.14 is based on the assumption of energy equipartition among the different m components. Implicit in the form of Eq. 6.91 is, therefore, the assumption that each component has the same FWHM linewidth Γ_{nl}.

The contribution of all detectable modes to the observed frequency spectrum is then usually modeled as a linear combination, that is,

$$\mathcal{O}(\nu) = \mathrm{sinc}^2 \left[\frac{\pi}{2} \left(\frac{\nu}{\nu_{\mathrm{Nyq}}} \right) \right] \sum_{nl} \mathcal{M}_{nl}(\nu), \qquad (6.93)$$

where we have not forgotten to correct for the sinc-function attenuation (Eq. 5.9 in Section 5.1.3) due to the finite duration of the data sampling in the time domain.

To the summed oscillation contribution are then added the background terms. The shot noise contribution, b_{shot}, is simply a flat offset (recall Eq. 5.19 in Section 5.1.4). Contributions due to the evolution of granulation (and possibly at still lower frequencies contributions due to starspots and magnetic active regions) are modeled using expressions of the form introduced in Section 5.5, each having a characteristic power b and timescale τ. Here, we use the general expression defined by Eq. 5.101 as the basic building block to describe each component, and sum over the contributions of the n_B modeled components. The model for the total background power spectral density may then be written as

$$\mathcal{B}(\nu) = \text{sinc}^2 \left[\frac{\pi}{2} \left(\frac{\nu}{\nu_{\text{Nyq}}} \right) \right] \left(\sum_{j=1}^{n_{\mathcal{B}}} \frac{b_j}{1 + (2\pi \nu \tau_j)^{\beta_j}} \right) + b_{\text{shot}}. \tag{6.94}$$

Note how the sinc-function attenuation is not applied to the shot-noise component, because its response in frequency is unaffected by the sampling.

Usually one component, but sometimes two, may be needed to model the observed power spectral density that is attributed to the granulation. It should be noted that there is no physical justification underlying the use of a second component, based at least on the example of the Sun (where no evidence suggests that more than one dominant granulation timescale exists). Unless the Sun is atypical, this suggests that the usual approach to building the background model is missing something. One possibility is potential changes to the shape in frequency of the power spectral density that might arise from the correlation of the granulation and the modes (see Section 5.3.3). Contributions from active-region signal can often be ignored, since they typically have only a very small contribution across the frequency range occupied by the modes.

To summarize, the full fitting model takes the form

$$\mathbf{M}(\nu) = \mathcal{O}(\nu) + \mathcal{B}(\nu). \tag{6.95}$$

6.4.2 Global, Pseudo-Global or Local?

In high-quality data on solar-type stars, we would expect to be able to fit more than ten overtones of the spectrum. In the *global* approach to peak-bagging, we attempt to fit the entire spectrum simultaneously. If we fold in the number of overtones that are detectable at each degree, l—and in the best photometric data this may even include a few $l = 3$ modes—and then consider the potential number of free parameters in the fitting model (not forgetting the background terms), it is easy to see that this may require the simultaneous optimization of well over 100 parameters. Fitting a spectrum made from multiyear data may then require many hours using MLE or MAP, usually longer for a sampling-based approach, even with a powerful computer cluster.

This then raises the obvious question: why not just fit the spectrum one (or two) radial orders at a time, thereby reducing significantly the number of free parameters varied at any one time? Although this *local* approach undoubtedly leads to dramatic improvements in speed and also gives better stability in the fitting—with fewer parameters to explore, one is less likely to get stuck in subsidiary maxima in multiparameter likelihood space—one may then be compromising the accuracy of the best-fitting outputs. When we restrict ourselves to a frequency window spanning, for example, one radial order, power will also be present in that window from the slowly decaying Lorentzian tails of the modes whose frequencies lie outside the window. If the fitting model only accounts for the power spectral density of the modes that lie inside the window, it follows that we will in principle have an inaccurate fitting model. The impact depends of course on the frequency separation of the overtones (i.e., on $\Delta\nu$) and on the peak linewidths Γ. Careful analysis of multiyear Sun-as-a-star helioseismic data has shown that the impact is just about detectable for the Sun. Since most of the available asteroseismic data are on stars that have a smaller $\Delta\nu$ and larger Γ than does the Sun, caution is advised regarding the use of this local approach.

Significant improvements in speed may be obtained by using the *pseudo-global* approach to MAP or MLE. This approach is able to preserve the accuracy of the global fitting model while reducing fitting times for the entire spectrum to a few minutes. Here, a model for the entire spectrum is used, but only the parameters in one (or two) orders are varied at any one time. One works systematically through the spectrum, updating the stored parameters of the global model as each order is optimized. Convergence of the fitting parameters is usually achieved after just a few passes through the entire spectrum.

However, speed must be balanced against robustness of the outputs, including not only the best-fitting parameters but also the associated uncertainties. Implicit in the MLE or MAP approach is the assumption that the likelihood functions are multidimensional Gaussians in the vicinity of the best-fitting parameters (Section 6.3.1). This requirement is usually met under conditions of high SNRs, given a sensible choice of fitting model. However, lower-quality data, an inappropriate choice of model (trying to squeeze too much out of the data), or both can lead to very badly behaved likelihood functions that are far from Gaussian shaped. The spot-value, best-fitting estimates and associated uncertainties provided by MLE or MAP may then provide a poor description of the actual state of knowledge of the parameters. Under such circumstances, a sampling approach is clearly the right one, since it allows a full exploration and mapping of the posterior distributions of the parameters, irrespective of the shapes of the distributions.

Borderline quality data may also point to the need to apply some regularization in the model, to, for example, constrain a smooth variation in frequency of some of the best-fitting parameters to secure stability of the fit. If an MLE-like approach is adopted, this implies MLE with use of strong priors (i.e., MAP). Clearly, important and tricky choices need to be made in the fitting model, which will depend on the quality and length of the data and the intrinsic stellar characteristics. For example, will the combination of the observed SNR, frequency resolution, and stellar rotation and magnetic fields allow one to set up a model in which free parameters are used to describe the frequency splitting of each individual mode? Or is the quality such that one is only likely to get robust results from a model that has just one free parameter to describe an average splitting over all the nonradial modes (and which ignores any potential asymmetry of the splittings due to magnetic fields)? Also, peaks due to some of the modes may not be fully resolved in spectra made from shorter lightcurves: is it then still appropriate to fit a simple Lorentzian model? And what about the fact that we might expect the mode peaks to show small, asymmetric departures from a pure Lorentzian form, as is the case for the solar p modes? In Sections 6.4.4 and 6.4.6 we shall discuss how the fitting model can be suitably augmented to incorporate these additional complexities. But before that, we make a few comments on the first-guess parameters and priors.

6.4.3 First-Guess Parameters and Priors

Let us begin with the l-tagged, first-guess frequencies. How they are used depends on the choice of optimization algorithm. For MLE or MAP they serve as the initial values used in the fitting model (i.e., starting points for multiparameter optimization). The same goes for the other parameters that are varied in the fitting model. For sampling (e.g., MCMC), and possibly also MAP (depending on the choice of priors), they fix a central value around which an allowed range of frequencies is explored, implying we must set a suitable range for (i.e., a prior on) each frequency. Again, the same follows for the other parameters.

We may estimate suitable first-guess parameters or priors for the heights and widths of the modes using the scaling relations in Section 5.4; and set those for the granulation parameters using the scaling relations in Sections 4.3.1 and 5.5. We can also base first-guess parameter estimates on a preliminary analysis of the spectrum (as per the frequencies): for example, an estimate of the shot-noise background is given by an average of the power-spectral densities of frequencies close to the Nyquist frequency. For the heights, we can apply some light smoothing to the spectrum (e.g., a 1 μHz-wide boxcar) and assess the maximum of the power excess that lies above a preliminary estimate of the broad-band background.

Priors may be set in a variety of ways, and they take various forms. For example, the prior might be nonuniform in the parameter space (e.g., a Gaussian-shaped prior, reflecting the expected form of the posterior probability) or simply a flat prior. The widths of the priors might be set based on the structure of the oscillation spectrum, on prior expectation of the level of uncertainty in the parameter being fitted, or on both.

With regard to using the structure of the spectrum, the frequency spacings between overtones provide one obvious set of priors (i.e., one does not want frequencies hopping between orders during the optimization process). The prior should also be wide enough to capture the full range of the posterior probability distribution. We can obtain some guidance given estimates of the expected precision achievable in each parameter. These expected uncertainties can be calculated given approximate estimates of (or limits on) the mode height and width, the background in the vicinity of the mode, and the duration of the observations.

The following expressions are written with the formal symbols for each parameter; it is worth bearing in mind that in practice, the values we would use in the formulas would be approximate. However, exploring a suitable range of values—including, for example, extreme or upper limits on the parameters—would provide useful guidance.

First, we define the inverse height-to-background ratio at the frequency of the mode:

$$\beta_{nl} = \frac{B(\nu_{nl})}{H_{nl}}. \tag{6.96}$$

The expected frequency precision of a single, fully isolated mode component is then given by

$$\sigma_{nl} \simeq \left(\frac{\Gamma_{nl} \sqrt{1 + \beta(\nu_{nl})} \left[\sqrt{1 + \beta(\nu_{nl})} + \sqrt{\beta(\nu_{nl})} \right]^3}{4\pi T} \right)^{1/2}, \tag{6.97}$$

where T is the duration of the observations.

Equation 6.97 suffices to provide an approximate estimate of the expected precision in the radial modes. (It should of course be borne in mind that only under some circumstances are the radial mode peaks likely to be fully isolated from neighboring peaks of the $l = 2$ modes.) Estimation of the frequency precision in the nonradial modes depends not only on the nonradial mode visibility—which changes $\beta(\nu_{nl})$ relative to the radial mode case—but also on the actual number of observed nonradial components and their observed heights (which depend on the angle of inclination offered by the star), and on how well resolved the individual components are (which depends on the ratio between the frequency splitting and the peak linewidth).

While accounting for the change in $\beta(\nu_{nl})$ is trivial, correcting for the other factors is somewhat more complicated. Here, we instead offer some approximate guidance using an empirical correction factor e_l for each angular degree l, based on results from fits to the oscillation spectra of several tens of *Kepler* solar-type stars that span a range of quality and dataset duration, the minimum being around 6 months. These factors may be regarded as being representative; values for individual stars will vary, depending on the specific combination of individual stellar and seismic parameters and the inclination angle of the star. If the model-computed eigenfrequencies of the star are ν_{nl}, approximate estimates of the formal uncertainties at the other l are given by

$$\sigma_{nl} \simeq e_l \, \sigma(\nu_{nl}), \tag{6.98}$$

where $e_0 = 1.0$, $e_1 = 0.85$, $e_2 = 1.60$, and $e_3 = 6.25$.

The estimation of the expected uncertainties in the height and width parameters is beset by similar complications (i.e., they again depend on the degree of the mode, the angle of inclination of the star, and the values of the parameters themselves). Approximate guideline estimates for the radial modes are given by

$$\sigma(\Gamma_{nl})/\Gamma_{nl} \simeq \left(\frac{\left[\sqrt{1 + \beta(\nu_{nl})} + \sqrt{\beta(\nu_{nl})} \right]^4}{\pi \Gamma_{nl} T} \right)^{1/2}, \tag{6.99}$$

and

$$\sigma(H_{nl})/H_{nl} \simeq \left(\frac{4 \left[\sqrt{1 + \beta(\nu_{nl})} \right]^3 \left[\sqrt{1 + \beta(\nu_{nl})} + \sqrt{\beta(\nu_{nl})} \right]}{\pi \Gamma_{nl} T} \right)^{1/2}. \tag{6.100}$$

6.4.4 Choices Regarding the Model and Fitting

As the quality and duration of the data as well as the intrinsic stellar properties vary, so different demands are placed on making the appropriate choices for the nature and complexity of the fitting model. How many parameters will actually be varied and optimized? Should the model be simplified? Or do the data demand the inclusion of extra complexity?

Choices must be made regarding the frequency and splitting parameters. The solar-type stars for which we usually get good asteroseismic data typically show rather modest rates of absolute and differential (in latitude and radius) rotation. The rotational frequency splittings δv_{nlm} of observable high-n, low-l p modes will then have very similar values. We may then introduce a representative splitting, δv_s, and write the frequencies of a given multiplet in the form:

$$v_{nlm} = v_{nl} + m\delta v_{nlm} \simeq v_{nl} + m\delta v_s. \tag{6.101}$$

In the entry-level scenario, we therefore fit an individual frequency v_{nl} for every mode, but only *one* rotational splitting δv_s for all nonradial modes in the spectrum:

$$\delta v_{nlm} = \delta v_s. \tag{6.102}$$

The next level of complexity would be to fit an individual frequency splitting δv_{nl} to each mode, preserving the assumption that the m components are arranged symmetrically in frequency:

$$\delta v_{nlm} = \delta v_{nl}. \tag{6.103}$$

When rotation periods are as short as a few days—as might be expected for young, hot solar-type stars—these simplifications break down. Second-order terms become important, and a reasonable description for the multiplet frequencies is then

$$v_{nlm} \simeq v_{nl} + m\delta v_s + (C_1 + m^2 C_2)\delta v_s^2, \tag{6.104}$$

where C_1 and C_2 are appropriate coefficients that describe the additional nonlinear contributions to the observed splittings (strictly, both depend on n and l). Typically, they might be expected to modify the predicted frequencies relative to the slow-rotation case by up to a few micro-Hertz.

Near-surface magnetic activity can also affect the distribution of the multiplet frequencies. When the near-surface activity is distributed in a nonspherically symmetric manner (i.e., into active bands of latitude as on the Sun), it will give rise to *acoustic asphericity*, so that the magnitudes of the frequency offsets in the multiplet will then depend on the angular degree l and the azimuthal order m of the mode.

To describe the resulting arrangement, let us consider a description of the full set of m frequencies of a mode in a polynomial expansion of the form

$$v_{nlm} = v_{nl} + \sum_{j=1}^{j=2l} a_j(n, l) l \mathcal{P}_l^j(m), \tag{6.105}$$

where v_{nl} corresponds to frequency centroid of the mode, and $\mathcal{P}_l^j(m)$ are polynomials related to Clebsch-Gordon coefficients. An expansion of this type is commonly used to fit the frequencies observed in data collected by highly spatially resolved observations of the Sun, where all m components are available. The even-a_j coefficients describe perturbations that are nonspherically symmetric in nature, (e.g., the acoustic asphericity from activity), while the odd-a_j coefficients describe spherically symmetric contributions to the frequency splittings (rotation). Here, for simplicity, we assume that for the low-l modes in question, the a_2 term dominates other terms in the description of the asphericity, and that the a_1 term dominates other terms in the description of the rotation. The required $\mathcal{P}_l^j(m)$ are then

$$\mathcal{P}_l^1(m) = m/l, \tag{6.106}$$

and

$$\mathcal{P}_l^2(m) = \frac{6m^2 - 2l(l+1)}{6l^2 - 2l(l+1)}, \tag{6.107}$$

so that

$$\nu_{nlm} = \nu_{nl} + a_1 m + a_2 l \left[\frac{6m^2 - 2l(l+1)}{6l^2 - 2l(l+1)}\right]. \tag{6.108}$$

For completeness, we may write the frequencies of each of the components of modes of degree $l = 1$ and 2 explicitly as follows:

$$l = 1: \quad \begin{aligned} \nu_{n10} &= \nu_{nl} - 2a_2, \\ \nu_{n11} &= \nu_{nl} + a_1 + a_2, \\ \nu_{n1-1} &= \nu_{nl} - a_1 + a_2; \end{aligned} \tag{6.109}$$

and

$$l = 2: \quad \begin{aligned} \nu_{n20} &= \nu_{nl} - 2a_2, \\ \nu_{n21} &= \nu_{nl} + a_1 - a_2, \\ \nu_{n2-1} &= \nu_{nl} - a_1 - a_2, \\ \nu_{n22} &= \nu_{nl} + 2a_1 + 2a_2, \\ \nu_{n2-2} &= \nu_{nl} - 2a_1 + 2a_2. \end{aligned} \tag{6.110}$$

So instead of fitting one frequency-splitting parameter (e.g., $\delta\nu_s$), we would fit two parameters, the coefficients a_1 and a_2. The relevant part of the fitting model becomes

$$\delta\nu_{nlm} = a_1 m + a_2 l \left[\frac{6m^2 - 2l(l+1)}{6l^2 - 2l(l+1)}\right]. \tag{6.111}$$

With no asphericity, $a_2 = 0$ and we have the slow-rotation approximation $a_1 \equiv \delta\nu_s$. At times of high activity on the Sun, the solar asphericity coefficient has a value $a_2 \simeq 0.1\,\mu\mathrm{Hz}$, that is, a not insignificant fraction of the rotational splitting $a_1 \simeq 0.4\,\mu\mathrm{Hz}$. This effect is therefore potentially important when analyzing long, high-quality spectra on other stars that have levels of activity similar to, or higher than, the Sun.

Next we must make choices regarding the height and linewidth parameters. In the entry-level approach, one fits a single height and width parameter to the modes in a given order n. The description of the heights must also account for the l-dependent mode visibilities S_l, which we introduced in Section 4.1. They fix the radial-mode normalized visibilities in power, which we define here explicitly as

$$V_l = (S_l/S_0)^2. \tag{6.112}$$

The heights and linewidths in the fitting model then become

$$H_{nl} = V_l H_n,$$ (6.113)

and

$$\Gamma_{nl} = \Gamma_n.$$ (6.114)

Helioseismic observations have demonstrated that near-surface magnetic activity affects the amplitudes, heights, and widths of the solar p modes. Elevated levels of activity tend to suppress the amplitudes and the heights while increasing the linewidths. As per our discussion of the frequencies above, these effects depend on the l and m of the mode but are rather modest for the Sun, and hence are unlikely to be a cause for concern in typical asteroseismic targets. Fitting individual heights and widths to each component would also be very likely to introduce instability in the fits.

Our entry-level model has a Lorentzian profile as its basic building block. However, we noted in Section 5.3.3 that the peaks of the solar p modes are asymmetric. This asymmetry is small, typically of the order of a few percent or less, and so demands high SNR and high frequency resolution data to extract robust estimates. Fits to *Kepler* and CoRoT data on solar-type stars have to date usually omitted peak asymmetry in the peak-bagging models. The simplest approach is to incorporate in the model an asymmetric Lorentzian-like function of the type introduced in Section 5.3.3 (e.g., Eq. 5.63), which includes an explicit asymmetry parameter, α. The profile then becomes

$$\mathcal{M}_{nl}(\nu) = \sum_{m=-l}^{l} \mathcal{E}_{lm}(i_s) H_{nl} \left(\frac{1 + 2\alpha_{nl}\xi_{nlm}(\nu)}{1 + \xi_{nlm}^2(\nu)} \right).$$ (6.115)

6.4.5 Some Real Examples

The top panel of Figure 6.14 again shows the frequency-power spectrum of *Kepler*-410, but this time with a best-fitting peak-bagging model overlaid. This fit used an entry-level model of the type discussed in Section 6.4.4, with a single splitting parameter $\delta\nu_s$ to describe the frequency splitting of the nonradial modes, but with no peak asymmetry. The fitting was performed using an MCMC sampler, with uniform priors on the various fitting parameters. The plotted best-fitting model was constructed by using as input for each parameter the median of its posterior distribution. Annotations mark the locations of modes of different angular degrees l. The bottom panel shows the échelle diagram of best-fitting frequencies (again, in each case the median of the relevant posterior distribution).

Figure 6.15 shows a zoom of the central part of the oscillation spectrum of *Kepler*-410. The finite width of the mode peaks is clearly evident. The nonradial modes also show clear evidence of frequency splitting due to rotation.

To help make sense of the appearance of these rotationally split modes, we must recall Eq. 4.14, which describes how the observed power in the modes is affected by the angle of inclination offered by the star. Figure 6.16 shows how the visibilities in power $\mathcal{E}_{lm}(i_s)$ of the different components of the $l = 1$ and $l = 2$ modes vary with

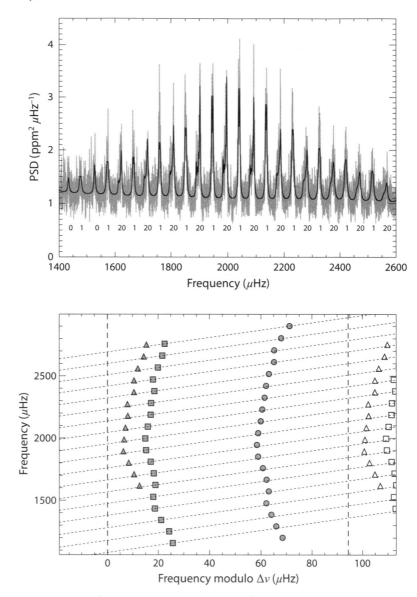

Figure 6.14. Top panel: Frequency-power spectrum of *Kepler*-410 (gray, after smoothing with a boxcar filter of width 0.5 μHz), and the best-fitting peak-bagging model (black). Annotations indicate the locations of modes of different degree l. Bottom panel: Échelle diagram of the best-fitting frequencies, with $l = 0$ modes plotted as squares, $l = 1$ modes as circles, and $l = 2$ modes as triangles. Modes are plotted with open symbols in the ranges where the échelle repeats.

the inclination i_s. The horizontal axis shows the component splittings normalized by the peak linewidth Γ. We also define a reduced splitting, which is the ratio of the frequency splitting to the peak linewidth:

$$\delta\nu_r = \frac{\delta\nu_s}{\Gamma}. \tag{6.116}$$

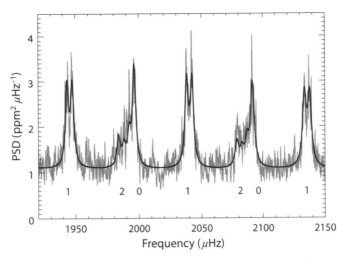

Figure 6.15. Zoom of the central part of the frequency-power spectrum of *Kepler*-410.

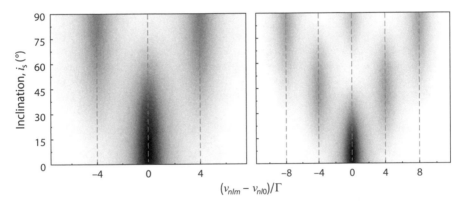

Figure 6.16. visibilities in power $\mathcal{E}_{lm}(i_s)$ of the different components of the $l = 1$ and $l = 2$ modes as a function of the angle of inclination i_s for a reduced splitting of $\delta v_r = 5$. Darker shading equates to higher visibility.

Here, we assume Lorentzian profiles for every component, and a ratio of $\delta v_r = 5$, which ensures that the individual components are clearly resolved in the figure (of which more below).

Notice that when i_s is close to zero, only the zonal ($m = 0$) components show prominent power. At intermediate angles we see power distributed among the various components. When i_s is close to 90 degrees (i.e., when the star is oriented such that its rotation axis lies close to the plane of the sky), only components that have $l + m$ even show significant power. At $l = 1$ we therefore see only the outer, sectoral ($m = \pm 1$) components, whereas at $l = 2$ we see power at $m = 0$ and $m = \pm 2$, although it is the power in the latter (sectoral) components that dominate.

The rotationally split $l = 1$ modes of *Kepler*-410 in Figure 6.15 look like doublets. Figure 6.17 shows schematic representations of $l = 1$ modes at different angles, assuming frequency splittings $\delta v_s \simeq 2\,\mu$Hz and peak linewidths $\Gamma \simeq 2\,\mu$Hz

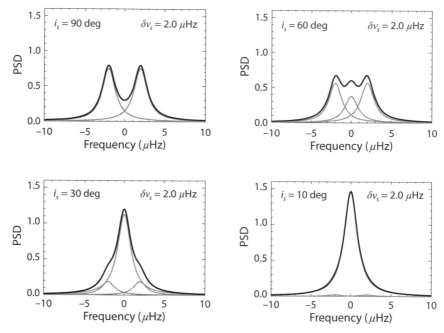

Figure 6.17. Schematic representations of $l = 1$ modes at different angles of inclination i_s, assuming an underlying frequency splitting of $\delta\nu_s \simeq 2\,\mu$Hz and peak linewidths $\Gamma \simeq 2\,\mu$Hz. Power due to the individual m components is plotted in gray, and the combined (composite) profile is plotted in black.

similar to those displayed by the $l = 1$ modes of the star. The rotation axis of *Kepler*-410 clearly lies close to the plane of the sky.

High-i_s cases like this should be much more common than those with a low i_s if we assume that the angular momentum vectors (rotation axes) of target stars are oriented randomly in space. The probability of observing a star with inclination i_s then follows the distribution

$$p(i_s) = \sin i_s. \tag{6.117}$$

The expected fraction of targets having angles in the range $(i_s)_1$ to $(i_s)_2$ is then simply

$$f[(i_s)_1, (i_s)_2] = \int_{(i_s)_1}^{(i_s)_2} \sin i_s di_s. \tag{6.118}$$

The above suggests the use of an isotropic prior on the angle when fitting the modes (i.e., one that respects the $\sin i_s$ distribution).

When the individual components of the nonradial modes are resolved and the oscillation spectrum is available at high SNRs, it is then possible to constrain both the splitting $\delta\nu_s$ and the angle i_s. *Kepler*-410 is one such example. Figure 6.18 shows the posterior probability surface in the $\delta\nu_s$-i_s plane for its spectrum. The figure also shows the marginalized posteriors of each parameter (i.e., as given by

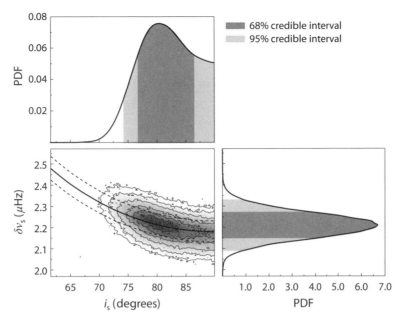

Figure 6.18. Estimated posterior distribution (PDF) in the $\delta\nu_s$–i_s plane for *Kepler*-410. Also shown are the marginalized posteriors in each parameter.

collapsing the two-dimensional surface in the relevant direction). We see a clear, preferred solution at a high inclination angle and a rotational frequency splitting of $\delta\nu_s \simeq 2.2\,\mu$Hz. The $l = 1$ modes in this star typically have values of $\delta\nu_r \approx 1$. However, when the rotationally split components of an oscillation spectrum are less well resolved, the SNR in the modes is lower, or both, it is no longer possible to obtain a unique solution, and degeneracies appear in the posterior probabilities. Figure 6.19 shows an example. This case comes from the analysis of KIC 3456181. The posterior probability surface now shows a ridge, which defines a curve of constant $\delta\nu_s \sin i_s$; thus it is still possible to constrain the *projected* rotational splitting, defined as the product of the splitting and the sine of the inclination:

$$\delta\nu_s^* = \delta\nu_s \sin i_s, \tag{6.119}$$

but not the splitting itself.

Figure 6.20 shows the impact of changing the natural frequency resolution of the spectrum of *Kepler*-410. Plotted here is the raw (unsmoothed) spectrum made from $T = 1{,}052$ days of data on *Kepler*-410 and the peak-bagging fit to the spectrum. This is the same spectrum that was plotted in Figures 6.14 and 6.15 (there the raw spectrum was lightly smoothed). Also plotted in Figure 6.20 is the raw spectrum made from a short 30-day subset of the lightcurve. Although peak-bagging the longer spectrum returns robust and precise estimates of the $l = 1$ mode splittings, attempting to constrain the splittings given only the short spectrum would clearly be much more challenging (the resolution being inferior by a factor of 35).

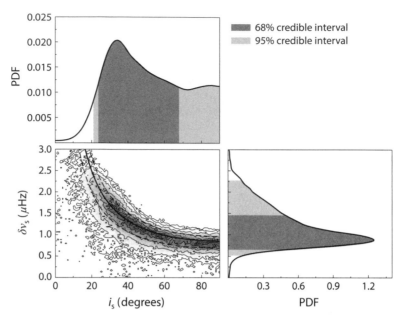

Figure 6.19. Estimated posterior distribution (PDF) in the $\delta\nu_s$–i_s plane for KIC 3456181. Also shown are the marginalized posteriors in each parameter.

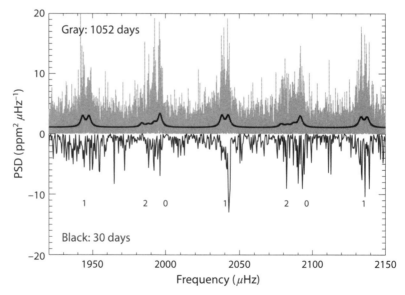

Figure 6.20. Raw (unsmoothed) spectrum made from $T = 1,052$ days of data from *Kepler-410* (gray) and the peak-bagging fit to the spectrum (heavy black line). Also plotted is the raw spectrum made from a short 30-day subset of the full lightcurve (bottom half of the figure).

The complexity of the fitting model clearly needs to be altered depending on the intrinsic quality of the data: subtle features that are clearly detectable in a high SNR spectrum made from a long lightcurve—implying also excellent resolution in frequency—may simply not be measurable in an inferior quality, lower-resolution spectrum. Figure 6.21 shows the spectrum of another planet-hosting star,

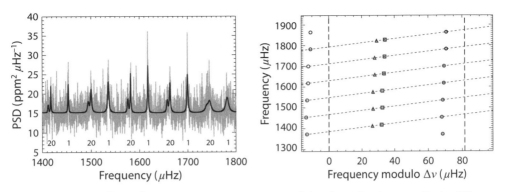

Figure 6.21. Left panel: Frequency-power spectrum of the planet-hosting star *Kepler*-103. Right panel: Échelle diagram, with $l = 0$ modes plotted as squares, $l = 1$ modes as circles and $l = 2$ modes as triangles.

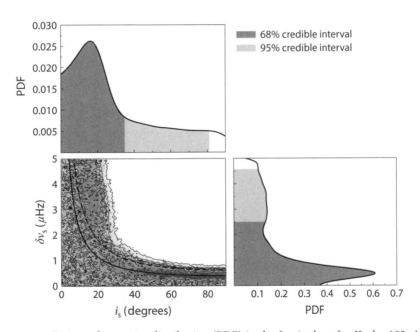

Figure 6.22. Estimated posterior distribution (PDF) in the $\delta\nu_s$-i_s plane for *Kepler*-103. Also shown are the marginalized posteriors in each parameter.

Kepler-103. This is a star whose spectrum is on the threshold of being of sufficient quality to extract individual frequencies. Only very weak constraints can be placed on the rotational frequency splitting of its detected nonradial modes (Figure 6.22).

6.4.6 More Subtleties

As noted already, implicit in the use of MLE or MAP is the assumption that the likelihood functions are Gaussian in the vicinity of the best-fitting parameter values. The distributions for some parameters may actually be log-normal (e.g., this usually turns out to be the case for the height and width parameters). An optimal approach is therefore one in which the logarithms of the parameters are varied and optimized, not the actual parameters themselves:

$$h_{nl} = \ln(H_{nl}), \tag{6.120}$$

and

$$\gamma_{nl} = \ln(\Gamma_{nl}). \tag{6.121}$$

Notice that we take the natural logarithms. Provided the parameters are well constrained—giving small fractional uncertainties—this means that (taking height as an example) the uncertainties of the logarithmic (varied) parameters and absolute parameters are related by

$$\sigma(h_{nl}) \simeq \sigma(H_{nl})/H_{nl}. \tag{6.122}$$

The existence of strong correlations between any parameters that are varied during optimization is undesirable, since it can lead to bias in the best-fitting outputs. The impact of these correlations can sometimes be mitigated by making careful choices regarding which parameters to vary. One example for the asteroseismic peak-bagging is the correlation between the rotational frequency splitting and the stellar angle of inclination. In borderline cases—for which it is hard to constrain both parameters—the impact of this correlation can be mitigated to some extent by varying the projected splitting $\delta\nu_s^*$ and i_s (or $\sin i_s$), as opposed to the more obvious approach of varying $\delta\nu_s$ and the angle. The frequency part of the Lorentzian function in the fitting model then becomes

$$\xi_{nlm}(\nu) = \frac{\nu - \nu_{nl} - \left(m\delta\nu_s^*/\sin i_s\right)}{\Gamma_{nl}/2}. \tag{6.123}$$

Another example is the strong anticorrelation shown between the height and linewidth parameters. A way to mitigate the impact of this correlation is to instead fit the mode amplitude and linewidth, which are somewhat less correlated. This effect is illustrated in Figure 6.23. Here, we show the two-dimensional posterior probability distributions for fits to a single order of p modes in *Kepler*-103. The left panel shows the results when the amplitude is fitted (no noticeable correlation); while the right-hand panel shows what happens when we instead fit for the peak height (strong anticorrelation).

When we fit directly for the mode amplitude, the fitting model for a given mode multiplet then becomes

$$\mathcal{M}_{nl}(\nu) = \sum_{m=-l}^{l} \mathcal{E}_{lm}(i_s) \left(\frac{A_{nl}^2}{\pi\Gamma_{nl}}\right) \left(\frac{1}{1 + \xi_{nlm}^2(\nu)}\right). \tag{6.124}$$

To round out this section, our final tweak concerns what to do when a mode peak is neither fully resolved nor unresolved (in effect, a coherent spike). Recall that we discussed this issue in Section 5.3.2, and presented an equation for the mode height

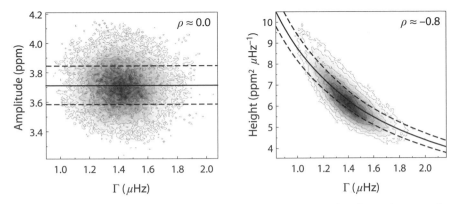

Figure 6.23. Estimated posterior distributions for fits to a single order of p modes in *Kepler*-103. Left panel: Results when the amplitude is fitted. Right panel: Results when the height is fitted. Note annotations that give the implied correlation coefficients.

that handles all possible regimes (Eq. 5.48). Use of this expression further modifies the multiplet fitting model to

$$\mathcal{M}_{nl}(\nu) = \sum_{m=-l}^{l} \mathcal{E}_{lm}(i_s) \left(\frac{A_{nl}^2 T}{\pi \Gamma_{nl} T + 2} \right) \left(\frac{1}{1 + \xi_{nlm}^2(\nu)} \right). \tag{6.125}$$

Formally, the best-fitting Lorentzian profile should also be convolved with the sinc-squared profile of the window function to which the observed profile approaches as T tends to τ.

6.4.7 A Peak-Bagging Model with Correlated Excitation

In Section 5.3.3 we discussed the concept of correlations giving rise to asymmetry of the mode peaks. First, there is the correlation between the convective granulation—which excites and damps the solar-like oscillations—and the signal due to the modes themselves. Second, we hypothesized that the excitation of some modes will also be correlated.

With regard to the correlated excitation, recall that here we assume that the excitation function of a mode of angular degree l, azimuthal degree m, and frequency ν_{nlm} can be described in terms of the component of the granulation noise that has the same spherical harmonic projection Y_l^m (in the corresponding range in frequency in the Fourier domain). An important implication is that overtones of the same (l, m) should have excitation functions that are correlated in time. Note that the Y_l^m for (l, m) and $(l, -m)$ are formally orthogonal and are therefore assumed to have independent (i.e., uncorrelated) excitation. It is important to note that correlation of the excitation in time does not imply correlation of the mode amplitudes in time. This can be understood by considering the damped, stochastically forced oscillator analogy. Modes of different frequencies will be kicked by the common excitation at different phases in their oscillation cycles, and provided the frequencies differ by more than a few linewidths—a condition easily met by consecutive overtones of the low-l modes—there will be significant differences in how the amplitudes vary in time.

Here we propose a strategy to introduce this correlated excitation, and the correlation with the granulation background, into the peak-bagging model. In the previous sections we built the fitting model by first constructing individual multiplets of a given n, l, [i.e., $\mathcal{M}_{nl}(\nu)$], which each contain up to $2l + 1$ different m components. We then summed the individual multiplet contributions over all n and l to give the model of the full oscillation spectrum. Here we must adopt a different approach. At the heart of our model now lies the assumption that modes of a given (l, m) are correlated. Our first step must now be to construct, in the complex domain, the overtone spectrum given by each (l, m) in a way that includes explicitly the correlated excitation. Next we must add to this, again in the complex domain, a suitable contribution from the granulation component. Only after completing these first two steps can we sum over the different (l, m)—which we assume are uncorrelated—to give the final model for the oscillation spectrum.

Let $\mathcal{A}_{lm}(\nu)$ be the complex spectrum (in amplitude) given by all overtones n of a given (l, m). With reference to Eq. 5.55, the real and imaginary parts can be written as, respectively,

$$\Re[\mathcal{A}_{lm}(\nu)] = \sum_n \left(\frac{\mathcal{E}_{lm}(i_s) H_{nl}}{2} \right)^{1/2} \left(\frac{\xi_{nlm}(\nu)}{1 + \xi_{nlm}^2(\nu)} \right), \tag{6.126}$$

and

$$\Im[\mathcal{A}_{lm}(\nu)] = \sum_n \left(\frac{\mathcal{E}_{lm}(i_s) H_{nl}}{2} \right)^{1/2} \left(\frac{1}{1 + \xi_{nlm}^2(\nu)} \right). \tag{6.127}$$

Let us assume that the correlation between different components having the same (l, m) is described by some coefficient ρ. The correlated part of the power spectrum for a given (l, m) is therefore

$$\mathcal{C}_{c;lm}(\nu) = |\rho| \, \mathcal{A}_{lm}(\nu) \mathcal{A}_{lm}^*(\nu), \tag{6.128}$$

where * denotes complex conjugate. Summing over all (l, m) gives the composite, correlated part of the spectrum:

$$\mathcal{C}_c(\nu) = \sum_{lm} |\rho| \, \mathcal{A}_{lm}(\nu) \mathcal{A}_{lm}^*(\nu). \tag{6.129}$$

The uncorrelated part of the spectrum is then just

$$\mathcal{C}_u(\nu) = (1 - |\rho|) \sum_{nl} \mathcal{M}_{nl}(\nu), \tag{6.130}$$

where we use the original formulation to construct its part.

We have accounted for correlations *between* modes. But remember that we must also account for the fact that the observed data includes a contribution from the granulation background that is assumed to be correlated with the modes (recall our discussion at the start of this section). Let us model the granulation background with a single component. The contribution at each (l, m) is given by

$$g_{lm}(\nu) = \frac{b_{lm}}{1 + (2\pi \nu \tau)^\beta}, \tag{6.131}$$

where the individual granulation powers are scaled according to

$$b_{lm} = b \times \frac{\mathcal{E}_{lm}(i_s) V_l^2}{\displaystyle\sum_{lm} \mathcal{E}_{lm}(i_s) V_l^2}, \tag{6.132}$$

with b being the full power of the composite background that we see in the frequency-power spectrum, and V_l is as defined in Eq. 6.112. This scaling step is a necessary fix to ensure that each (l, m) sees the same ratio in power of granulation to modes, correcting for the effect of the spatial filtering inherent in the observations (which the star knows nothing about).

From there we may build the correlated and uncorrelated parts of the spectrum, including the granulation contribution. Here we shall assume that the correlation between the modes and granulation is described by the same coefficient, ρ, as in Eq. 6.128. We then have

$$\mathcal{C}_c(\nu) = \sum_{(lm)} |\rho|\, \psi_{lm}(\nu) \psi_{lm}^*(\nu), \tag{6.133}$$

where

$$\psi_{lm}(\nu) = \Re[\mathcal{A}_{lm}(\nu)] + i\, \Im[\mathcal{A}_{lm}(\nu)] \pm g_{lm}^{1/2}, \tag{6.134}$$

and

$$\mathcal{C}_u(\nu) = (1 - |\rho|) \left(\sum_{nl} \mathcal{M}_{nl}(\nu) + \frac{b}{1 + (2\pi \nu \tau)^\beta} \right). \tag{6.135}$$

The sign of the complex contribution $g_{lm}^{1/2}$ fixes whether the asymmetry from the correlation with the background is positive or negative. Putting this all together, the full peak-bagging model—including the shot noise and sinc-function attenuation—is then

$$\mathbf{M}(\nu) = \text{sinc}^2 \left[\frac{\pi}{2} \left(\frac{\nu}{\nu_{\text{Nyq}}} \right) \right] [\mathcal{C}_c(\nu) + \mathcal{C}_u(\nu)] + b_{\text{shot}}. \tag{6.136}$$

6.5 CHALLENGES FOR PEAK-BAGGING POSED BY MIXED MODES

Once stars leave the main sequence and enter the subgiant phase, they begin to show detectable signatures of mixed modes, which poses additional challenges for the peak-bagging analysis.

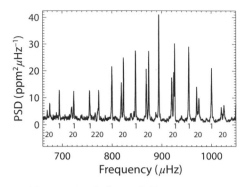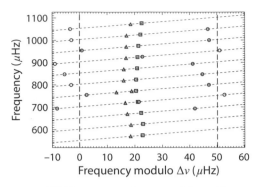

Figure 6.24. Left panel: Frequency-power spectrum of KIC 11026764 after smoothing with a boxcar filter of width 1.0 μHz. Annotations indicate the locations of modes of different degree l. Right panel: Échelle diagram, with $l = 0$ modes plotted as squares, $l = 1$ modes as circles, and $l = 2$ modes as triangles.

We saw in Chapter 3 that as the stars evolve, the frequency limit up to which g modes can be sustained increases in response to evolutionary changes in the deep stellar interior. Once this upper limit has extended into the range occupied by the detectable, high-overtone p modes, we may begin to detect the effects of g modes coupling to p modes. In Figure 1.6 we saw an example of a subgiant (HD 183159) where a single g mode frequency lay in the middle of the spectrum of the detectable p modes. This gave rise to a clear avoided crossing in the $l = 1$ modes (i.e., from several p modes coupling to the g mode). Here we introduce another similar example, this time a star that shows two clear avoided $l = 1$ crossings in the detectable p-mode range, in addition to an avoided crossing in its $l = 2$ modes. Figure 6.24 shows the frequency-power spectrum (left panel), and échelle diagram (right panel) of KIC 11026764. The two $l = 1$ g mode frequencies lie at approximately $\simeq 720\,\mu$Hz and $920\,\mu$Hz. We see significant disruptions in the $l = 1$ p mode ridge, with frequencies shifted significantly from the putative, undisturbed $l = 1$ ridge. Note that there are two $l = 1$ frequencies that lie almost on top of the $l = 0$ ridge. The use of a copied or replicated échelle diagram will reveal the above pattern even more clearly, as we shall go on to demonstrate in Section 6.5.2.

Let us look next at an even more evolved star. Figure 6.25 shows the oscillation spectrum and échelle diagram of the *Kepler* target KIC 11771760. This target has intrinsic properties that place it at the base of the RGB. About eight $l = 1$ g mode frequencies lie in the range where p modes are detected. The resulting frequency pattern—where each p mode now couples to several g modes—is at first glance more complicated and seemingly more confused.

We may estimate roughly how many pure g mode frequencies we would expect in each order $\Delta\nu$ (here for $l = 1$ modes) using

$$\delta\nu = -\Delta\Pi_1\nu^2, \tag{6.137}$$

with $\Delta\Pi_1$ being the period spacing and $\delta\nu$ the corresponding spacing in frequency, at frequency ν. The number of modes $(N_g)_1$ per order, in the vicinity of $\nu_{\max}$, is then simply

$$(N_g)_1 \simeq 1/(\Delta\Pi_1)\left(\Delta\nu/\nu_{\max}^2\right). \tag{6.138}$$

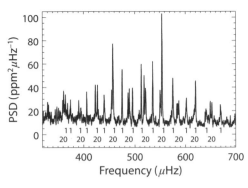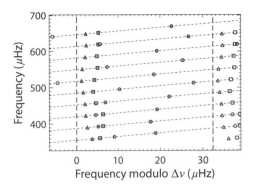

Figure 6.25. Left panel: Frequency-power spectrum of KIC 11771760 after smoothing with a boxcar filter of width $1.0\,\mu$Hz. Annotations indicate the locations of modes of different degree l. Right panel: Échelle diagram, with $l = 0$ modes plotted as squares, $l = 1$ modes as circles, and $l = 2$ modes as triangles.

Figure 6.26 shows measured $l = 1$ g mode period spacings from *Kepler* data (top panel) and the corresponding estimates of $(N_g)_1$ (bottom panel) for $l = 1$ modes. Here we use data on measured period spacings, ΔP, which are designed to provide estimates of $\Delta \Pi_1$. Subgiants and RGB stars are plotted in gray; core helium-burning stars are plotted with open symbols, including both red-clump (RC) stars that have gone through the helium flash, and higher-mass, secondary-clump stars that avoided the flash. A few example stars are plotted with filled black symbols, including HD 183159 and KIC 11026764. Both these stars lie in the very sparse subgiant regime, meaning their spectra are reasonably straightforward to understand visually. We add a caveat: the presence of mixed $l = 1$ modes lying on top of the $l = 0$ ridge in both stars is a feature that might not be so obvious from a visual check alone, a point we shall return to in Section 6.5.2. KIC 11771760 is in the region where $(N_g)_1 \simeq 1$, which leads to oscillation spectra and échelle diagrams that can be very hard to parse visually. As stars evolve up the RGB and the density of g-mode frequencies increases (i.e., $\Delta \Pi_1$ gets shorter), the spectra again become a bit more straightforward to unpick, provided one has sufficiently good resolution in frequency to resolve clearly the various mixed components (rotational splitting can present a challenge). However, once stars reach the core-helium burning phase, the g mode period spacings increase, and, coupled with the effect of wider peak linewidths, this can give the spectra a more messy appearance.

Before we look at the spectra in more detail, we must first understand how the observed mode powers and damping rates are affected by the coupling between the p mode and g mode cavities.

6.5.1 Mode Inertia, Nonradial Modes, and Mixed Modes

The impact of the coupling on the observed amplitudes may be understood in terms of the mode inertia, which can be thought of as a measure of the fraction of stellar mass that is engaged in pulsation. The g-dominated modes have high mode inertia because their perturbations are largest in the core, where the density is high. This reduces their observed surface amplitudes relative to those expected for pure (radial)

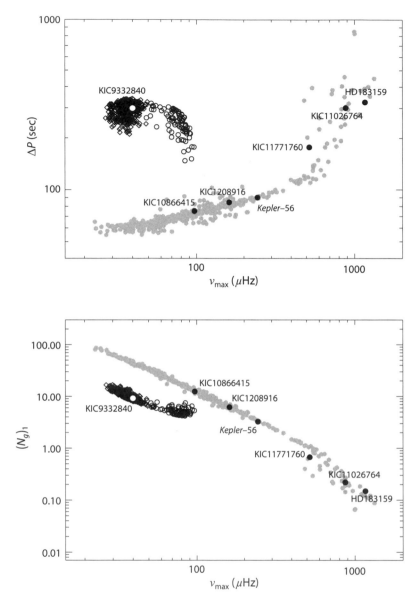

Figure 6.26. Top panel: Measured g mode period spacings as a function of ν_{max}, from *Kepler* observations. Results on subgiants and RGB stars are plotted in gray; and those on RC and secondary-clump stars are plotted with open symbols. Bottom panel: Estimate of $(N_g)_1$, the approximate number of $l = 1$ g mode frequencies per order $\Delta\nu$, in the vicinity of ν_{max}. Data are from Mosser *et al.*, 2014, *Astron. Astrophys.*, **572**, 5.

p modes at the same frequencies. Perturbations due to the p-dominated modes are, in contrast, highest in the envelope; hence these modes are more easily detected.

The damping rate, and hence the implied peak width of a mixed mode, is reduced by the ratio of the (increased) inertia of that mixed mode relative to the radial mode inertia expected at the same frequency. Provided the peaks are well resolved, the peak height is not affected by the coupling. However, when the mode peak is unresolved, the height is reduced by the same factor as the peak width. The g-dominated modes, which have very high inertia and hence very narrow peak widths, are therefore much harder to detect than are p-dominated modes of lower inertia. This is yet another factor that needs to be accounted for in the peak-bagging.

Let us develop our discussion using the equations introduced in Section 5.3.1. We may combine Eqs. 5.43 and 5.44 to express the mode amplitude A in terms of the rate at which energy is supplied to the modes, $\mathcal{E}$, and the associated mode mass (i.e., the product of the mode inertia I and the mass of the star M):

$$A^2 = \frac{\mathcal{E}}{2\eta I M}. \tag{6.139}$$

We wish to explore how the amplitude of a nonradial mixed mode compares to that of a pure (radial) p mode at the same frequency. Here we consider just the $l = 1$ modes. Using a suffix to tag the degree l, we assume that the forcing of modes of similar frequency is unchanged at the different degrees, so that we can write

$$\mathcal{E}_1 I_1 \simeq \mathcal{E}_0 I_0. \tag{6.140}$$

Recall that the mode linewidth is linearly proportional to the damping constant η. It may be shown that the damping is expected to show an inverse dependence on inertia:

$$\Gamma \propto \eta \propto I^{-1}. \tag{6.141}$$

Combining Eqs. 6.139–6.141 therefore implies that

$$A \propto I^{-1/2}. \tag{6.142}$$

What impact do these dependences on I have on the predicted heights H? To answer this question, we revisit the concept of the inertia ratio Q: the inertia of a nonradial mode relative that of a radial mode at (notionally) the same frequency. We then have

$$Q_1 = \frac{I_1}{I_0(\nu_1)}. \tag{6.143}$$

If I_1 is the total inertia of our mixed $l = 1$ mode, we assume that to reasonable approximation we can write this as a simple linear combination of the inertia in the g mode cavity and the p mode cavity:

$$I_1 \simeq (I_1)_g + (I_1)_p. \tag{6.144}$$

Moreover, if we also assume that

$$(I_1)_p \simeq I_0, \tag{6.145}$$

it then follows that

$$Q_1 \simeq \frac{(I_1)_g + (I_1)_p}{(I_1)_p}. \tag{6.146}$$

We may think of the impact of the inertia in terms of a modification to the radial-mode dependencies. Equations 6.141 and 6.142 imply that the mixed-mode amplitudes will be modified by the factor $Q_1^{-1/2}$ relative to the notional radial-mode values at the same frequency, whereas the mixed-mode linewidths are modified by the factor Q_1^{-1}. Putting this all together, we have

$$H_1 = \begin{cases} [V_1 A_0(\nu_1)^2]/[\pi \Gamma_0(\nu_1)], & \text{for } T \gg 2\tau_0(\nu_1) \\ \\ [V_1 T A_0(\nu_1)^2/2]/Q_1, & \text{for } T \ll 2\tau_0(\nu_1). \end{cases} \tag{6.147}$$

In Eq. 6.147, $A_0(\nu_1)$ and $\Gamma_0(\nu_1)$ are the notional radial-mode amplitude and linewidth values, respectively, at the same frequency as the mixed mode, and V_1 is the $l = 1$ visibility in power. For the more general (nonlimiting) case (recall Eq. 5.48), we have

$$H_1 = \frac{V_1 T A_0(\nu_1)^2}{\pi T \Gamma_0(\nu_1) + 2Q_1}. \tag{6.148}$$

We draw some important conclusions from these equations. If a mixed mode is resolved, then we do not need to make any allowance in our peak-bagging fitting model for the impact of the coupling on the mode height. The height does not depend on the inertia ratio. However, the linewidth will be narrower, by the factor Q_1^{-1}, than would be expected for a pure p mode at the same frequency. As such, fitting the same linewidth as the closest $l = 0$ mode (as per the entry-level model in Section 6.4.4) may then be a poor choice (although how much so will depend on whether the modification to the linewidth is significant at the level of precision of the data).

If instead the mixed mode is g-dominated in character—implying a high inertia ratio—the linewidth may be sufficiently narrow as to render the mode peak unresolved in the frequency spectrum. The peak height will be then suppressed by the factor Q_1^{-1}. Again, the entry-level model will require modification.

Figure 6.27 shows theoretically predicted échelle diagrams for a selection of stellar evolutionary models, ranging from the subgiant phase all the way to the RGB. The symbol sizes are scaled by predictions of the observed amplitudes. Those predictions make allowance for the l-dependent visibilities V_l, and the frequency-dependent Gaussian oscillation envelope, but crucially, they also include the $A \propto I^{-1/2}$ dependence from Eq. 6.142. The top-left panel is close to the observed case HD 183159; the top-right panel is not dissimilar to KIC 11026764, with two avoided crossings lying comfortably within the detectable range; the second left panel down approximates KIC 11771760. In Section 6.5.2, we shall look at observed cases that are close to the predictions shown in the bottom two panels, both stars ascending the RGB.

An important take-away message from this figure is that the amount of coupling can have a significant impact on the detectability of individual mixed modes. On the one hand this presents challenges for simply detecting modes, but on the other it may

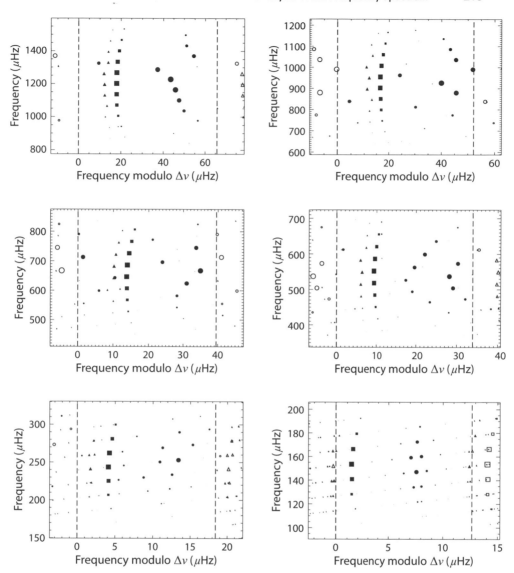

Figure 6.27. Theoretically predicted échelle diagrams for a selection of stellar evolutionary models of the same mass and metallicity in the subgiant phase to the red-giant branch. The symbol sizes are scaled by predictions of the observed amplitudes, with $l = 0$ modes plotted as squares, $l = 1$ modes as circles, and $l = 2$ modes as triangles.

help clean up the appearance of the spectrum, making it easier to parse. This is what we go to discuss next.

6.5.2 Parsing the Mixed-Mode Spectrum

Our ability to make sense of the spectrum is crucial, since without knowing which modes are which, we will not be able to fit an appropriate model to extract the mode parameters.

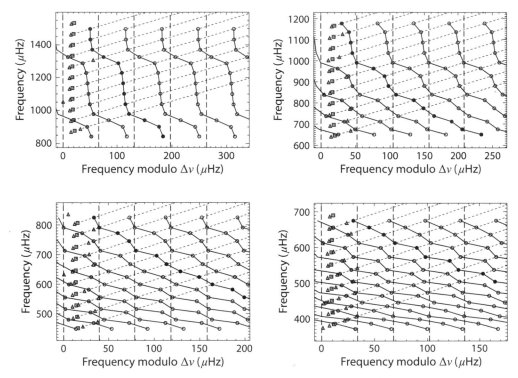

Figure 6.28. Copied or replicated échelle diagrams for the top four theoretically predicted échelle diagrams in Figure 6.27, with $l = 0$ modes plotted as squares, $l = 1$ modes as circles, and $l = 2$ modes as triangles. As usual, dotted lines follow the sloping orders, while the vertical dashed lines appear at multiples of $\Delta\nu$.

Let us first consider the subgiant regime (e.g., HD 183159 and KIC 11026764) and stars in the vicinity of the base of the RGB (e.g., KIC 11771760). We have either a single g mode coupling to several p modes; or a few g modes coupling to multiple p modes (i.e., each p mode is affected by more than one g mode). A copied or replicated échelle diagram is then a useful tool for parsing the spectrum. Figure 6.28 shows replicated échelles for the top four theoretically predicted échelle diagrams in Figure 6.27.

Each diagram is made by copying the base échelle (which covers the range $\{0, \Delta\nu\}$ on the abscissa) a multiple number of times, modulo the large separation. We concentrate our discussion on the $l = 1$ mixed-mode frequencies and so for clarity only those frequencies have been copied in Figure 6.28.

The échelles plotted throughout this chapter again show $[\nu \bmod(\Delta\nu)]$ on the horizontal axis, so that each order slopes slightly upward (to the right), as indicated by the dotted lines. Remember that we get an extra $l = 1$ mode at each avoided crossing. In the base échelle diagram we therefore expect to see one $l = 1$ mixed mode in each horizontal row of modes. In the replicated échelle we may therefore construct the connecting lines in the figure—which pass through one $l = 1$ mode in each horizontal row—allowing us to follow the mixed modes through the copied échelles. This provides a simple way to identify whether there is a missing detection in the sequence. It also allows us to see whether a mixed mode is indeed expected to

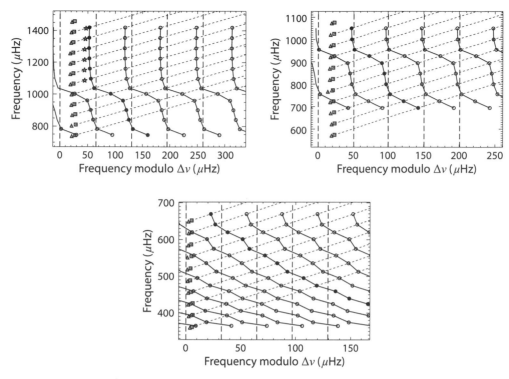

Figure 6.29. Replicated échelle diagrams of HD 183159 (top-left panel), KIC 11026764 (top right panel), and KIC 11771760 (bottom panel).

lie close to the $l = 0$ or $l = 2$ ridges, without the need to resort to predictions from a model. Figure 6.29 shows the replicated échelles for HD 183159, KIC 11026764, and KIC 11771760.

Finally, recalling that KIC 11026764 also has detected mixed $l = 2$ modes, Figure 6.30 shows a replicated échelle diagram where only the $l = 2$ frequencies have been copied from the base échelle. The single, $l = 2$ avoided crossing is readily apparent. The coupling is weaker at $l = 2$ than at $l = 1$.

As stars evolve up the RGB and the density of g mode states increases, there will be several (up to many) g modes coupling to each p mode. The detectable mixed modes bunch more closely around the pure $l = 1$ p mode frequency. This is shown in the bottom two panels of Figure 6.27. The increased density of states now makes the replicated diagrams less useful. However, we may instead use the asymptotic description for mixed modes introduced in Chapter 3 to help guide our understanding of the observed patterns of peaks. We rewrite Eq. 3.95 for $l = 1$ mixed modes:

$$\nu = (\nu_p)_{n,1} + \frac{\Delta \nu}{\pi} \arctan \left[q \tan \left(\frac{\pi}{\nu \Delta \Pi_1} - \epsilon_g \right) \right], \qquad (6.149)$$

where $(\nu_p)_{n,1}$ corresponds to the pure $l = 1$ p mode frequency, $\Delta \Pi_1$ is the asymptotic period spacing, and q is the coupling constant. Solving for the roots of ν in Eq. 6.149 gives the predicted mixed-mode frequencies relative to $(\nu_p)_{n,1}$ for the assumed $\Delta \nu$,

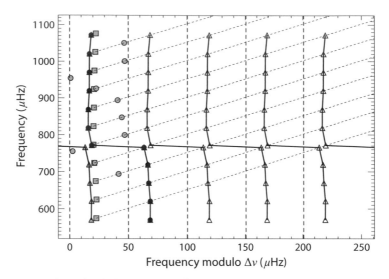

Figure 6.30. Replicated échelle diagram of KIC 11026764, where only $l = 2$ modes have been copied.

$\Delta\Pi_1$, and q. We make the approximation that any mixed modes in a given order are the result of a set of g modes coupling to a single p mode, and that the influence of other, neighboring p modes is negligible.

It is more useful in practical terms to express the roots in ν relative to the $l = 0$ p mode frequencies, which are easy to measure in the spectrum. Using the asymptotic expression for p modes, we have

$$(\nu_p)_{n,1} = \nu_{n,0} + \Delta\nu/2 - \delta\nu_{01}. \tag{6.150}$$

One may then envisage performing a search in $\{\Delta\nu, \Delta\Pi_1, \delta\nu_{01}, q\}$ to find the combination of parameters whose roots best match the observed frequencies. This then serves as a starting point for a peak-bagging fit (i.e., we have a set of first-guess frequencies and constraints on some global seismic parameters).

The top panel of Figure 6.31 shows the observed oscillation spectrum of *Kepler*-56, a low-luminosity planet-hosting star (also marked on Figure 6.26). This star is quite similar to *Kepler*-432, which we met in Chapter 1. The spectra of both stars show clusters of $l = 1$ mixed modes. The bottom-left panel of Figure 6.31 shows the échelle diagram of the observed frequencies of *Kepler*-56. The bottom-right panel is a zoom of the spectrum containing three closely spaced $l = 1$ mixed modes. These modes show clear rotational frequency splitting. The lines connect rotationally split components of the same mixed mode. Power is present in all three m components in each mode, indicating that the inclination of the star, i_s, is intermediate in angle (around 45 degrees).

These modes also show clearly the impact of the coupling on the damping rates, discussed in Section 6.5.1. The central mode in the figure is p-dominated in character (i.e., it lies closest to the notional pure p mode frequency). In contrast, the two outer modes are g-dominated in character and hence have narrower linewidths compared to the central mode.

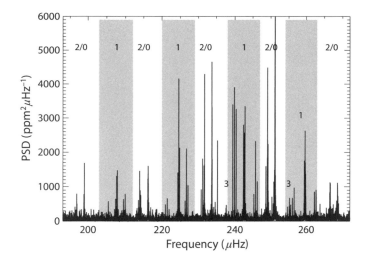

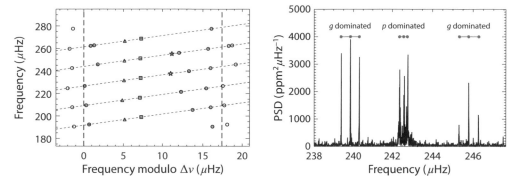

Figure 6.31. Top panel: Observed frequency spectrum of *Kepler*-56. The gray shaded regions contain $l = 1$ modes, while unshaded regions contain $l = 0$ and 2 modes. Bottom-left panel: Échelle diagram of its frequencies, with $l = 0$ modes plotted as squares, $l = 1$ modes as circles, $l = 2$ modes as triangles, and $l = 3$ modes as stars. Bottom-right panel: Zoom of part of the spectrum showing three closely spaced $l = 1$ mixed modes. The lines and filled circles link the rotationally split components of each mode.

Figure 6.32 shows the spectrum of a slightly more evolved RGB star, KIC 12008916. The right panel is again a zoom showing three closely spaced $l = 1$ mixed modes. Unlike those for *Kepler*-56, these modes show detectable power only in the outer (sectoral) components, indicating a high ($i_s \simeq 90$ degrees) inclination. We again see a variation of the peak linewidths across the three modes.

Figure 6.33 shows the oscillation spectrum of an even more evolved RGB star. KIC 10866415 lies in a slightly tricky parameter regime, where the spacings in frequency of its mixed $l = 1$ modes are very close to the magnitudes of the rotational frequency splittings (see the zoom in the right panel). Our final example in Figure 6.34 is an RC star, KIC 9332840. The period spacings for RC stars (see Figure 6.26) are much larger than for RGB stars of similar ν_{max}. The coupling between the p and g mode cavities is also stronger in the RC stars. Across a given cluster of mixed modes, the contrast between the inertia ratios of modes close to and further

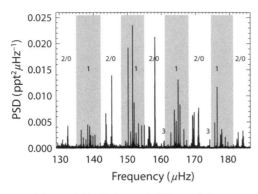
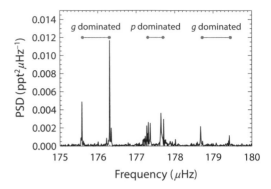

Figure 6.32. Left panel: Observed frequency spectrum of KIC 12008916. Right panel: Zoom of part of the spectrum, showing three closely spaced $l = 1$ mixed modes. The lines and filled circles link the rotationally split components of each mode.

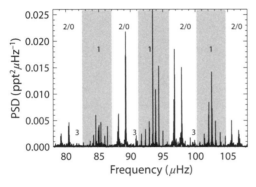
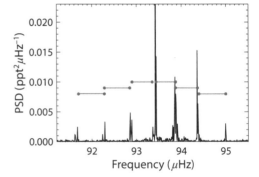

Figure 6.33. Left panel: Observed frequency spectrum of KIC 10866415. Right panel: Zoom of part of the spectrum, showing several closely spaced $l = 1$ mixed modes. The lines and filled circles link the rotationally split components of each mode.

from the notional pure p mode frequency is not as pronounced as for RGB stars at the same $\nu_{\max}$. We see detectable mixed modes in the RC phase that reach all the way to, and indeed overlap with, the $l = 0$ and $l = 2$ modes. This is why, for reasons of clarity, we have shaded the regions occupied by the even-l modes in Figure 6.34; aside from the marked $l = 3$ modes, all other prominent peaks in the spectrum are due to mixed $l = 1$ modes.

The peak linewidths tend to be wider than in RGB stars. The overall result is a messier looking oscillation spectrum. Disentangling the rotational frequency splittings—which are much smaller than for stars showing a similar $\nu_{\max}$ on the RGB—is much more challenging (see the right panel of Figure 6.34).

The need to disentangle the frequency splittings of the nonradial modes clearly adds another challenge to finding the optimal fitting model. In addition, the increased density in frequency of detectable mixed modes means that assigning a height and linewidth parameter to every mode may make the fit unstable and lead to poor constraints on the best-fitting values (and decidedly non-Gaussian posteriors). Matters may be less problematic if the spectrum is fitted an order at a time. Nevertheless, we can look again to asymptotic theory to offer a way of parametrizing

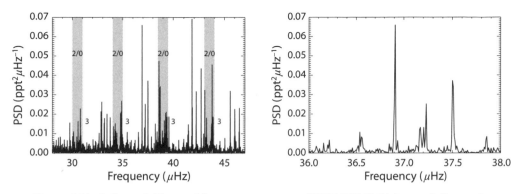

Figure 6.34. Left panel: Observed frequency spectrum of KIC 9332840. Right panel: Zoom of part of the spectrum showing, three closely spaced $l = 1$ mixed modes.

the variation of the amplitude, linewidth, and splitting across the observable mixed $l = 1$ modes in a given order, so that we can in principle reduce the number of free parameters in the fitting model.

Let us introduce a new quantity, ζ, which measures the ratio of the inertia of the g mode part of the cavity to the inertia of the entire cavity (i.e., recalling Eqs. 6.144 and 6.145):

$$\zeta = \frac{(I_g)_1}{I_1} \simeq \frac{(I_g)_1}{(I_g)_1 + (I_p)_1}. \tag{6.151}$$

This can be approximated in terms of observable quantities already discussed (see also Section 9.3.1), so that

$$\zeta \simeq \left[1 + \frac{f_1(\nu)}{q f_2(\nu)}\right]^{-1}, \tag{6.152}$$

with

$$f_1(\nu) = \cos^2\left[\pi\left(\frac{1}{\nu \Delta \Pi_1} - \epsilon_g\right)\right]\nu^2 \Delta \Pi_1, \tag{6.153}$$

and

$$f_2(\nu) = \Delta \nu \cos^2\left[\pi\left(\frac{\nu - (\nu_p)_{n,1}}{\Delta \nu}\right)\right]. \tag{6.154}$$

We may therefore estimate ζ for each mixed mode, using prior constraints on the global seismic parameters that we can extract from the observed spectrum (plus we must also adopt a suitable value for the coupling constant q, on which more in Section 9.3.1). From Eq. 6.146 it follows that

$$\zeta \simeq \left(\frac{Q_1 - 1}{Q_1}\right). \tag{6.155}$$

We therefore have the means to constrain the amplitude and linewidth of each mode during the fitting process. Specifically, we fit a single amplitude and linewidth parameter for each set of mixed $l = 1$ modes and then modify the parameters used to construct each individual mode according to

$$\Gamma \propto Q^{-1} \propto (1 - \zeta), \tag{6.156}$$

and

$$A \propto Q^{-1/2} \propto (1 - \zeta)^{1/2}. \tag{6.157}$$

The mode splittings are also affected by the mode trapping. The g-dominated modes, which lie farther away from the pristine p mode frequency, tend to show larger frequency splittings, since they have greater sensitivity to the more rapidly rotating core. This effect is apparent in the splittings of the mixed modes shown in the panels of Figures 6.31 (*Kepler*-56) and 6.32 (KIC 12008916). In both cases the central mixed mode is a p-dominated mode. The outer two mixed modes are more g-dominated in character, and hence show the larger splittings. One may write the splitting in terms of two parameters, one related to the splitting from the g mode part of the cavity ($\delta \nu_g$) and one related to the splitting from the p mode part of the cavity ($\delta \nu_p$). We then have

$$\delta \nu_s \simeq \zeta \delta \nu_g + (1 - \zeta) \delta \nu_p. \tag{6.158}$$

Hence we also have a way to parameterize the splitting across the mixed modes in the order, using two free parameters (the above g and p mode splittings) and Eq. 6.158 to modify the magnitude of the splitting for each mode.

6.6 FURTHER READING

Several recent reviews cover various aspects of the analysis of asteroseismic data on solar-like oscillators:

- Appourchaux, T., 2014, *A Crash Course on Data Analysis in Asteroseismology*, in *Asteroseismology, 22nd Canary Islands Winter School of Astrophysics*, eds. Pallé, P. L., and Esteban, C., Cambridge University Press.
- Bedding, T. R., 2014, *Solar-Like Oscillations: An Observational Perspective*, in *Asteroseismology, 22nd Canary Islands Winter School of Astrophysics*, eds. Pallé, P. L., and Esteban, C., Cambridge University Press.
- Chaplin, W. J., and Miglio, A., 2013, *Asteroseismology of Solar-Type and Red-Giant Stars*, Ann. Rev. Astron. Astrophys., **51**, 353.
- Chaplin, W. J., 2014, *Sounding the Solar Cycle with Helioseismology: Implications for Asteroseismology*, Asteroseismology, 22nd Canary Islands Winter School of Astrophysics, eds. Pallé, P. L., and Esteban, C., Cambridge University Press.
- García., R. A., 2015, *Observational Techniques to Measure Solar and Stellar Oscillations, Ecole Evry Schatzman 2014: Asteroseismology and Next Generation Stellar Models*, EAS Publications Series, **73**, 193, eds. Michel, E., Charbonnel, C., and Dintrans, B., EDP Sciences.

- Mosser, B., 2015, *Stellar Oscillations I. The Adiabatic Case, Ecole Evry Schatzman 2014: Asteroseismology and Next Generation Stellar Models*, EAS Publications Series, **73**, 3, eds. Michel, E., Charbonnel, C., and Dintrans, B., EDP Sciences.

The following list contains recent examples of the application of state-of-the-art approaches to peak-bagging data on solar-like oscillators:

- Appourchaux, T., 2008, *Bayesian Approach for g-Mode Detection, or How to Restrict Our Imagination, Astron. Nachr.*, **329**, 485.
- Handberg, R., and Campante, T. L., 2011, *Bayesian Peak-Bagging of Solar-Like Oscillators Using MCMC: A Comprehensive Guide, Astron. Astrophys.*, **527**, 56.
- Appourchaux, T., et al., 2012, *Oscillation Mode Frequencies of 61 Main-Sequence and Sub-Giant Stars Observed by* Kepler, *Mon. Not. R. Astron. Soc.*, **543**, 54.
- Corsaro, E., and De Ridder, J., 2014, *DIAMONDS: A New Bayesian Nested Sampling Tool. Application to Peak Bagging of Solar-Like Oscillations, Astron. Astrophys.*, **571**, 71.
- Corsaro, E., De Ridder, J., and García, R. A., 2015, *Bayesian Peak Bagging Analysis of 19 Low-Mass Low-Luminosity Red Giants Observed with* Kepler, *Astron. Astrophys.*, **578**, 76.
- Davies, G. R., et al., 2016, *Oscillation Frequencies for 35* Kepler *Solar-Type Planet-Hosting Stars Using Bayesian Techniques and Machine Learning, Mon. Not. R. Astron. Soc.*, **456**, 2183.

These papers discuss analysis of the oscillation spectra of evolved solar-like oscillators. The data plotted in Figure 6.26 come from Mosser et al. (2014):

- Mosser, B., et al., 2012, *Probing the Core Structure and Evolution of Red Giants Using Gravity-Dominated Mixed Modes Observed with* Kepler, *Astron. Astrophys.*, **540**, 143.
- Mosser, B., et al., 2012, *Spin Down of the Core Rotation in Red Giants, Astron. Astrophys.*, **548**, 10.
- Benomar, O., et al., 2013, *Properties of Oscillation Modes in Sub-Giant Stars Observed by* Kepler, *Astrophys. J.*, **767**, 158.
- Benomar, O., et al., 2014, *Asteroseismology of Evolved Stars with* Kepler: *A New Way to Constrain Stellar Interiors Using Mode Inertias, Astrophys. J.*, **782**, L29.
- Mosser, B., et al., 2014, *Mixed Modes in Red Giants: A Window on Stellar Evolution, Astron. Astrophys.*, **572**, 5.
- Deheuvels, S., et al., 2015, *Seismic Evidence for a Weak Radial Differential Rotation in Intermediate-Mass Core Helium Burning Stars, Astron. Astrophys.*, **580**, 96.
- Davies, G. R., and Miglio, A., 2016, *Asteroseismology of Red Giants: From Analysing Light Curves to Estimating Ages, Astron. Nachr.*, **337**, 774 (eprint arXiv:1601.02802).

6.7 EXERCISES

The archive file `artstars.tar.gz` contains simulated *Kepler* lightcurves and associated metadata for three artificial stars. All files are ASCII, with headers that explain the file formats and content. The metadata files contain comprehensive information on the fundamental properties of the artificial stars and the input parameters that describe the simulated oscillation spectra, granulation, activity, and shot-noise signatures. These simulated lightcurves may be used to test codes that inspect and analyze asteroseismic data. We suggest a few simple exercises below to get the reader started.

1. Compute the frequency-power spectrum of each lightcurve. Calibrate the spectra so that they respect Parseval's Theorem with single-sided scaling (see Section 5.1.4). The mean power spectral density at very high frequencies—close to the Nyquist frequency of these simulated short-cadence *Kepler* data—should be about $0.15\,\mathrm{ppm}^2\,\mu\mathrm{Hz}^{-1}$ in every case.

2. Implement the moving-median background filter (see Eq. 6.34 in Section 6.2), and apply it to the spectra. Explore how altering the width of the filter affects how well it describes the underlying background power spectral density.

3. Implement one or more of the methods used to detect the oscillation spectra (see Sections 6.2.1 and 6.2.2). Explore the limiting detection thresholds by adding varying levels of random Gaussian noise to the lightcurves, to increase the shot-noise contribution to the background power spectral density.

4. Investigate how the frequency range of the spectrum selected for analysis affects the estimation of the average large frequency separation, $\Delta\nu$ (e.g., using the PSPS and matched-filter methods). To what extent can one get good constraints on the frequency dependence of $\Delta\nu$, and how robust are those constraints for different intrinsic SNR levels in the spectrum?

5. Try fitting models to the background power spectral density. Explore different strategies (see Section 6.2): one where the detected oscillation spectrum is omitted from the range of fitted frequencies (so that it does not need to be modeled); and another where it is included, which requires inclusion of the simplified model of the power spectral density due to the oscillations (e.g., a Gaussian power envelope). How well do the fitted granulation parameters match those of the underlying spectrum? Is it possible to reach firm conclusions about the number of such components included in the data? (Note that all lightcurves have only one granulation component.)

6. Implement some of the false-alarm and odds-ratio tests used to flag the locations of candidate individual modes (Section 6.1). Explore how the number and robustness of the flagged detections changes with increasing shot-noise levels and with alterations made to the test parameters (e.g., threshold levels and bin ranges).

7. Write your own peak-bagging code, and apply it to the artificial data. Try something simple to begin with, such as code that will fit one radial order of the spectrum at a time.

7 Drawing Inferences from Average Seismic Parameters

Now that we have learned how to extract seismic data from observations, we turn our focus to determining stellar properties from those data.

As discussed earlier, the average seismic parameters, $\Delta\nu$ and $\nu_{\max}$, are relatively easy to determine. Very often these parameters are the only robust seismic data that can be obtained from low SNR power spectra. While this can be limiting, it still allows us to determine the global properties of the stars in question. In this chapter we discuss how these seismic parameters are used in determining the mass and radius of a star.

7.1 THE DIRECT METHOD

The scaling relations we first met in Chapter 1 allow us to determine the mass and the radius of a star using $\Delta\nu$ and $\nu_{\max}$ if the effective temperature is known. Equations 3.103 and 3.106 link $\Delta\nu$ and $\nu_{\max}$ to the mass, radius, and T_{eff} of a star, and the equations can be rearranged to get

$$\frac{M}{M_\odot} \simeq \left(\frac{\nu_{\max}}{\nu_{\max,\odot}}\right)^3 \left(\frac{\Delta\nu}{\Delta\nu_\odot}\right)^{-4} \left(\frac{T_{\mathrm{eff}}}{T_{\mathrm{eff},\odot}}\right)^{3/2}, \tag{7.1}$$

and

$$\frac{R}{R_\odot} \simeq \left(\frac{\nu_{\max}}{\nu_{\max,\odot}}\right) \left(\frac{\Delta\nu}{\Delta\nu_\odot}\right)^{-2} \left(\frac{T_{\mathrm{eff}}}{T_{\mathrm{eff},\odot}}\right)^{1/2}. \tag{7.2}$$

Using these equations is usually called the *direct method* of determining stellar properties. The results obtained this way are model independent, though approximate. Since the equations give us radius, we can also determine luminosity from T_{eff} and the estimated value of the radius. The mean density can be obtained from $\Delta\nu$ itself, and the value of $\log g$ can be determined from $\nu_{\max}$ and T_{eff}. Thus other than age, all global properties of the star can be obtained quite easily. The ease of using this

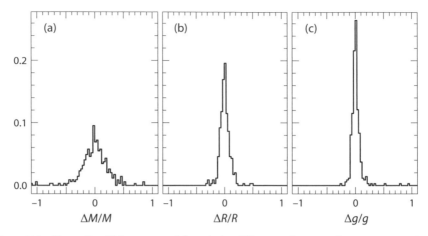

Figure 7.1. Normalized histograms of the relative differences between the exact properties of stellar models and those obtained using Eqs. 7.1 and 7.2. Note that the best results are obtained for radius and surface gravity; the mass estimates have a large spread.

method, particularly the fact that one does not need access to any stellar structure or stellar pulsation codes, has made this a popular way of determining stellar properties. While this direct method is attractive because of its simplicity and because it gives model-independent results, some issues make this method insufficient for a serious analysis of stellar properties.

One can see from Eqs. 7.1 and 7.2 that the errors in the estimates of mass and radius can be quite high. A 2.5% uncertainty in $\Delta\nu$ and 5% uncertainty in ν_{max}, values quite typical of what has been found with the *Kepler* data, combined with a 1.5% uncertainty in T_{eff} for a star about as hot as the Sun results in a $\simeq 20\%$ uncertainty in mass and a 10% uncertainty in radius. These estimates assume that Eqs. 3.103 Eq. 3.106 are exact, which (as we shall see a little later) is not the case.

To illustrate the properties of the direct method and others described later, we use a set of models constructed to closely mimic the properties of dwarf and subgiant stars observed by *Kepler*. Each of the models used here was drawn from a large grid of models and matches the metallicity, effective temperature, and luminosity of individual stars in the *Kepler* sample. For each model, $\Delta\nu$ was calculated using $l = 0$ frequencies, and ν_{max} was calculated using the scaling relation. We added one realization of the observed uncertainties to each of [Fe/H], T_{eff}, $\Delta\nu$, and ν_{max}—the observables that are used as inputs. The results obtained with the direct method for this set of models are shown in Figure 7.1. The masses estimated this way can be quite uncertain. And since both mass and radius are determined from the same data, their errors are correlated, as can be seen in Figure 7.2.

A separate problem that arises when using Eqs. 7.1 and 7.2 to determine mass and radius is that the originating equations (Eqs. 3.103 and 3.106) that link $\Delta\nu$ and ν_{max} to the mass, radius, and temperature of a star assume that all values of T_{eff} are possible for a star of a given mass and radius. However, the equations of stellar structure and evolution tell us otherwise—we know that for a given mass and radius, only a narrow range of temperatures is allowed. Additionally, we know that the mass-radius-temperature relationship depends on the metallicity of a star; the scaling

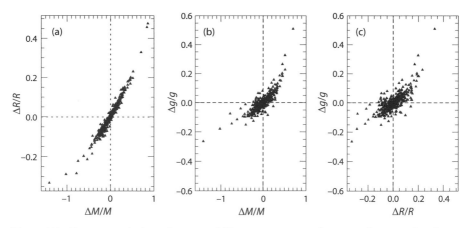

Figure 7.2. Error correlations between different properties determined using the direct method. The correlations are illustrated by plotting, for each model, the relative differences between the exact and inferred values of one property against that of another. Note that the plot of errors in radius plotted against errors in mass shows that the errors in the two quantities are completely correlated. Errors on surface gravity estimates are also correlated with the errors in mass and radius.

relations do not account for that. In the presence of data errors, these discrepancies can, in extreme cases, lead to results that do not satisfy physical constraints.

However, the biggest problem with the direct method is the fact that the $\Delta\nu$ and ν_{max} scaling relations are not exact. $\Delta\nu$ for stellar models can be calculated as the average spacing between its oscillation frequencies. Usually radial modes (i.e., $l = 0$ modes) are used for this. Since the power envelope of modes is essentially a Gaussian, this weighting may be used to calculate the average $\Delta\nu$ of a model to ensure that the calculated $\Delta\nu$ corresponds more closely to the observed value. The $\Delta\nu$ calculated in this manner can then be compared with $\Delta\nu$ calculated from the scaling relation. When we do this, we find that while the $\Delta\nu$ scaling relation is good enough to make rough estimates of masses, the deviations can be as high as 5%. This can be seen in Figure 7.3. If the $\Delta\nu$ scaling relation were exact, the plots would show flat lines at unity. Instead, we see that the ratio is a boomerang-shaped function of temperature. The deviation also has a metallicity dependence. Additionally, at the low-temperature end for ν_{max} values corresponding to red giants, the relation between $\Delta\nu$ and density is also mass dependent. The deviation also depends on the atmospheric physics of the models (see the right panel of Figure 7.3). The dependence of atmospheric parameters is another manifestation of the *surface term*, which is discussed in Section 8.3.

The deviation of $\Delta\nu$ from the scaling relation affects the mass and radius estimates obtained using Eqs. 7.1 and 7.2. This is illustrated in Figure 7.4, which shows how the masses and radii computed from $\Delta\nu$ obtained from the scaling relations compare with masses and radii obtained from frequencies. Along the red-giant branch, the error in the derived masses is a function of mass, metallicity, and ν_{max}; using the direct method leads to an overestimation of the mass until very close to the tip of the giant branch. The systematic error in mass estimates can lead to further incorrect deductions of stellar properties. Stars spend so little time in the

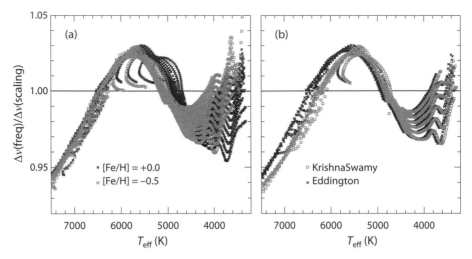

Figure 7.3. Ratio of the average large separations calculated using individual frequencies to those estimated with the scaling relations. Left panel: Results for models of different masses constructed with two different metallicities. Right panel: Results for models with [Fe/H] = 0 constructed with two different model atmospheres.

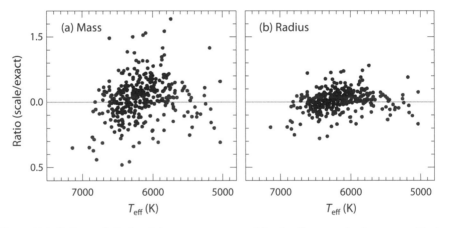

Figure 7.4. Left panel: Ratio of the masses computed by the direct method: masses with $\Delta\nu$ obtained from the scaling relations divided by masses with $\Delta\nu$ obtained from frequencies (Eqs. 7.1 and 7.2). Right panel: Same, but for the ratio of the radii.

red-giant phase that the mass of a star can be used to estimate its age regardless of where it is on the red-giant branch. The systematic overestimation of red-giant masses leads to a systematic underestimation of red-giant ages. This can be seen in Figure 7.5.

The parameter $\log g$ is not affected by the $\Delta\nu$ scaling relation, but depends on the $\nu_{\max}$ scaling relation instead. The accuracy of the $\nu_{\max}$ scaling relation can only be tested indirectly. This has been done in a number of ways, the most direct way being by comparing radius estimates obtained using $\Delta\nu$ and $\nu_{\max}$ with those obtained from interferometric radius estimates. These tests show that the $\nu_{\max}$ scaling relation should hold to better than 5%.

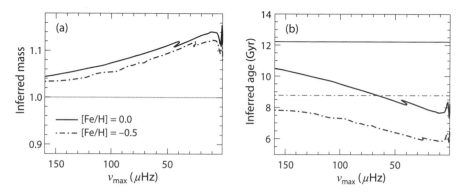

Figure 7.5. Left panel: Estimated mass of $1M_\odot$ RGB models derived assuming that the scaling relations are correct. Right panel: Derived ages shown as heavy lines; the corresponding horizontal lines show the real age.

The inexact nature of the scaling relations (which gives rise to systematic errors in the derived stellar properties) along with the large random errors caused by error propagation explain why stellar properties derived in this manner should be used only as approximate or guideline estimates.

7.2 GRID-BASED MODELING

The way out of some of the problems with the direct method is to use an approach that incorporates the relations between mass M, radius R, and $T_{\rm eff}$ implicitly by using grids of stellar models. Knowing the metallicity of the star in question helps immensely with this. At the price of making the results slightly model dependent, the method gives us results that are physical. Such an approach is usually called *grid-based modeling* or even sometimes "gridling," though to be more precise, these are *grid-based searches*. Another advantage of using grid-based modeling is that we can get an estimate of stellar ages; this is not possible using the direct method, because $\Delta\nu$ and $\nu_{\rm max}$ do not depend explicitly on age. Grid-based modeling can be used to overcome three of the major issues with the direct method: the problems with $\Delta\nu$ scaling, the lack of constraint on the scaling relations by the equations of stellar structure and evolution, and the failure to account for the effect of metallicity. Unfortunately, possible problems with the $\nu_{\rm max}$ scaling may still remain, although while doing grid-based searches, we do have the freedom not to use $\nu_{\rm max}$ as an observable.

The essential ingredient for grid-based modeling is of course a fairly dense grid of stellar models of different metallicities that cover a wide range of masses and evolutionary stages. It has been customary to use published evolutionary tracks, such as the Yale-Yonsei isochrones or the BASTI isochrones. The models in the grid have known global properties. The mass, radius, and $T_{\rm eff}$ for each star in the grid can be used to calculate $\Delta\nu$ and $\nu_{\rm max}$ using the scaling relations (Eqs. 3.103 and 3.106). However, because of the aforementioned issues relating to the exact scaling between $\Delta\nu$ and density, the trend is moving increasingly toward making custom grids with models whose $\Delta\nu$ are calculated using their model-calculated frequencies. The principle behind grid-based modeling is simple: for stars with the observed values of $\Delta\nu$, ν_{max}, [Fe/H], and T_{eff}, search through the grid for models that have properties that are "close" to the observations and then use the properties of those selected models to estimate the properties of the star.

7.2.1 A Simple Prescription for Grid-Based Modeling

While many grid-based pipelines have been developed to analyze data obtained with the *Kepler* mission, we shall use the pipeline that we developed, the Yale-Birmingham (YB) pipeline, to discuss how one can do a grid-based search and also to elucidate some important issues relating to grid-based modeling results.

The process starts with a given set of observations and their associated uncertainties for a star. We shall assume that the observations consist of $\Delta\nu$, ν_{max}, [Fe/H], and T_{eff}. As we shall see soon, we can easily add or remove observables.

The first step is to search for all models in the grid with properties close to the observed properties. This is usually defined in terms of observational uncertainties, and a 3–6σ search radius is usually enough. The set of models selected is the intersection of models that satisfy each of the properties within the search cycle (i.e., models that have $\Delta\nu$ within 3–6σ of the observed one *and* also have ν_{max} within 3–6σ of the observed ν_{max}, and so on). It is important to note that the search radius should be large enough to account for the fact that the grid is made of discrete models—too small a search radius could mean that no models will be selected. The selected models are then used to define a likelihood function for each selected model:

$$\mathcal{L}(\Delta\nu, \nu_{max}, T_{eff}, [Fe/H]) = \mathcal{L}(\Delta\nu)\mathcal{L}(\nu_{max})\mathcal{L}(T_{eff})\mathcal{L}([Fe/H]), \qquad (7.3)$$

where $\mathcal{L}(\Delta\nu)$, $\mathcal{L}([Fe/H])$, and so forth are the likelihoods for each variable. We define

$$\mathcal{L}(\Delta\nu) = \frac{1}{\sqrt{2\pi}\sigma(\Delta\nu)} \exp -\left(\frac{(\Delta\nu_{obs} - \Delta\nu_{model})^2}{2\sigma^2(\Delta\nu)} \right), \qquad (7.4)$$

$\sigma(\Delta\nu)$ being the uncertainty on $\Delta\nu_{obs}$, which is the $\Delta\nu$ observed for the star. The likelihood function for each of the other variables can be written in an analogous manner, for example, for [Fe/H]:

$$\mathcal{L}([Fe/H]) = \frac{1}{\sqrt{2\pi}\sigma([Fe/H])} \exp -\left(\frac{([Fe/H]_{obs} - [Fe/H]_{model})^2}{2\sigma^2([Fe/H])} \right). \qquad (7.5)$$

Note that the total likelihood defined in Eq. 7.3 assumes that the errors of each observable are uncorrelated (i.e., the error-covariance matrix is diagonal). If this is not the case, we have to define the likelihood as

$$\mathcal{L}(\Delta\nu, \nu_{max}, T_{eff}, [Fe/H]) = \frac{1}{(2\pi)^{n/2}\sqrt{|C|}} \exp(-\chi^2/2), \qquad (7.6)$$

where

$$\chi^2 = (\bar{o}_{\text{obs}} - \bar{o}_{\text{model}})^T \mathbf{C}^{-1}(\bar{o}_{\text{obs}} - \bar{o}_{\text{model}}), \tag{7.7}$$

with $\bar{o}_{\text{obs}}$ a vector defining all n observations, and $\bar{o}_{\text{model}}$ describing the corresponding quantities of the model. The quantity $\mathbf{C}$ is the complete error covariance matrix. In the limit that the errors are uncorrelated, $\mathbf{C}$ reduces to the diagonal form, and we recover Eq. 7.3.

The 3–6σ search radius is purely a matter of convenience: it can be as large as we wish; however, including only those models with properties close to the observed ones cuts down on the time required to compute the likelihood function. But it is advisable to define a minimum search radius independent of data uncertainties to ensure that the grid density does not affect the results too much. Moreover, if we select too few models from the grid to compare with the observations, we might significantly bias the estimated properties and their uncertainties.

The discrete nature of the likelihood function and the fact that it is not a monotonic function of any one property (e.g., radius, mass, or density) make it difficult to determine the parameters for which the likelihood function is a maximum using the usual method of fitting a continuous function to the points and determining where the maximum lies. As a result, different strategies are used by different researchers. The property under question—for instance, radius—can be determined as the likelihood-weighted average of the radii of the models used to calculate the likelihood function. Other schemes described in the literature include taking the average of models that are in the upper 90th percentile of the likelihood function. Yet others involve summing the likelihood over a suitable range of bins of the property under question, then by associating the center of each bin with the number of data in it, a probability density function is created, from which the properties are derived. While the likelihood function is generally sharply peaked, it is quite possible to get likelihood functions that are double-peaked in the case of stellar ages. In such situations, one might need a different weighting scheme.

From the definition of the likelihood in Eq. 7.3, it is clear that we can include other observations or discard some. For instance, if the grids had information about the average small-frequency separation $\delta\nu_{02}$ and we observed that quantity for a star, we could easily include this information by multiplying Eq. 7.3 by $\mathcal{L}(\delta\nu_{02})$. $\mathcal{L}(\delta\nu_{02})$, like that for most other observations, can be written in a manner analogous to Eq. 7.4 or 7.5. The exception would be an observation such as the parallax of the star. Stellar grids do not assume distances; all they list are luminosities and absolute magnitudes of the stars. Thus, we need to search for models in absolute-magnitude space, and the likelihood needs to be defined in terms of absolute magnitude. For instance, if V is the extinction-corrected apparent magnitude of a star with parallax π, then we have

$$\mathcal{L}(\pi) = \frac{1}{\sqrt{2\pi}\sigma(\pi)} \exp -\left(\frac{\left[\pi_{\text{obs}} - 10^{\left(\frac{M_{V,\text{model}} - V_{\text{obs}}}{5} - 1 \right)} \right]^2}{2\sigma^2(\pi)} \right). \tag{7.8}$$

Stellar properties (e.g., mass and radius) are related in a highly nonlinear fashion. There is a complex dependence on metallicity, age, and other physics. The relationship between the average seismic parameters and the properties of a star are, as a result, nonlinear too. This is particularly relevant for such stellar properties as age, which do not affect $\Delta \nu$ and $\nu_{\max}$ explicitly. Small errors in the measured metallicity, temperature, and seismic parameters of a star can therefore lead to large errors in the inferred properties of a star. Since a small change in the inputs can often cause large changes in the results, we have found that relying on the likelihood function derived from the central values of the observables does not give us very robust estimates of stellar properties or their associated uncertainties (i.e., one has to keep in mind that the central values are only one realization of the underlying true properties). Additional steps are therefore needed.

We have found that one way to get robust results is to rely on simulations. The first, key step is to generate many input parameter sets by adding different random realizations of Gaussian noise to the actual (central) observational input parameter set. We have found 10,000 to be a reasonable number. For each realization, as well as the central value, we determine the likelihood function and hence the most likely value of the properties as described earlier in this section. The ensemble of most likely values of a given stellar property, such as radius, constitutes the probability distribution function of that property. YB uses the the median of the distribution as the estimated value of the property, and it uses the equivalent 1σ limits from the median (i.e., that capture 68% of the values) as a measure of the uncertainties. The mean could be used too, though in cases with an asymmetric distribution function, the median of course gives more robust results. We illustrate the process of grid modeling in Figure 7.6.

7.2.2 Some Properties of Grid-Based Modeling

Metallicity Z (or equivalently, [Fe/H]) does not play any role in determining properties of stars using the direct method; however, it makes an enormous difference in grid-based results. In Figure 7.7 we compare the results obtained using $\Delta \nu$, $\nu_{\max}$, and $T_{\rm eff}$ with those obtained when metallicity is added to the inputs. Note that metallicity improves all results, but particularly those for mass and age. This is expected— the age of a star of a given mass depends on metallicity. Also, low-metallicity stars are generally bluer than their high-metallicity counterparts, changing the mass-temperature relation. For comparison we also show the results obtained with the direct method, and one without metallicity. The grid-based results are better, except for gravity, where the direct method can give marginally better results (however, that depends critically on how precise the $\nu_{\max}$ and $T_{\rm eff}$ estimates are).

That grid-based modeling uses preconstructed grids of models means that the results are somewhat model dependent. In Figure 7.8 we show the quality of grid-based results when two different grids, YY and BASTI, are used. The proxy stars are the same as those shown in Figure 7.1. As can be seen, mass estimates change slightly, radius and gravity change little, but age changes substantially when we use different grids. This in itself is not surprising. The age of a star depends sensitively on the parameters used to construct models. The inclusion of diffusion, overshoot, or both can change the age of a model of a given mass, temperature, and luminosity. The seismic properties of a star do not have any direct information about the age of

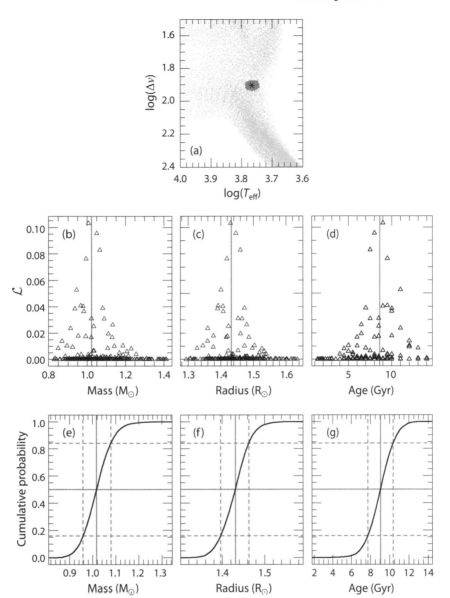

Figure 7.6. The different steps involved in the grid-modeling process. Top panel: A portion of the grid (light gray points), with the observed position marked with a star. The dark gray points mark the 10,000 realizations of the observations. Middle panels: Likelihood function as a function of mass, radius, or age for the observed central position. Bottom panels: Cumulative distribution functions in mass, radius, or age obtained for the likelihood weighted average of the observed star and the 10,000 realizations. The solid horizontal lines mark the medians of the distributions; the vertical lines mark the value of mass, radius, or age corresponding to the median. The dashed lines mark the 1σ uncertainties.

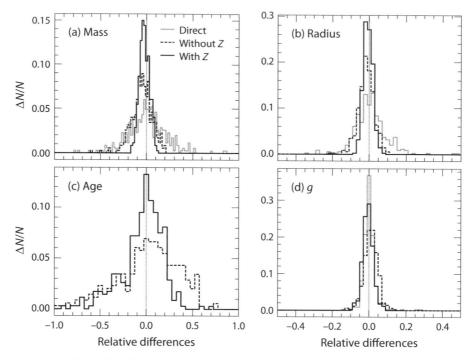

Figure 7.7. Normalized histograms of the relative differences in mass, radius, age, and surface gravity obtained with grid-based modeling with or without metallicity (Z) as an input. Results are for simulated data that closely mimic data obtained by *Kepler*. For comparison, the results of the direct method are shown in gray. Note that age cannot be obtained with the direct method.

a star, and hence they only act as indicators, not the determining factors, of age. As a result, we cannot determine ages in a manner that minimizes model dependence.

The uncertainties in some estimated stellar properties can be substantially correlated, as can be seen in Figure 7.9. These correlations are not caused by the grid-based modeling searches among a set of models; instead they arise for a more fundamental reason. Recall that $\Delta\nu \propto \sqrt{M/R^3}$, and thus if we overestimate mass, we will overestimate radius as well, as seen in Figure 7.9. Mass and age are correlated simply because stars with higher masses evolve faster; thus if we overestimate mass, we will underestimate age. Surface gravity is the least correlated: this is because $g \propto M/R^2$, and the correlations between M and R work to minimize the correlations with g.

7.2.3 Including Priors

The form of the likelihood function in Eq. 7.3 allows us to include priors (e.g., an assumed initial mass function) straightforwardly, and grid-based modeling can be done very easily in a Bayesian framework. If **v** is the set of parameters of a model (e.g., such mass, radius), and if $\mathcal{O}$ is the observed data, then from Bayes' rule, the

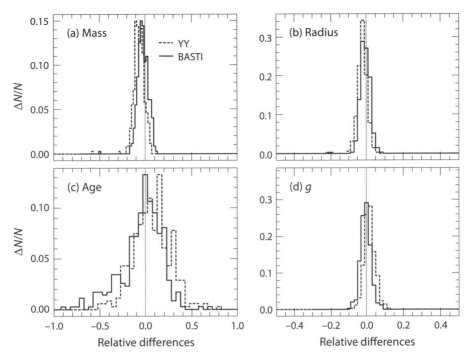

Figure 7.8. An illustration of the model-dependence of grid-modeling results. The panels show normalized histograms of the relative difference between the exact results and the results obtained with grid modeling based on two different grids, YY and BASTI. Note that age is affected the most.

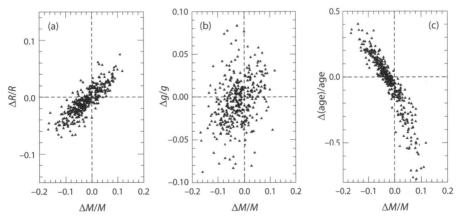

Figure 7.9. Error correlations between different model properties. Plots of the relative differences in radius, surface gravity, and age are shown as a function of the relative differences in mass. Note that both radius and age are highly correlated with mass, but interestingly, unlike the case of the direct-method calculations shown in Figure 7.2, errors in surface gravity have very low correlation with errors in mass. Although not shown here, errors in surface gravity are also not highly correlated with errors in radius.

probability distribution function of **v** given the observation $\mathcal{O}$ is

$$p(\mathbf{v}|\mathcal{O}) \propto p(\mathbf{v})\mathcal{L}(\mathcal{O}|\mathbf{v}), \tag{7.9}$$

where $\mathcal{L}(\mathcal{O}|\mathbf{v})$ is the likelihood of $\mathcal{O}$ given **v**; and $p(\mathbf{v})$ is the prior probability distribution of **v**, which represents prior knowledge of the variables. The likelihood represented by Eq. 7.3 or 7.6 is nothing but $\mathcal{L}(\mathcal{O}|\mathbf{v})$. In the case of grid modeling, $p(\mathbf{v}|\mathcal{O})$, like $\mathcal{L}(\mathcal{O}|\mathbf{v})$, is a discrete function. Thus the marginalized posterior distribution function for any parameter x can be written as

$$p(x|\mathcal{O}) \propto \int \delta(x(\mathbf{v}) - x)\,p(\mathbf{v}|\mathcal{O})\mathrm{d}\mathbf{v}, \tag{7.10}$$

where d**v** denotes integration over all the parameters in vector **v**.

7.3 FURTHER READING

The set of models used to illustrate the properties of the different methods is from

- Coelho, H. R. et al., 2015, *A Test of the Asteroseismic ν_{max} Scaling Relation for Solar-Like Oscillations in Main-Sequence and Sub-Giant Stars*, Mon. Not. R. Astron. Soc., **451**, 3011.

The models are based on the properties of *Kepler* stars analyzed by

- Chaplin, W. J. et al., 2014, *Asteroseismic Fundamental Properties of Solar-Type Stars Observed by the NASA* Kepler *Mission*, Astrophys. J. Suppl., **210**, 1.

The source of the Yale-Yonsei isochrones is:

- Demarque, P., et al., 2004, *Y^2 Isochrones with an Improved Core Overshoot Treatment*, Astrophys. J. Suppl., **155**, 667.

The source of BASTI isochrones is:

- Pietrinferni, A. et al., 2004, *A Large Stellar Evolution Database for Population Synthesis Studies. I. Scaled Solar Models and Isochrones*, Astrophys. J. Suppl., **612**, 168.

Various grid modeling pipelines have been developed; they differ in the details of how they select the parameters from the model search and how they estimate uncertainties in the results. To read about the different pipelines, readers are referred to the following papers. Some use Bayesian schemes, others do not.

- Yale-Birmingham pipeline: Basu, S., Chaplin, W. J., and Elsworth, Y., 2010, *Determination of Stellar Radii from Asteroseismic Data*, Astrophys. J., **710**, 1596; Gai, N. et al., 2011, *An In-Depth Study of Grid-Based Asteroseismic Analysis*, Astrophys. J., **730**, 63 and Basu, S. et al., 2012, *Effect of Uncertainties in Stellar Model Parameters on Estimated Masses and Radii of Single Stars*, Astrophys. J., **746**, 76.

- BASTA: Silva Aguirre, V. et al., 2015, *Ages and Fundamental Properties of* Kepler *Exoplanet Host Stars from Asteroseismology*, Mon. Not. R. Astron. Soc., **452**, 2127.
- RadEx10: Creevey, O. L. et al., 2013, *A Large Sample of Calibration Stars for Gaia: Log g from* Kepler *and CoRoT Fields*, Mon. Not. R. Astron. Soc., **431**, 2419.
- AMS: Hekker, S. et al., 2013, *Asteroseismic Surface Gravity for Evolved Stars*, Astron. Astrophys., **556**, 59.
- RADIUS: Stello, D. et al., 2009, *Radius Determination of Solar-type Stars Using Asteroseismology: What to Expect from the* Kepler *Mission*, Astrophys. J., **700**, 1589.
- SEEK: Quirion, P. O., Christensen-Dalsgaard, J., and Arentoft, T., 2010, *Automatic Determination of Stellar Parameters via Asteroseismology of Stochastically Oscillating Stars: Comparison with Direct Measurements*, Astrophys. J., **725**, 2176.
- SFP: Kallinger, T. et al., 2010, *Asteroseismology of Red Giants from the First Four Months of* Kepler *Data: Fundamental Stellar Parameters*, Astron. Astrophys., **522**, 1.
- PARAM: Da Silva, L. et al., 2006, *Basic Physical Parameters of a Selected Sample of Evolved Stars*, Astron. Astrophys., **458**, 609; Miglio, A. et al., 2013, *Galactic Archaeology: Mapping and Dating Stellar Populations with Asteroseismology of Red-Giant Stars*, Mon. Not. R. Astron. Soc., **429**, 423.

A comprehensive study of the properties of grid-based modeling (and comparison with results obtained with the direct method) can be found in the following papers:

- Gai, N. et al., 2011, *An In-Depth Study of Grid-Based Asteroseismic Analysis*, Astrophys. J., **730**, 63.
- Basu, S. et al., 2012, *Effect of Uncertainties in Stellar Model Parameters on Estimated Masses and Radii of Single Stars*, Astrophys. J., **746**, 76.

The idea of grid-based searches is not original to asteroseismology. Grid-based searches have of course been utilized for stellar parameter studies using nonasteroseismic data. Readers are referred to the following to see the variety of schemes that have been used. These also demonstrate how to use Bayesian schemes to perform grid-based searches:

- Jørgensen, B. R., and Lindegren, L., 2005, *Determination of Stellar Ages from Isochrones: Bayesian Estimation Versus Isochrone Fitting*, Astron. Astrophys., **436**, 127.
- Takeda, G. et al., 2007, *Structure and Evolution of Nearby Stars with Planets. II. Physical Properties of* ∼1000 *Cool Stars from the SPOCS Catalog*, Astrophys. J. Suppl., **168**, 297.

7.4 EXERCISES

1. The archive `frequencies.tar.gz` contains files with frequencies of different models. The global properties of the models are listed in file `stellar_properties.txt`.

(a) Calculate $\Delta\nu$ and ν_{max} for the models using the scaling relations. You may use $\Delta\nu_\odot = 135\ \mu Hz$ and $\nu_{max,\odot} = 3{,}090\ \mu Hz$.

(b) Calculate $\Delta\nu$ of the models using the mode frequencies around ν_{max}. How different are these values from those obtained in part (a)?

(c) Explore what happens to the value of $\Delta\nu$ as different frequency ranges and weighting schemes are used to calculate the quantity.

2. File grid_YY.txt has a subset of the YY grid.

(a) The Astero Fitting at Low Angular degree Group (asteroFlag) is an international collaboration of asteroseismologists. The group produced synthetic data for testing different methods of determining stellar parameters. Use the synthetic stellar properties in the paper by Stello et al., 2009, *Astrophys. J.*, **700**, 1589 to see how well you can use the YY grid to determine the properties of the stars.

(b) Table 7.1 lists $\Delta\nu$, ν_{max}, metallicity, and temperature for some synthetic stars. Determine the mass, radius, and age of these stars using both the direct method and a grid-based one using the YY grid. How do the results change depending on whether you use only ($\Delta\nu$, ν_{max}) or ($\Delta\nu$, ν_{max}, [Fe/H], T_{eff})? Note that the data have one realization of the error added.

TABLE 7.1. DATA FOR PROBLEM 2(B)

Star	$\Delta\nu$ μHz	ν_{max} μHz	[Fe/H] dex	T_{eff} K
1	178.1 ± 4.5	$5{,}123 \pm 256$	-0.43 ± 0.10	$5{,}407 \pm 100$
2	169.7 ± 4.2	$5{,}622 \pm 281$	0.17 ± 0.10	$4{,}830 \pm 100$
3	151.8 ± 3.8	$4{,}128 \pm 206$	-0.24 ± 0.10	$5{,}388 \pm 100$
4	133.8 ± 3.3	$3{,}675 \pm 184$	-0.11 ± 0.10	$6{,}045 \pm 100$
5	129.7 ± 3.2	$3{,}433 \pm 172$	0.04 ± 0.10	$5{,}735 \pm 100$
6	126.1 ± 3.2	$3{,}317 \pm 166$	-0.35 ± 0.10	$6{,}574 \pm 100$
7	96.6 ± 2.4	$2{,}474 \pm 124$	-0.30 ± 0.10	$8{,}352 \pm 100$
8	88.9 ± 2.2	$2{,}083 \pm 104$	-0.33 ± 0.10	$6{,}405 \pm 100$
9	83.9 ± 2.1	$2{,}000 \pm 100$	-0.18 ± 0.10	$5{,}923 \pm 100$
10	64.1 ± 1.6	$1{,}421 \pm 71$	-0.12 ± 0.10	$5{,}882 \pm 100$
11	71.4 ± 1.8	$1{,}592 \pm 80$	-0.57 ± 0.10	$6{,}529 \pm 100$
12	35.2 ± 0.9	647 ± 32	-0.48 ± 0.10	$7{,}104 \pm 100$
13	9.6 ± 0.2	123 ± 6	-0.37 ± 0.10	$4{,}913 \pm 100$
14	5.9 ± 0.2	56 ± 3	0.09 ± 0.10	$4{,}414 \pm 100$
15	3.6 ± 0.1	44 ± 2	-0.41 ± 0.10	$4{,}986 \pm 100$

8 Interpreting Frequencies of Individual Modes: Comparing Frequencies

The frequencies of individual modes can be measured for stars with high SNR power spectra. A careful analysis of the frequencies allows us to learn much more about the star than merely its mass and radius.

Stellar oscillation frequencies can be analyzed in a simple manner: model the star of known metallicity and effective temperature, calculate the frequencies of the models, compare these frequencies with the observed frequencies, and in some manner get the "best fit" model. The properties of the model then represent the properties of the star. While this may sound simple, the large number of model parameters that one can play with makes the exercise quite time consuming.

In this chapter we discuss different ways in which one can use the frequencies to determine stellar properties. We discuss straightforward comparisons of observed frequencies with those of stellar models, and we also discuss the limitations and challenges of such comparisons. Different approaches are used to model stars whose frequencies have been observed. Some methods are similar to what is done if only the nonseismic properties of a star (e.g., $T_{\rm eff}$, [Fe/H], and perhaps luminosity) are known: make a model, calculate the observables (including frequencies), compare with the data, change the model parameters, and repeat until a satisfactory fit is obtained. Other methods make use of the fact that the average asteroseismic data give a good indication of the mass and radius of the star being modeled, and so these methods construct models around that mass and radius. Yet other methods involve optimization schemes. However, even the most complex optimization schemes used thus far do not explore the effects of all inputs and parameters that can be varied while constructing stellar models.

8.1 CHOOSING THE PHYSICS OF MODELS

A very large number of inputs are required to construct stellar models. The first step in modeling is to decide what inputs to use. The major microphysics inputs—nuclear reaction rates, radiative opacities and the equation of state—are usually chosen once and then kept fixed. Other inputs, such as the mixing length parameter or initial abundances, are kept fixed in some calculations, but allowed to vary in others. The major inputs that need to be fixed (or varied, depending on the way the modeling is done) are:

(1) The metallicity scale. This defines Z/X for [Fe/H] = 0. It is usual to use solar scaling, but the solar metallicity scale is somewhat uncertain. The common choices are usually the Grevesse and Sauval value of $(Z/X)_\odot = 0.023$.[1] or the Asplund et al. value of $(Z/X)_\odot = 0.018$.[2] The relative abundances of different elements in the two cases are also different. This means that a choice of one or the other requires a change of opacity tables being used. If detailed abundance measurements are available, one can use them; such details are often not available for most stars. For low-metallicity stars one also needs to consider whether the relative abundance of α elements (e.g., oxygen) with respect to iron is larger than that in the Sun. In the case of the Sun, the ratio [α/Fe] is zero by definition. Lower metallicity stars generally have higher values of [α/Fe], which affects opacity. Assuming a value of [α/Fe] is only important when we do not have abundances of individual elements. In all cases, radiative opacities that are consistent with the abundances should be used.

(2) Diffusion and gravitational settling. Another issue to consider is whether the diffusion and gravitational settling of helium and heavy elements will be included. The inclusion of diffusion increases computation time. Sometimes choices are mixed For example, in the case of YY isochrones[3] the diffusion of helium was included, but diffusion of other elements was not. This choice is not uncommon[4] and avoids issues related to excessive loss of heavy elements from the thin convection zones of relatively hot stars. Once the choice over including diffusion is taken, the next step is deciding which set of diffusion coefficients to choose. The set of coefficients is usually not changed in the modeling process.

(3) Initial abundances. Stars are usually modeled by evolving them from the pre-main sequence or the zero-age main sequence (ZAMS). In either case one needs to decide what the initial helium abundance Y_0 of the model is. Y_0 affects the luminosity of a star. It is usual to assume that the initial helium abundance is a function of metallicity (i.e., to use a simple chemical evolution model, as is done when constructing isochrones). For instance, the YY isochrones assume that $Y = 0.23 + 2Z$, while the models from the Dartmouth group[5] assume $Y = 0.245 + 1.54Z$ for their normal models. However, it should be noted that in some of the iterative optimization methods, Y_0 is kept as a free parameter and allowed to vary. A complication arises when diffusion is included: since the matching models must have the *current* surface metallicity of the star, they need to be modeled with a higher initial metallicity and helium abundance than in a model without diffusion. Often some trial-and-error is involved in this stage.

---

[1] Grevesse and Sauval, 1998, *Sp. Sci. Rev.*, **85**, 161.

[2] Asplund, et al., 2009, *Ann. Rev. Astron. Astrophys.*, **47**, 481.

[3] Demarque et al., 2004, *Atrophys. Sp. Sci.*, **155**, 667.

[4] See for example, Gilliland et al., 2011, *Astrophys. J.*, **726**, 2 as well as Mathur et al., 2012, *Astrophys. J.*, **749**, 152.

[5] Dotter et al., 2008, *Atrophys. Sp. Sci.*, **178**, 89.

(4) The mixing length parameter. The mixing length parameter α affects the radius of a low-mass star, and hence this is potentially one of the most important parameters in modeling. However, its value is uncertain because there is no clear physics-based way of choosing it. It is common to use the solar-calibrated value of α, as has been done for all published isochrones. However, some methods of constructing models to fit asteroseismic data do allow α to change in order to get the best fit. Simulations of stellar convection indicate that properties of convection change with metallicity, $T_{\rm eff}$, and $\log g$; hence, allowing α to deviate from the solar-calibrated value is probably the better choice.

(5) Core overshoot. For stars that may have a convective core, one must decide whether to include convective overshoot. If overshoot is included, one needs to decide on the value of $\alpha_{\rm ov}$.

This list of parameters is not exhaustive. One could, for example, include undershoot below envelope convection zones, calculate the effects of rotation on the stellar structure, or examine the effects of chemical mixing due to rotation-induced flows. Once decisions have been made about the different inputs, one can construct models, calculate the frequencies, and compare them with observations.

8.2 EFFECTS OF DIFFERENT MODEL PARAMETERS

How different model parameters affect evolutionary sequences of models is a well-studied problem; the effects of metallicity, helium abundance, diffusion, and so forth, on models are well known. Here we look at this from a different point of view: in Chapter 7 we showed how the average asteroseismic parameters can provide good estimates of the mass and radius of a star, thus either implicitly or explicitly, our aim is to construct models of a given mass and radius by some means. We therefore examine here how the different free parameters affect models that have a given mass and radius.

To illustrate the effects of the different choices, we show results obtained for a set of $1M_\odot$ models with different effective temperatures (between 5,370 and 6,170 K), and metallicity ([Fe/H], between -0.1 and $+0.1$); all models have a radius of $1R_\odot$. The models were constructed in two different ways. In the first case the mixing length parameter α was kept fixed, and the initial helium abundance Y_0 was changed to obtain a model with the required radius and temperature; in the second case Y_0 was fixed, and α was changed to construct the models. The variations of Y_0 and α gave models with different ages, even when their mass, radius, $T_{\rm eff}$, and metallicity were the same.

Figure 8.1 shows how Y_0 varies on the $T_{\rm eff}$-[Fe/H] plane for these $1M_\odot$, $1R_\odot$ models when they are constructed with the solar-calibrated value of the mixing-length parameter α. What can be seen is that if we were to draw contours of constant Y_0, they would lie diagonally in this plane, showing that for a given value of $T_{\rm eff}$, we need higher values of Y_0 to construct the models as we increase the metallicity. We also see a large region where the value of Y_0 is below primordial levels. If we believe that the amount of helium in the universe increases continuously because of hydrogen fusion, we should not expect to see stars that are in this region of the plane.

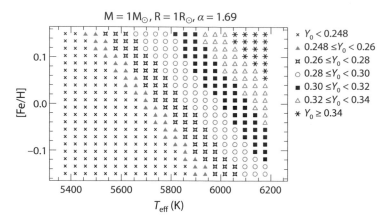

Figure 8.1. The initial helium abundance Y_0 needed to construct $1M_\odot$ models that have a radius of $1R_\odot$ using a solar-calibrated mixing-length parameter α for a range of effective temperatures and metallicities. The models were constructed without diffusion and with $\alpha = 1.69$. Small crosses mark Y_0 values that are unphysical, in the sense that Y_0 is lower than the primordial helium abundance.

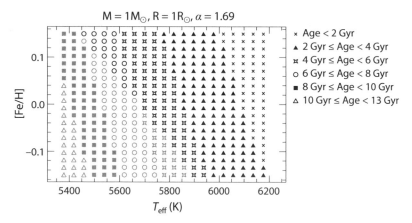

Figure 8.2. The same as Figure 8.1, but showing age. Gray points indicate models with subprimordial (i.e., unphysical) initial helium abundance.

Figure 8.2 shows the age distribution of the models, and again the age contours are almost diagonal. If these models represented real stars, the initial helium abundance would provide limits to their ages for a given T_{eff} and metallicity. The typical uncertainty in stellar effective temperatures is about 80–100 K, and that in metallicity is of the order of 0.1 dex; thus even in the absence of uncertainties in mass and radius, we will get models with a spread in age and initial helium abundance.

However, Figures 8.1 and 8.2 only show results for one particular value of the mixing-length parameter. The demarcation between the allowed and the forbidden region changes as α changes, and this can be seen in Figure 8.3. The figure shows age and Y_0 as a function of T_{eff} for different values of the mixing length and metallicity. The Y_0–T_{eff} relation is almost linear, but age is not. For a given value of

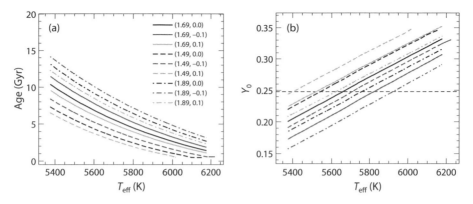

Figure 8.3. Left panel: Age as a function of $T_{\rm eff}$ for $1{\rm M}_\odot$, $1{\rm R}_\odot$ models for three different values of α and metallicity. The line styles correspond to α and gray scale to metallicity. The pairs of numbers in the legend denote $(\alpha, [{\rm Fe/H}])$. Right panel: Y_0 as a function of $T_{\rm eff}$. The legend is the same as in the left panel. The horizontal dashed line marks the primordial helium abundance.

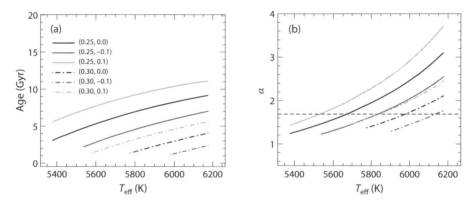

Figure 8.4. Left panel: Age as a function of $T_{\rm eff}$ for $1{\rm M}_\odot$, $1{\rm R}_\odot$ models for two different values of Y_0 and three values of metallicity. The line styles correspond to Y_0 and gray scale to metallicity. The pairs of numbers in the legend denote $(Y_0, [{\rm Fe/H}])$. Right panel: α as a function of $T_{\rm eff}$. The legend is the same as in the left panel. The horizontal dashed line marks the solar-calibrated value of α.

metallicity, age increases as α increases, as does Y_0. Thus the "forbidden" region in the $T_{\rm eff}$–[Fe/H] plane changes with α. Since α is essentially unconstrained, the forbidden region is not as strong a constraint as it would be if we had independent constraints on α. The behavior of the $1{\rm M}_\odot$, $1{\rm R}_\odot$ models constructed with a given value of Y_0 are shown in Figure 8.4. Increasing Y_0 decreases age at a given metallicity and temperature; and increasing Y_0 requires us to decreases α to make a model with the required radius.

Of course what we are concerned with is the behavior of the mode frequencies of these models. Since the models have the same mass and radius, all have the same $\Delta\nu$ to first order. The scaling relation would imply that they have exactly the same $\Delta\nu$, though variations in $T_{\rm eff}$ and [Fe/H] do cause small deviations. It is far more interesting to look at the frequencies themselves.

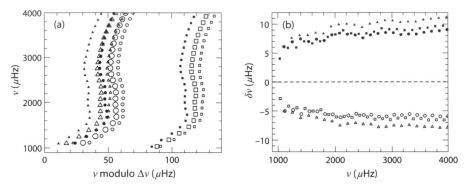

Figure 8.5. Left panel: Échelle diagram for three models with the same mass, radius, $T_{\rm eff}$, and metallicity but constructed with three different values of α. The large, unfilled symbols are for the solar calibrated $\alpha = 1.69$; small, unfilled symbols for $\alpha = 1.89$; and small filled symbols for $\alpha = 1.49$. The circles, squares, and triangles represent modes of $l = 0$, 1, and 2, respectively. The models have ages 4.29 Gyr, 6.20 Gyr, and 2.29 Gyr, and Y_0 of 0.274, 0.257, and 0.293 for $\alpha = 1.69$, 1.49, and 1.89, respectively. Right panel: Frequency differences between the $\alpha = 1.69$ and $\alpha = 1.89$ models plotted as open symbols and those between the $\alpha = 1.69$ and $\alpha = 1.49$ models plotted as filled symbols. The oscillatory behavior is a result of the different helium abundances in the models.

The left panel of Figure 8.5 shows the échelle diagram of three models with the same temperature (5,817 K), metallicity ([Fe/H] $= 0$), mass (1M$_\odot$), and radius (1R$_\odot$). The models were constructed with different mixing length parameters and consequently have different initial helium abundances and ages. What we can see is that although the large separation of the models is the same to better than 1%, the frequencies are not. The right panel of Figure 8.5 shows the frequency differences between one of the models and the other two; the differences are large, and if data on individual frequencies were available, we should, in principle, be able to detect these differences.

What happens when we have models of the same mass, radius, $T_{\rm eff}$, but different metallicities is more interesting. We show an échelle diagram and frequency differences for such a case in Figure 8.6. All have 1M$_\odot$, 1R$_\odot$, $\alpha = 1.69$, and $T_{\rm eff}$=5,817 K. They have different metallicities ([Fe/H] $= -0.1$, 0, and 0.1) and consequently different ages (5.03 Gyr, 4.29 Gyr, and 3.76 Gyr, respectively) and different Y_0 (0.247, 0.274, and 0.296, respectively). Unlike in Figure 8.5, the frequencies are very similar, and we can barely see the differences in the échelle diagram. The frequency-difference figure shows that the differences are indeed very small. While the differences are larger than usual levels of uncertainty in the observed frequencies, they are small enough to cause difficulties in fitting models unless the metallicity of the star is known precisely. Thus precise metallicity measurements are required to carry out reliable asteroseismic analyses.

Changing mass at a given radius, or radius at a given mass, of course will result in large changes in the properties of the models. The large separation will be different, and substantial changes will occur in the mode frequencies. At a given radius, temperature, and metallicity, a higher-mass star will have a lower age than a lower-mass star. A higher-mass star will also have a lower initial helium abundance. The forbidden region (i.e., the region where the initial helium abundance is less than

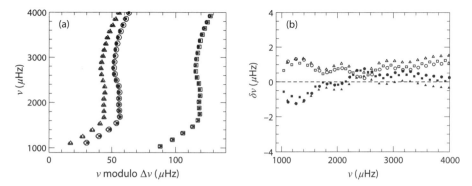

Figure 8.6. Left panel: Échelle diagram for three models with the same mass, radius, T_{eff}, and α but constructed with three different values of [Fe/H]. The large, unfilled symbols are for [Fe/H] = 0; small, unfilled symbols for [Fe/H] = +0.1; and small, filled symbols for [Fe/H] = −0.1. The circles, squares, and triangles represents modes of $l = 0$, 1 and 2, respectively. Right panel: Frequency differences between the [Fe/H] = 0 model and [Fe/H] = +0.1 models are plotted as open symbols, and those between the [Fe/H] = 0 and [Fe/H] = −0.1 models are plotted as filled symbols.

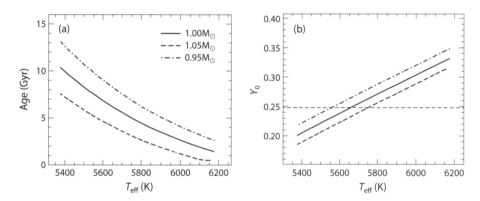

Figure 8.7. Left panel: Age as a function of T_{eff} for models with different masses but the same radius ($1R_{\odot}$), metallicity ([Fe/H] = 0), and mixing-length parameter ($\alpha = 1.69$). Right panel: Y_0 as a function of T_{eff} for the models shown in the left panel. The legend is the same as in the left panel. The horizontal dashed line marks the primordial helium abundance.

the primordial helium abundance) occurs at a higher temperature. This is illustrated in Figure 8.7. For models of the same mass, T_{eff}, [Fe/H], and α, age increases with increasing radius, while Y_0 decreases. The edge of the forbidden region shifts to higher temperatures. This is illustrated in Figure 8.8.

8.3 A COMPLICATING FACTOR: THE SURFACE TERM

The description in the previous section may have given the impression that modeling a star to fit its oscillation frequencies is very similar to modeling a star that does not have any seismic data, and that the only difference is that we have access to

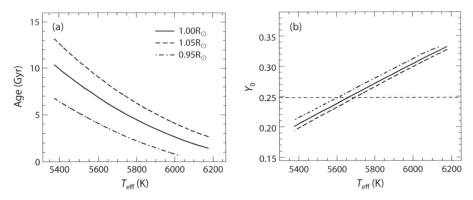

Figure 8.8. Left panel: Age as a function of T_{eff} for models with different radii but the same mass ($1M_\odot$), metallicity ([Fe/H] = 0), and mixing-length parameter ($\alpha = 1.69$). Right panel: Y_0 as a function of T_{eff} for the models in the left panel. The legend is the same as in the left panel. The horizontal dashed line marks the primordial helium abundance.

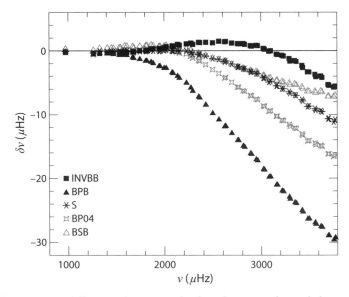

Figure 8.9. Frequency differences between solar low-degree modes and those of different published solar models. INVBB is a seismic model; models BPB, S, BP04, and BSB are standard models. Note that the frequency differences are small at low frequencies and increase with frequency. Model source: Model INVBB is from Antia et al., 1996, *Astron. Astrophys.*, **307**, 609; Model BPB from Basu et al., 2000, *Astrophys. J.*, **529**, 1084; Model S Christensen-Dalsgaard et al., 1996, *Science*, **272**, 1286; Model BP04 from Bahcall et al., 2005, *Astrophys. J.*, **618**, 1049; Model BSB(GS98) from Bahcall et al., 2006, *Astrophys. J. Suppl.*, **165**, 400.

data that probe the interior of the star. However, comparing the frequencies of models with those of real stars is complicated by our ignorance of how to model properly the near-surface layers of a star. This difficulty gives rise to a frequency-dependent frequency offset between models and observed frequencies. It can be seen in Figure 8.9, which shows the frequency differences between the Sun and several published solar models. Such a frequency-dependent frequency difference is also seen between different models of the Sun. These surface-related differences can easily hide frequency differences caused by differences in the internal structure.

We cannot properly model the observed frequencies of a star for two reasons. The first is that it is usual to calculate the frequencies of models assuming that the modes are fully adiabatic, when in reality adiabaticity breaks down close to the surface. This can of course be rectified by calculating fully nonadiabatic frequencies; however, the inability to model the interaction of convection with pulsations will still result in systematic errors. The second reason is that there are large uncertainties in modeling the near-surface layers of stars. The sources of these uncertainties include the way convection is treated in stellar models. Models are usually constructed using the mixing-length approximation or its variants, and these model the region of inefficient convection close to the surface in a rather crude manner. Additionally, most models do not include the dynamical effects of convection and pressure support due to turbulence; these are again important in the near-surface layers. The treatment of stellar atmospheres is also crude, and one generally uses simple atmospheric models, such as the Eddington T-τ relation defined in Eq. 2.32, or others of a similar nature. These atmospheric models are often fully radiative gray atmospheres and do not include convective overshoot from the interior. Some of the microphysics inputs that affect the near-surface regions also can be uncertain, in particular low-temperature opacities because of difficulties in including all relevant molecular lines and the impact of grains.[6] Since all these factors are relevant in the near-surface regions, their combined effect is usually called the *surface effect*, and the differences in frequencies they introduce is referred to as the *surface term*. Figure 8.10 shows the frequency differences between solar models constructed with different surface physics but identical interior physics, and we can see that the frequency differences are a function of frequency alone.

For low- and intermediate-degree modes, the surface term depends only on the frequency of the mode once mode inertia is taken into account. And usually, the surface term is small at low frequencies and increases rapidly with frequency. It is also a smooth function of frequency—the smoothness can be explained in terms of the signature of acoustic glitches described in Section 3.4. Because the surface term can be thought of as the effect of a glitch at shallow depths (i.e., at small values of τ_g), the signature of this glitch is an oscillatory function with a long wavelength, which in the observed frequency range can be represented as a low-degree polynomial in frequency. This can be seen clearly in the top panel of Figure 3.17.

The surface term makes frequency comparisons difficult. Figure 8.11 shows what happens to the frequency differences in the right panels of Figure 8.5 and Figure 8.6 when we simulate a surface term in the frequencies of one of the models. The frequency differences now look very different. The steep frequency-dependent component hides some of the shorter-scale frequency modulation caused by differences in the helium abundance. Any criterion, such as a χ^2 metric, that quantifies the difference between the models becomes much larger, but worse, no changes in model parameters will allow us to fit the data. To make meaningful

---

[6]Astrophysical grains, like other dust grains, are aggregates of molecules that exist in relatively cool environments and can have sizes in the submicron to several microns range. Grains scatter light and thus add to opacities.

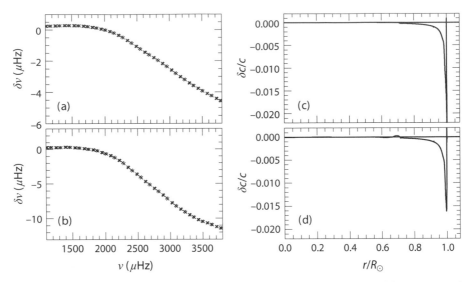

Figure 8.10. Top-left panel: Frequency differences between two solar models constructed with different model atmospheres but otherwise identical physics. One model was constructed with the Eddington T-τ relationship and the other with the Krishna Swamy relationship. The differences are in displayed as Eddington − Krishna Swamy. Top-right panel: Relative sound-speed difference between the two models shown in the top-left panel. Bottom-left panel: Frequency differences between two solar models constructed with different prescriptions of convection but otherwise identical physics. One model was constructed with the standard mixing length-formalism, the other with the Canuto and Mazzitelli formalism. The differences are displayed as mixing-length–Canuto Mazzitelli. (MLT − CM). Bottom-right panel: Relative sound-speed difference between the two models in the bottom-left panel. Sources: Krishna Swamy relationship, Krishna Swamy, K. S., 1966, *Astrophys. J.*, **145**, 174; Canuto and Mazzitelli Formalism, Canuto, V. M., and I. Mazzitelli, 1991, *Astrophys. J.*, **370**, 295.

comparisons between the frequencies of a model and that of a star, the frequencies of the model need to be suitably corrected to remove the surface term, and the corrected frequencies used to calculate a goodness-of-fit statistic. Uncorrected frequencies could result in accepting models that are not correct.

There is, however, no first principles way of calculating this correction—if there were one, we would not have been talking of correcting the frequencies, we would have made the correction an integral part of the frequency calculations. Representing the surface term as any low-degree polynomial is not enough—that could remove frequency differences that arise as a result of real differences in structure deeper in the star. When we have data on low-frequency modes, the correction is easier to do, but most often, we only have frequencies around ν_{max}. When we have frequency differences of the kind shown in the right panel in Figure 8.11, and frequencies only around ν_{max}, trying to remove the surface term usually leads to the result that both models, the one represented with open symbols and the other with the filled symbols, give equally good fits. This is particularly true if we were to fit a polynomial to the surface term. As a result, some specific methods have been developed to correct for the surface term. Some methods rely on first determining the surface term for solar models and assuming that the term can be scaled and applied to other stars.

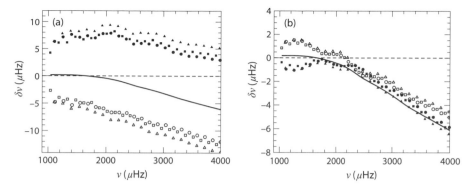

Figure 8.11. Left panel: The same as the right panel in Figure 8.5 but with a simulated surface term (shown as the continuous curve) in the frequencies of the reference model. Right panel: The same as the right panel in Figure 8.6 but again with a surface term in frequencies of the reference model.

One of the more common ways of correcting the surface term is the formulation of Kjeldsen et al. (2008; see Further Reading at the end of this chapter). This correction was motivated by the fact that the surface term for model S (see Figure 8.9) is negligible at low frequencies, but around ν_{max} it can be expressed as a power law in ν. The assumption made is that the same power law can be applied to the surface terms between other stars and their models. The surface term between any star and its model is written as

$$\nu_{nl}^{\mathrm{obs}} - \nu_{nl}^{\mathrm{mod}} = a \left[\frac{\nu_{nl}^{\mathrm{obs}}}{\nu_0} \right]^b ; \tag{8.1}$$

where the left-hand side is the surface term. Thus, the "corrected" frequency is given by

$$\nu_{nl}^{\mathrm{corr}} = \nu_{nl}^{\mathrm{mod}} + a \left[\frac{\nu_{nl}^{\mathrm{obs}}}{\nu_0} \right]^b . \tag{8.2}$$

The exponent b in Eqs. 8.1 and 8.2 is determined from the solar surface term. The reference frequency ν_0 is usually assumed to be ν_{max}. The factor a, though formally the frequency difference at $\nu_{\mathrm{obs}} = \nu_0$, is often determined in an average manner.

Another method uses a direct scaling of the solar surface term. The solar surface term can be represented as a $\nu_{nl,\odot}$-$\delta\nu_{nl,\odot}$ relation, where $\delta\nu_{nl,\odot}$ is the solar surface term. To use this relation as a surface term correction applicable to models of other stars, both $\nu_{nl,\odot}$ and $\delta\nu_{nl,\odot}$ are scaled to the mass and radius of the stellar model under consideration using the homology scaling

$$r = \frac{\langle \Delta\nu^{\mathrm{mod}} \rangle}{\langle \Delta\nu_\odot \rangle}, \tag{8.3}$$

where the angular brackets denote the average values of the quantities. The resulting $r\nu_{nl,\odot}$-$r\delta\nu_{nl,\odot}$ are then used to correct the stellar model for the surface term. The corrected frequencies are given by

$$\nu_{nl}^{\text{corr}} = \nu_{nl}^{\text{mod}} + \beta r \delta \nu_{nl,\odot}, \tag{8.4}$$

with $r \delta \nu_{nl,\odot}$ evaluated at $r \nu_{nl,\odot} = \nu_{nl}^{\text{obs}}$. The factor β is selected to minimize

$$\chi_{\text{sur}}^2 = \sum_{nl} \frac{\left(\nu_{nl}^{\text{obs}} - \nu_{nl}^{\text{corr}}\right)^2}{\sigma^2(\nu_{nl})}. \tag{8.5}$$

A variant of this method with r defined as $\nu_{\text{max}}^{\text{mod}}/\nu_{\text{max},\odot}$ is also used.

A subtlety that is missing in Eqs. 8.1 and 8.4 is that these equations, strictly speaking, apply only to radial modes, though they also work well for other low-degree p modes. Intermediate-degree p modes cannot be corrected in this manner. Since we only observe low-degree modes for stars other than the Sun, this is not an issue. Another problem is that the equations cannot be directly applied to mixed modes either. Mixed modes have high mode inertia and are thus not affected as much by near-surface uncertainties as p modes. In the case of the Sun, we know that the surface term appears to be different for intermediate-degree (low inertia) modes than for the low-degree (high inertia) modes. However, the degree dependence of the surface term can be corrected by scaling the frequency differences between the Sun and solar models by the inertia ratio, Q_{nl} (see Eq. 6.143):

$$Q_{nl} = \frac{I_{nl}(\nu)}{I_0(\nu)}, \tag{8.6}$$

where $I_{nl}(\nu)$ is the mode inertia of a mode of degree l, radial order n, and frequency ν, as defined in Eq. 3.36; and $I_0(\nu)$ is the mode inertia of a radial mode at the same frequency, obtained by interpolating between the mode inertia of radial modes of different orders (and hence different frequencies). Mixed modes can be corrected in the same manner. Thus Eq. 8.2 has to be modified to

$$\nu_{nl}^{(\text{corr})} = \nu_{nl}^{\text{mod}} + Q_{nl}^{-1} a \left[\frac{\nu_{nl}^{\text{obs}}}{\nu_{nl}}\right]^b, \tag{8.7}$$

and Eq. 8.4 needs to be modified to

$$\nu_{nl}^{\text{corr}} = \nu_{nl}^{\text{mod}} + Q_{nl}^{-1} \beta r \delta \nu_{nl,\odot}, \tag{8.8}$$

to make them applicable to all modes. One needs to remember that the solar surface term is not a well-defined quantity, but that it changes with the surface physics used in models, as shown in Figure 8.9. Thus scaling the solar surface term to correct for the surface term of other stars requires us to first construct a solar model with the same physics that is used to construct models of other stars.

More recently two new forms for the two surface term corrections have been suggested:

$$\nu_{nl}^{\rm obs} - \nu_{nl}^{\rm mod} = \frac{c}{I_{nl}} \left(\frac{\nu^{\rm mod}}{\nu_{\rm ac}^{\rm mod}} \right)^3 , \qquad (8.9)$$

and

$$\nu_{nl}^{\rm obs} - \nu_{nl}^{\rm mod} = \frac{1}{I_{nl}} \left[c_3 \left(\frac{\nu^{\rm mod}}{\nu_{\rm ac}^{\rm mod}} \right)^3 + c_{-1} \left(\frac{\nu^{\rm mod}}{\nu_{\rm ac}^{\rm mod}} \right)^{-1} \right] , \qquad (8.10)$$

where I_{nl} is the mode inertia; $\nu_{\rm ac}$ is the acoustic cut-off frequency; and c, c_3, and c_{-1} are constants obtained by fitting the frequency differences. Tests show that Eq. 8.10 works very well over a large portion of the HR diagram. The most useful feature of these forms is that they do not require a solar model to determine any of the parameters. Another advantage is that the mode inertia scaling is already present.

A problem often faced when applying the surface term is that we can obtain two models with differing properties that give the same quality of fit to observations. Thus we need to use other diagnostics to separate out the model that really explains the data from the model that does not. For this we use specific combinations of frequencies. The frequency combinations probe the interior of the star in different ways; for instance, some combinations are most sensitive to conditions in the core, others to the interface between radiative and convective regions. The combinations therefore provide diagnostics of different stellar properties. The one we have already encountered is the average large-frequency separation $\Delta\nu$, which tells us about the average density of a star. Some of these combinations are relatively insensitive to the surface term and can be used to verify that the model we think fits the data actually does so. We discuss some of these combinations in Section 9.1.

8.4 COMMON WAYS OF CONSTRUCTING MODELS TO FIT FREQUENCIES

The parameter space of models is large, and as a result, many methods to obtain a model that could match the observed frequencies have been developed. Some of the most commonly used ones are discussed here.

8.4.1 Method A: Constructing Evolutionary Sequences of Models

Using evolutionary sequences of models is not new in stellar astrophysics, and its use in asteroseismology to fit frequencies is merely an extension of what has been done before.

The process starts with making decisions on the physics and parameters, such as mixing length and initial helium abundances. An evolutionary sequence of models is constructed for a reasonably fine grid of masses over a range of metallicities covered by spectroscopic metallicity estimates. For each model along an evolutionary sequence, one calculates frequencies, corrects them for the surface term using one of the methods described in Section 8.3, and then calculates a χ^2 for the goodness of fit.

Thus we calculate

$$\chi_\nu^2 = \frac{1}{N-1} \sum_{nl} \left(\frac{\nu_{nl}^{\text{mod}} - \nu_{nl}^{\text{obs}}}{\sigma(\nu_{nl})} \right)^2, \tag{8.11}$$

where ν_{nl}^{mod} are the *surface-term corrected* frequencies of the model, ν_{nl}^{obs} the observed frequencies, and N the total number of frequencies. In this way, one can determine the model $\mathcal{M}_{\text{min}}$ in the evolutionary sequence for a given mass and metallicity that has the smallest value of χ_ν^2. Models along an evolutionary track are usually constructed at a discrete set of ages. It is therefore quite possible that smallest χ_ν^2 will actually be between two adjacent models. To determine the properties of the best-fitting missing model, one can use the frequencies of model $\mathcal{M}_{\text{min}}$ by scaling

$$\nu_{nl}^{\text{mod}}(\text{best}) = r \nu_{nl}^{\text{mod}}(\mathcal{M}_{\text{min}}), \tag{8.12}$$

where the factor r is determined so that χ_ν^2 becomes even lower than that obtained for model $\mathcal{M}_{\text{min}}$. Recall from Chapter 3 that the frequencies of a model scale as $\sqrt{M/R^3}$. Since we are dealing with models along an evolutionary track (i.e., models with the same mass and metallicity), Eq. 8.12 implies that the radius of the best-fitting model can be obtained as

$$R(\text{best}) = r^{-2/3} R(\mathcal{M}_{\text{min}}). \tag{8.13}$$

The other properties of the best-fitting model, such as T_{eff} and age, can be determined by interpolating in radius to this value. If the grid of models along an evolutionary track is dense enough, a linear interpolation suffices.

Identifying the model along each evolutionary track that best fits the frequencies is not enough, since the model may not have the correct temperature. Hence the next step is to calculate the total χ^2. Assuming that both T_{eff} and [Fe/H] are observed, the total χ^2 can be defined as

$$\chi^2 = \chi_\nu^2 + \left(\frac{T_{\text{eff}}^{\text{mod}} - T_{\text{eff}}^{\text{obs}}}{\sigma(T_{\text{eff}})} \right)^2 + \left(\frac{[\text{Fe/H}]^{\text{mod}} - [\text{Fe/H}]^{\text{obs}}}{\sigma([\text{Fe/H}])} \right)^2. \tag{8.14}$$

The relative weights of the different terms in Eq. 8.14 can be varied. Often the spectroscopic parameters are taken as a group to define χ_{spec}^2, in which case the total χ^2 is just $\chi_\nu^2 + \chi_{\text{spec}}^2$.

Once the total χ^2 of the best-fitting model, for each mass sequence and for all metallicities under consideration is calculated, one can look for the global minimum in χ^2. Usually one plots the value of χ^2 as a function of different properties (e.g., mass and age) to determine the best-fitting properties. One can now look to obtain a global minimum. The average values of the model properties are usually determined as likelihood-weighted averages. Recall that the likelihood $\mathcal{L}$ is related to χ^2 by

$$\mathcal{L} \propto \exp(-\chi^2/2). \tag{8.15}$$

The uncertainties are also defined as the likelihood-weighted standard deviation.

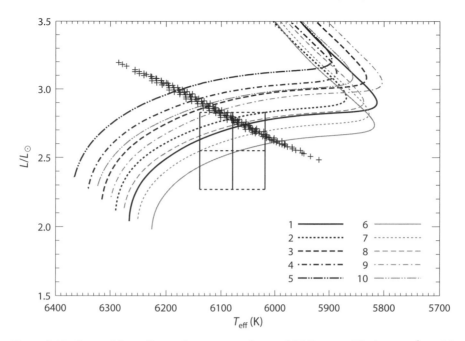

Figure 8.12. Some of the stellar tracks constructed to model HD 17156. The lowest χ_ν^2 models from each evolutionary sequence are marked with crosses. The box is centered on the L and T_{eff} estimates of the star. Reproduced with permission from Gilliland et al., 2011, *Asteroseismology of the Transiting Exoplanet Host HD 17156 with Hubble Space Telescope Fine Guidance Sensor,* Astrophys. J., **726**, 2, copyright American Astronomical Society.

The method applied to the star HD 17156 is shown in Figures 8.12 and 8.13. The goal of the analysis was to estimate the mean density of the star, though mass, radius and age were also estimated. A grid of models was constructed by changing the mass between 1.26 and 1.33$M_\odot$ in steps of 0.01$M_\odot$ with [Fe/H] covering the range of 0.18–0.30 dex in steps of 0.02 dex. Three values of core overshoot—0, 0.05 H_P, and 0.1 H_P—were used. The model with the minimum χ_ν^2 for each evolutionary sequence was used in the subsequent analysis. The analysis showed that the star has a mean density of 0.5301 ± 0.0044 g cm^{-3}, age of 3.2 ± 0.3 Gyr, mass of $1.285 \pm 0.026 M_\odot$, and radius of $1.507 \pm 0.012 R_\odot$.

This method of obtaining the best-fitting model is in a way quite similar to the grid-based method discussed in Section 7.2. One starts with a grid of models of known physics and searches for a minimum in χ^2 between them. In that sense, this is an efficient way of modeling—the models can be constructed ahead of time and their frequencies stored. The drawback is that decisions about the values of key parameters, such as the mixing length and the initial helium abundance, have to be taken ahead of time. While the stellar-physics community is comfortable doing that (after all, that is the basis of constructing isochrones and using them), this approach does not make it easy to study how combinations of different input parameters affect the frequency spectrum of a model. Thus, often the method is extended to make multiple sequences of stellar models with different values of the mixing-length parameter, Y_0, or both.

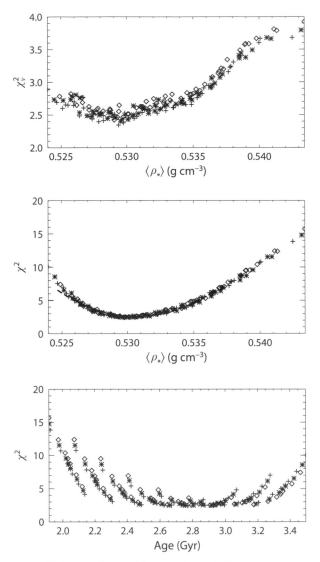

Figure 8.13. Results of fitting the observed frequencies and effective temperature of HD 17156 to the grid of stellar models. Crosses, stars, and diamonds correspond to overshooting of 0, $0.05 H_P$, and $0.1 H_P$, respectively. Top panel: Variation of χ_ν^2 as a function of the mean density of the model. Middle panel: Variation of sum $\chi_\nu^2 + \chi_{\rm spec}^2$ as a function of mean density. Bottom panel: $\chi_\nu^2 + \chi_{\rm spec}^2$ as a function of age. Reproduced with permission from Gilliland et al., 2011, *Asteroseismology of the Transiting Exoplanet Host HD 17156 with Hubble Space Telescope Fine Guidance Sensor, Astrophys. J.,* **726**, 2, copyright American Astronomical Society.

8.4.2 Method B: Monte-Carlo Simulation over Parameters

Method A relies on constructing evolutionary sequences of models of different masses and metallicities. The radius and temperature of the models that go through the error-box in temperature and luminosity (or its proxy) depend on the adopted values of the mixing-length parameter and initial helium abundance. A somewhat different approach to modeling and interpreting the individual frequencies of a star is to make models of a given mass, radius, temperature, and metallicity. The mass and radius ranges of the models are determined by grid-based modeling (see Chapter 7). Constructing a model with a given mass, radius and temperature requires us to iterate the mixing-length parameter or the initial helium abundance. By construction, the models have the temperature and metallicity of the observed star. The frequencies of the models can be calculated and compared with observations. This approach allows one to explore a finer grid of stellar properties and also to determine the effects of the chosen initial helium abundance and mixing-length parameter on the final results.

As with any modeling exercise, we have to decide ahead of time which microphysics we want to use, and whether we include diffusion and core overshoot. The next step to carry out a grid-based modeling using average seismic parameters and the spectroscopic estimates of $T_{\rm eff}$ and [Fe/H] to determine the mass of the star. The central values of mass, large separation, effective temperature, and metallicities and their uncertainties are used to make many realizations of the $\{M, \Delta\nu, T_{\rm eff}, [{\rm Fe/H}]\}$ quartet, and for each realization, the radius R is calculated using M and $\Delta\nu$. Models are then constructed to match each $\{M, R, T_{\rm eff}, [{\rm Fe/H}]\}$ set. Since grid-based results have a small model dependence, it is customary to assume that the uncertainties in the grid-based mass estimate are larger than the formal uncertainties returned by the grid-based modeling exercise. This assumption allows the exploration of a larger range of masses and prevents the final results from being biased. Exploring a larger range of masses also corrects for the fact that the $\Delta\nu$ scaling law deviates from the strict $\Delta\nu \propto \sqrt{\bar{\rho}}$ relation.

As discussed in Section 2.8, models with a given M and [Fe/H] that have a specified R and $T_{\rm eff}$ (or alternatively, specified luminosity and $T_{\rm eff}$) can only be constructed in an iterative manner. Modeling is usually done in two ways (see Section 8.2): in one, Y_0 is varied over a range of values, and for each Y_0 the model is constructed by varying α to get the desired radius and $T_{\rm eff}$; in the other, α is varied over a range of values, and for each α the model is constructed by changing Y_0 until the model with the desired radius and $T_{\rm eff}$ is obtained. We thus get a set of stars with a range of α and Y_0 and age for a given $\{M, R, T_{\rm eff}, [{\rm Fe/H}]\}$ set. The only way to distinguish these models is through their frequencies.

Once the models are constructed, their frequencies can be calculated, corrected for the surface term, and the value of χ_ν^2 can be calculated using Eq. 8.11. This way of constructing models automatically results in low values of $\chi_{\rm spec}^2$ (since the models are constructed to have the observed $T_{\rm eff}$ and [Fe/H] within uncertainties). However, it is still useful to calculate the total χ^2 using Eq. 8.14. The lowest χ^2 model can be determined easily. Average properties can again be determined as a likelihood-weighted average, and the uncertainties determined from the same statistic.

As in the case of constructing evolutionary sequences of models, deciding to make models with different values of overshoot generally implies making separate sets of models. Similarly, deciding to include models with and without diffusion would also lead to a separate set of models. Of course one could include overshoot as a parameter in the Monte Carlo exercise, but the number of realizations needed to cover the entire parameter space properly will increase as well. For models with diffusion, one needs to keep in mind that the initial metallicity and helium abundance needed to model the star will be higher than the final metallicity. Since an iterative

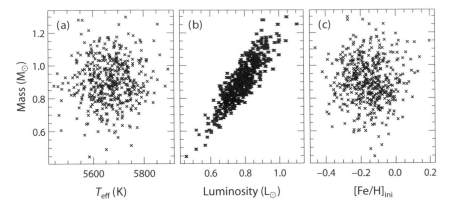

Figure 8.14. Distribution of the properties of the input models used to determine the properties of *Kepler*-93. Note that mass and luminosity are correlated—a result of the fact that mass and radius are correlated through the observed value of $\Delta\nu$.

process is used to construct models of a given mass and radius, one can include an iteration over the initial metallicity to achieve the required final metallicity. Such an iteration is done routinely for constructing models of the present-day Sun.

As has been shown in Section 8.2, one issue often seen in the case of models constructed iteratively to have a predetermined $\{M\text{-}R\text{-}T_{\text{eff}}, [\text{Fe/H}]\}$ combination is that for some values of the mixing-length parameter α, the Y_0 required to construct the models is less than the amount of helium produced by the Big Bang. Although nothing in the theory of stellar structure and evolution precludes models with low helium abundances, it is clearly unphysical to expect real stars to have subprimordial initial helium abundances. These models are either discarded or are given low weights, with the weights depending on how low the value Y_0 is compared with the primordial abundance of helium. It can, however, be argued that if subprimordial helium values are a result of uncertainties in the input parameters, we should not discard them since this could bias the results.

The results of the Monte Carlo method as applied to *Kepler*-93 are shown in Figures 8.14, 8.15, and 8.16. Figure 8.14 shows the distribution of the initial mass, T_{eff}, luminosity, and initial metallicity (the models include the diffusion of helium and heavy elements, and hence the initial metallicity has to be higher than the observed metallicity). The clear correlation between mass and luminosity is caused by the correlation between mass and radius through the observed value of $\Delta\nu$. Figure 8.15 shows χ^2 plotted as a function of mass, radius, age, and temperature of the models. Many of the models had unphysically low values of the initial helium abundance, and those models were discarded from the final analysis. From this exercise we find that the star has a mass of $0.937 \pm 0.045 M_\odot$, radius of $0.923 \pm 0.015 R_\odot$, and an age of 6.7 ± 0.9 Gyr. The final published values of the properties of this star are from statistical combinations of the properties obtained by this method and by others described in this chapter: $0.911 \pm 0.033 M_\odot$, $0.919 \pm 0.011 R_\odot$, and 6.6 ± 0.9 Gyr.

From Figure 8.15 we see that there is a limit of mass and radius beyond which no models have an acceptable initial helium abundance. This is shown more clearly in Figure 8.16. Note that in the temperature and metallicity ranges admitted by observations, only the low mass-radius end of the parameter space leads to physical results.

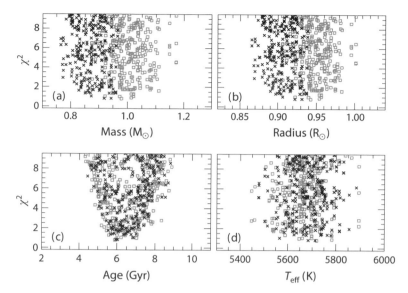

Figure 8.15. Variation of χ^2 with model properties. The χ^2 includes the contribution from the asteroseismic data as well as spectroscopic data on the $T_{\rm eff}$ and metallicity of the star. The points in gray are models that have initial helium abundances lower than the primordial helium abundance.

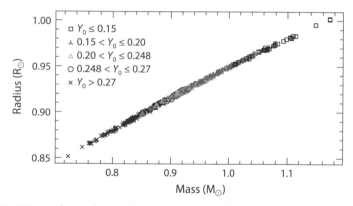

Figure 8.16. Mass-radius relation for the models of *Kepler*-93 that satisfy spectroscopic constraints as well as constraints on $\Delta\nu$. The points are coded by the initial helium abundance of the models. Note that higher the mass (and hence radius) of a model, the lower is the initial helium abundance required to construct the models; this is consistent with Figure 8.7 (right panel).

This method of determining the best-fitting properties of a star has the advantage that we work in the correct global parameter range from the very beginning. This allows us to explore regions of parameter space (such as nonsolar values of the mixing length, as well as initial helium abundances not tied to metallicity) that are often ignored when constructing evolutionary sequences of models. But this method is computationally intensive. Constructing models in an iterative manner to get the required set of final properties can be time consuming, particularly in the case of evolved stars. Additionally, a very large number of models needs to be constructed to cover the uncertainties in temperature, metallicity, mass, and radius, while trying to keep good coverage of the mixing-length parameter and Y_0.

8.4.3 Method C: Optimization Methods

Given that the aim of the modeling exercise is to get a best-fitting solution to the observed frequencies, it should come as no surprise that optimization schemes are also used. These generally combine a stellar structure code and a stellar pulsation code that are called by an optimization routine. Optimization methods are generally easy to implement. Decisions that need to be made ahead of time include which stellar properties and parameters (e.g., mass, mixing length, abundances, and age) will be varied to achieve a best fit. Also fixed ahead of time is the set of observations against which the models are tested.

An optimization routine like the Levenberg-Marquardt method works reasonably well. The goodness-of-fit measure is the total χ^2:

$$\chi^2 = \sum_i \frac{(O_i^{\text{obs}} - O_i^{\text{mod}})^2}{(\sigma_i^{\text{obs}})^2}, \tag{8.16}$$

where O_i^{obs}, σ_i^{obs} and O_i^{mod} are the observed value, uncertainty, and theoretical prediction of the observable i. The observable can be a seismic quantity, such as a frequency or a large separation, or a nonseismic one, such as effective temperature and metallicity at the surface. When frequencies are available, the χ^2 can be separated into an asteroseismic value and a spectroscopic value, and the asteroseismic value has to be calculated after surface-term corrections have been applied to the frequencies of the models. The Levenberg-Marquardt algorithm depends on derivatives of the fitting function with respect to the parameters, and these derivatives need to be calculated numerically. The convergence criterion for χ^2 also needs to be specified. Since the method depends on determining gradients numerically, it needs to begin by constructing several evolutionary tracks, each with one of the parameters changed slightly. This process can easily lead to a local minimum in the parameter space, and it is therefore usual to perform several runs starting at different initial guesses.

Other optimization algorithms can be used of course, and there are codes in the community that do so. For instance the asteroseismology module of the MESA code[7] offers a choice between two algorithms to minimize χ^2: the Hooke-Jeeves algorithm[8] and the Bound Optimization by Quadratic Approximation technique.[9]

As can be deduced, optimization algorithms allow us to vary parameters that need to be held constant when constructing grids of models. The disadvantage is that these algorithms can be extremely time consuming.

---

[7] Paxton et al., 2013, *Astrophys. J. Suppl.*, **208**, 4.

[8] Hooke and Jeeves, 1961, *J. Assoc. Comp. Machin.*, **8**, 212.

[9] Powell, 2009, *Technical report of the Dept. of Applied Mathematics and Theoretical Physics, University of Cambridge.*

8.5 DEALING WITH AVOIDED CROSSINGS IN SUBGIANTS

All the methods described so far work very well for main sequence stars. For evolved stars with mixed modes that give rise to avoided crossings, the techniques need to be modified. More efficient methods also can be used. One of the most important properties of mixed modes is that their frequencies change rapidly with time, as was shown in Chapter 2. While this means that we can get precise ages of the stars, it also means that it is easy to miss the correct age.

Modifying Method A is easy enough. What is needed is that models along the evolutionary tracks be made at very short age intervals during the evolved phases. This does increase the computational time, but the general principle of the method remains the same.

Method B is more difficult to adapt. One way is to iterate over the initial helium abundance for a much finer grid of mixing-length parameters. The other way is to find models close to the χ^2 minimum and then do separate simulations around those models. In either case, computation time increases immensely. Since most observed mixed modes are $l = 1$ mixed modes, a more practical way of modifying Method B is to first get the best fits to the $l = 0$ and $l = 2$ modes. The mass, mixing length, metallicity, and helium abundance of this model are then used to construct an evolutionary sequence of models, and in particular, to create a fine grid of models with ages slightly smaller and slightly larger than the best-fitting model. One then determines which model best fits the mixed modes. Since the evolution of the pure p modes is much slower than that of the mixed modes, one still obtains a model that matches the p modes. One issue though is that the fit to T_{eff} often becomes worse in this adaptation.

The computation time for Method C also increases if we want to model stars with avoided crossings. In the Levenberg-Marquardt type codes, one would have to use a small time step to be able to catch the best model. Similarly, the number of iterations in the other optimization methods increases as well. Again, a two-step process—first fitting the $l = 0$ and $l = 2$ modes, and then the mixed modes—cuts down the number of iterations.

The fast evolution of mixed-mode frequencies has led to the development of some specialized methods of dealing with these stars. Some of these methods depend on using a reduced set of observations to make the problem more tractable. One such method uses only two observables: the average $\Delta\nu$ calculated using the observed radial models; and ν_{cross}, which is the frequency at which we find the avoided crossing (defined to be the frequency of the most g-type mode). The idea behind the method is first to determine the model that satisfies both the $\Delta\nu$ and ν_{cross} constraints. This method gives very precise values of the mass and age of the star, since the values of $\Delta\nu$ and ν_{cross} are both monotonic functions of age. As a result, for a given set of physics, there is only one value of mass and age that can satisfy the two observational constraints simultaneously. For stars of a given mass, $\Delta\nu$ decreases with age, and thus is a proxy of age; however, at a given age, $\Delta\nu$ depends on mass. The frequency of an avoided crossing increases with age for a star of a given mass but decreases with increasing mass for a star with a given value of $\Delta\nu$. Since the observed $\Delta\nu$ and ν_{max} provide an initial estimate of the mass range, the process can thus begin with a constrained mass range. This method, as applied to the subgiant HD 49385, is illustrated in Figure 8.17.

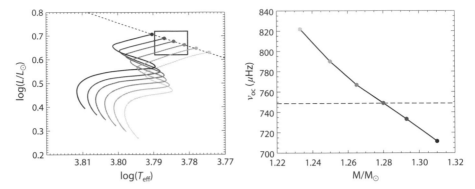

Figure 8.17. Results for the subgiant HD 49385. Left panel: Evolutionary tracks in the HR diagram for models with masses ranging from $1.23M_\odot$ (faintest line) to $1.31M_\odot$ (darkest line) with the circles denoting models that have $\Delta\nu$ matching the observed average separation. The dotted line shows the locus of all such points. The box marks the observed luminosity and effective temperature and their 1σ uncertainties. Right panel: The frequency of the avoided crossing that is closest to the observed one for models in the panel on the left. The dashed line is the observed frequency. Note that these two panels allow us to determine the mass of the model. The associated properties of the model can be used as estimates of the properties of the star. The models shown here are for a particular choice of physics. As the physics inputs (e.g., the amount of overshoot) are changed, the best-fitting mass will change as well. Reproduced with permission from Deheuvels and Michel, 2011, *Constraints on the Structure of the Core of Subgiants via Mixed Modes: The Case of HD 49385, Astron. Astrophys.,* **535**, A91, copyright european Southern Observatory.

Just because a given model fits the observed value of $\Delta\nu$ and the frequency of the avoided crossing does not mean that it also will fit the other frequencies. Thus the exercise needs to be repeated after changing other inputs, including (but not limited to) the amount of overshoot, initial abundances, and the mixing-length parameter. The best-fitting model is then chosen from the superset of the models. An optimization routine is generally used to determine the model that fits the observed $\Delta\nu$ and ν_{cross}.

8.6 MODELING RED GIANTS

The complicated nature of red-giant frequency spectra makes fitting all frequencies very difficult. As of the writing this chapter, no one has developed a method of fitting all $l = 1$ modes of a redgiant reliably. The most promising avenue seems to be to ignore the $l = 1$ modes completely at first and obtain models that fit the $l = 0$ and $l = 2$ models properly. These fits give good estimates of the mass and radius of the star. Since stars spend a very small fraction of their life on the red-giant branch, estimating the mass of a red giant is equivalent to estimating its age. Fits to the $l = 1$ frequencies, when possible, reveal details of the core structure.

The different strategies used in modeling that are described in Section 8.4 can be used easily in constructing models of red giants. However, Method A (Section 8.4.1) is definitely the fastest when it comes to these evolved stars. Since constructing red-giant models is time consuming, Methods B and C can be very slow.

One point to be aware of when calculating frequencies of red giants is that many of the modes have high-order g mode-like characteristics. To calculate the frequencies of these modes correctly requires models to have tens of thousands of mesh points. The models are not usually constructed with this resolution, and it is therefore advisable to increase the number of mesh points by interpolation. The interpolation needs to be done carefully, particularly since one needs to ensure that true discontinuities are not smoothed over.

Although the frequency spectra of red giants can be complicated, there is one more observable that can be determined from the data—the period spacing of $l = 1$ modes, which has additional diagnostic potential. We discuss this in more detail in Section 9.3.

8.7 ALTERNATIVE METHODS FOR FITTING FREQUENCIES

The usual ways of fitting frequencies as described in Section 8.4 rely on being able to make ad hoc corrections for the surface term. There are other ways of fitting the data that do not need explicit surface-term corrections to determine a goodness of fit. These techniques rely on the asymptotic theory of stellar pulsations. We now discuss two of the techniques; however, as we shall see later, the two methods are actually related.

8.7.1 Phase Matching

The Background

As we have seen in Chapter 3, the frequencies of a stellar model are eigenvalues of a fourth-order set of equations that govern stellar oscillations, subject to boundary conditions at the center and the surface of the star. For $l = 0$ modes, the fourth-order set of equations can be reduced to a second-order equation of the form

$$\frac{\mathrm{d}^2\psi_0}{\mathrm{d}t^2} - Q\frac{\mathrm{d}\psi_0}{\mathrm{d}t} + [\omega^2 - U]\psi_0 = 0, \tag{8.17}$$

where t is the acoustic radius defined by $t = \int_0^r \mathrm{d}r/c_{\mathrm{s}}$; and $\psi = P_1/\sqrt{c_{\mathrm{s}}\rho}$, P_1 being the usual pressure perturbation, and the subscript 0 of ψ indicates ψ for $l = 0$ modes. The acoustic potentials Q and U are defined by the following expressions:

$$U = A^2 + QA - \frac{\mathrm{d}A}{\mathrm{d}t} + N^2 - 4\pi G\rho, \tag{8.18}$$

$$Q = \frac{1}{F}\frac{\mathrm{d}F}{\mathrm{d}t}, \tag{8.19}$$

$$A = \frac{c_{\mathrm{s}}}{4\Gamma_1}\frac{\mathrm{d}\Gamma_1}{\mathrm{d}r} - \frac{c_{\mathrm{s}}N^2}{4g}\left(\frac{3 - \Gamma_1}{4}\right)\frac{g}{c_{\mathrm{s}}} - \frac{c_{\mathrm{s}}}{r}, \tag{8.20}$$

$$F = 1 + \frac{4\pi G\rho}{\omega} - \frac{N^2}{\omega^2}, \tag{8.21}$$

where as usual $\omega = 2\pi\nu$. G is, as usual, the gravitational constant; Γ_1 is the adiabatic index; c_s is the sound speed; ρ is the density; and N the buoyancy frequency defined in Eq. 3.32. Equation 8.17 can be reduced to a first-order equation,

$$\omega \frac{d\chi_0}{dt} - \omega^2 + Q\omega\chi_0 - (\omega^2 - U)\chi_0 = 0, \tag{8.22}$$

where

$$\chi_0 = \frac{\omega\psi_0}{d\psi_0/dt}. \tag{8.23}$$

Equation 8.23 has a solution of the form

$$\chi_0 = \frac{\omega\psi_0}{d\psi_0/dt} = \tan[\omega t + \delta_0(\nu, t)], \tag{8.24}$$

where δ_0 is an inner phase shift that satisfies

$$\frac{d\delta_0}{dt} = -\frac{U}{\omega}\sin^2(\omega + \delta_0) - \frac{Q}{2}\sin[2(\omega t + \delta_0)]. \tag{8.25}$$

The equation can also be solved by integrating inward from the surface, in which case there will be an outer phase shift $\alpha(\omega, \tau)$, where τ is now the acoustic depth, and $\tau = T_0 - t$, where $T_0 = \int_0^R dr/c_s$ is the acoustic radius of the star.

For $l \neq 0$, the inner and outer phase shifts can be expressed in the same way in terms of variables ψ_l and χ_l, except that $\pi l/2$ is subtracted from δ to get the standard ordering of n values. Thus in general we have

$$\chi_l(\nu, t) = \tan\left[\omega t + \frac{\pi l}{2}\right] + \delta_l(\nu, t), \tag{8.26}$$

and

$$\chi_l(\nu, \tau) = -\tan[\omega\tau - \alpha_l(\nu, \tau)]. \tag{8.27}$$

For $\omega = 2\pi\nu$ to be an eigenvalue of the equations, the inner and outer solutions must match at any intermediate fitting point t_f (or, in terms of acoustic depth, fitting point $\tau_f = T_0 - t_f$), and this gives rise to the condition

$$\omega_{nl} T_0 - \pi(n + l/2) - \alpha_{nl}(\nu_{nl}, \tau_f) + \delta_{nl}(\nu_{nl}, t_f) = n\pi. \tag{8.28}$$

One should note that because we are dealing with trigonometric functions, α_l and δ_l can only be determined to within an arbitrary multiple of π, and the value depends on the procedure used to determine α_l and δ_l. One simple way of determining the two shifts is by using the expressions

$$\alpha_l(\nu_l, \tau_f) = 2\pi\nu_l\tau_f - (j_\alpha + 1/2)\pi - \tan^{-1}(\chi_f), \tag{8.29}$$

and

$$\delta_l(\nu_l, t_f) = (j_\delta + l/2 + 1/2)\pi - 2\pi \nu_l t_f + \tan^{-1}(\chi_f), \tag{8.30}$$

where χ_f is χ at the fitting point f. With the above definition, $n = j_\alpha + j_\delta + 1$ is the number of nodes of the eigenfunction ψ_l. For any given model one can calculate ψ_l, ν_l, t, and τ and hence calculate α_l and δ_l.

Equation 8.28 is not particularly helpful when comparing with observations, because it has a term with the total acoustic radius T_0. However, it can be rewritten in terms of the observable $\Delta\nu$ and a phase function $\mathcal{G}(l, \nu)$ defined as

$$\mathcal{G}(l, \nu_{nl}) \equiv \frac{\nu_{nl}}{\Delta\nu} + \frac{\delta_l(\nu_{nl})}{\pi} - \frac{l}{2} - n = \alpha_l^\star(\nu_{nl}), \tag{8.31}$$

where

$$\alpha_l^\star(\nu_{nl}) = \alpha_l(\nu_{nl}) + \pi\nu_{nl}\left(\frac{1}{\Delta\nu} - 2T_0\right). \tag{8.32}$$

Recall that the observed $\Delta\nu$ is an average of the spacing between modes of the same order and therefore can differ from $1/(2T_0)$. The last term on the right-hand side of Eq. 8.32 is a linear function of frequency. If the phase shift were calculated at a fitting point close to the surface (e.g., $r_f > 0.95R$), α_l would be essentially independent of l and would be a function of frequency alone. This would mean that $\alpha_l^\star(\nu_{nl})$ is independent of degree l, and hence $\mathcal{G}(l, \nu)$ is independent of degree. This helps us determine the best-fitting model to a star.

If a model has the same internal structure as a star, then $\mathcal{G}(l, \nu_{nl})$ of the model calculated at the observed frequencies of the star should collapse into a single function of frequency; otherwise, $\mathcal{G}(l, \nu_{nl})$ will be l dependent. Thus the absence of an l dependence is the indicator of a good fit. The method of phase fitting has been applied successfully to the stars HD 49933 and HD 177153.

Obtaining the Fits

The fitting process begins with making models and solving the oscillation equations at the observed frequencies ν_{nl}^{obs}, without applying the outer boundary condition on pressure. Note that the inner phase shifts of a model can be calculated at any frequency if one does not impose an outer boundary condition on the pressure perturbation. Applying the outer boundary condition on the pressure will give us the eigenfrequencies of the model, which is not what we want here.

Solving the equations gives us the Eulerian pressure perturbation P_1, which is then used to calculate the scaled pressure perturbation $\psi(t) = rP_1/\sqrt{\rho c_s}$. This is used to calculate the inner phase function δ_l at the observed frequency using Eq. 8.30. This function is then used to determine $\mathcal{G}(l, \nu_{nl}^{\text{obs}})$ using Eq. 8.31, with the observed average $\Delta\nu$. The resulting $\mathcal{G}(l, \nu_{nl}^{\text{obs}})$ should be a function of frequency alone if the model is a good fit.

To determine the l independence of $\mathcal{G}(l, \nu_{nl}^{\text{obs}})$, we test whether it can be fitted with an arbitrary function of frequency. Such a function can be defined as $\mathcal{A}(\nu_{nl}^{\text{obs}}) = \sum_i^M a_i \phi(\nu_{nl}^{\text{obs}})$, where the ϕ are suitable basis functions in frequency (such as Chebyshev polynomials or B-splines), a_i are unknown coefficients, and M is the total number of basis functions. The number M should not be too large, and keeping M less than the number of $l = 0$ modes is a good guideline. To determine the goodness of fit, one needs the uncertainties on $\mathcal{G}(l, v_{nl}^{\text{obs}})$. One way to determine them is to calculate $\mathcal{G}(l, v_{nl}^{\text{obs}} \pm \sigma_{nl}^{\text{obs}})$, σ_{nl}^{obs} being the uncertainties on v_{nl}^{obs}, and subtract out $\mathcal{G}(l, v_{nl}^{\text{obs}})$ from the result. The next step is to minimize

$$\chi^2 = \frac{1}{N - M} \sum_{nl} \left(\frac{\mathcal{G}(l, v_{nl}^{\text{obs}}) - \mathcal{A}(v_{nl}^{\text{obs}})}{e_{nl}} \right)^2, \tag{8.33}$$

where e_{nl} are the uncertainties on $\mathcal{G}(l, v_{nl}^{\text{obs}})$, and N is the total number of modes. The sum is performed over all modes nl. The interior structure of the model is consistent with that of the star if $\chi^2 \leq 1$.

A problem with the above method is that it requires us to know the radial order n of the observed modes. While this is not too much of a problem with main sequence stars, evolved stars cause difficulties. Evolved stars have mixed modes, and assigning an order n to them is not easy; and even if we did, we cannot identify n with the number of nodes in the eigenfunction. As a result we need to modify the method somewhat. The modification relies on the fact that there are no $l = 0$ mixed modes. The frequencies of $l = 0$ modes follow a regular pattern and can be labeled in order of increasing frequency (e.g., with $k = 1, 2, \ldots$). We can now define $\mathcal{G}(l, v_{nl}^{\text{obs}})$ for the $l = 0$ frequencies as

$$\mathcal{G}_0(v_{k0}^{\text{obs}}) = \frac{v_{k0}}{\Delta v} + \frac{\delta_0(v_{k0})}{\pi} - k, \tag{8.34}$$

The function $\mathcal{G}_0$ can now be interpolated to any frequency v_{kl} with $l > 0$, to obtain values g_{kl}. This allows us to calculate

$$\mathcal{N}(v_{kl}) = \frac{v_{k0}}{\Delta v} - \frac{l}{2} + \frac{\delta_0(v_{k0})}{\pi} - g_{kl}. \tag{8.35}$$

If the observed modes of the frequency fit the model, $\mathcal{N}(v_{kl})$ should be an integer. Of course, we do not expect $\mathcal{N}(v_{kl})$ to be an integer for most models; thus we determine the integer closest to $\mathcal{N}(v_{kl})$, that is,

$$n_{kl} = \text{Int}[\mathcal{N}(v_{kl}) + 0.5] \tag{8.36}$$

and we define

$$\mathcal{G}(l, v_{kl}) \equiv \frac{v_{kl}}{\Delta v} + \frac{\delta_l(v_{kl})}{\pi} - \frac{l}{2} - n_{kl}, \tag{8.37}$$

and then use Eq. 8.33 to determine the goodness of fit.

8.7.2 ϵ Matching

The phase fitting method might seem rather complicated, and one of the drawbacks of the method is that pulsation codes need to be modified so as to solve the pulsation equations without the surface boundary condition on pressure. There are easier ways of implementing the method, and one of them is through the use of the phase term ϵ.

Recall that frequencies roughly satisfy the equation

$$\nu_{nl} = \Delta\nu\left(n + \frac{l}{2} + \epsilon_{nl}\right),\tag{8.38}$$

where ϵ_{nl} is a phase. In other words,

$$\epsilon_{nl} = \frac{\nu_{nl}}{\Delta\nu} - n - \frac{l}{2}.\tag{8.39}$$

For any given l, ϵ_{nl} is a function of frequency and measures the departure of the structure of a model from that of a uniform sphere; the function is determined by integrals over the structure of the star. ϵ has contributions from the interior as well as from the stellar surface. The dominant contribution to ϵ_{nl} from the interior comes from the inner phase shift $\delta_l(\nu, t)$; the contribution from the outer layers is because of the outer phase shift $\alpha_l(\nu, t)$, where as before, t is the acoustic radius of any point. Since eigenfrequencies of a model satisfy Eq. 8.28 at any arbitrary t (and hence $\tau = T_0 - t$), the combination $\alpha_l(\nu_{nl}, T_0 - t) - \delta_l(\nu_{nl}, t)$ is independent of the matching radius t_f, and hence the frequencies follow

$$\nu_{nl} = (2T_0)^{-1}\left(n + \frac{l}{2} + \epsilon_{nl}\right),\tag{8.40}$$

where

$$\epsilon_{nl} = \alpha_l(\nu_{nl}, T_0 - t_f) - \delta_l(\nu_{nl}, t_f).\tag{8.41}$$

For $l = 0$ and $l = 1$ modes, $\alpha_l(\nu_{nl}, T_0 - t_f)$ depends completely on the structure of layers above t_f and is independent of l. For modes with $l \geq 2$, α can be assumed to be independent of l if the fitting point is at $t_f > T_0/2$. Thus with a suitable choice of fitting point, $\alpha_l(\nu_{nl}, T_0 - t_f)$ is a function of frequency alone.

In Eq. 8.40, ϵ_{nl} is related to the acoustic radius T_0, though we are more used to seeing the equation in terms of the observed $\Delta\nu$, as in Eq. 8.38. We can rewrite Eq. 8.40 in terms of $\Delta\nu$ so that it is more familiar:

$$\nu_{nl} = \Delta\nu\left[n + \frac{l}{2} + \alpha_l(\nu_{nl}, T_0 - t_f) - \delta_l(\nu_{nl}, t_f)\right] + \nu_{nl}(1 - 2\Delta\nu T_0).\tag{8.42}$$

Since $\nu_{nl}[1 - 2\Delta\nu T_0]$ is a function of frequency alone, it can be absorbed into α. We can therefore define

$$\alpha^*(\nu_{nl}, T_0 - t) = \alpha(\nu_{nl}, T_0 - \tau) + \nu_{nl}\left(\Delta\nu^{-1} - 2T_0\right),\tag{8.43}$$

and thus we get back Eq. 8.38, but with

$$\epsilon_{nl}(\nu_{nl}) = \alpha^*(\nu_{nl}, T_0 - t) - \delta_l(\nu_{nl}, t).\tag{8.44}$$

The question is how Eq. 8.44 can help us determine a model that matches the observations. We want to find models whose interior structure matches the structure of the star, but its outer layers may be different (and we know this is inevitable, since we cannot model the near-surface layers properly). Assume we have two models, tagged 1 and 2, that have the same internal structure. Then we have

$$\epsilon_{nl}^1(\nu_{nl}^1) = \left[\frac{\nu_{nl}^1}{\Delta\nu^1} - n - \frac{l}{2}\right] = \alpha^1(\nu_{nl}^1) - \delta_l^1(\nu_{nl}^2), \tag{8.45}$$

and

$$\epsilon_{nl}^2(\nu_{nl}^2) = \left[\frac{\nu_{nl}^2}{\Delta\nu^2} - n - \frac{l}{2}\right] = \alpha^2(\nu_{nl}^2) - \delta_l^2(\nu_{nl}^2). \tag{8.46}$$

Since the models have the same internal structure,

$$\delta_l^2(\nu_{nl}^1) = \delta_l^1(\nu_{nl}^2). \tag{8.47}$$

Thus we have

$$\epsilon_{nl}^2(\nu_{nl}^1) - \epsilon_{nl}^1(\nu_{nl}^1) = \alpha^2(\nu_{nl}^1) - \alpha^1(\nu_{nl}^1) = F(\nu_{nl}^1), \tag{8.48}$$

where F is a function of frequency. Thus to get a model that fits the observations, we need to determine ϵ_{nl} for the model and the observations using Eq. 8.38. The ϵ_{nl} for the observations should then be subtracted from that of the model interpolated to the same frequency as the observations. If the residuals are independent of l, the models match the stars. To determine whether this is the case, we calculate

$$\Delta\epsilon(\nu_{nl}^{\text{obs}}) = \epsilon^{\text{model}}(\nu_{nl}^{\text{obs}}) - \epsilon^{\text{obs}}(\nu_{nl}^{\text{obs}}), \tag{8.49}$$

and define an arbitrary function of frequency $\mathcal{F}(\nu) = \sum_i^M a_i\phi_i(\nu)$, where the $\phi_i(\nu)$ are basis functions in frequency, and the a_i are unknown coefficients. We then minimize

$$\chi^2 = \frac{1}{N-M}\sum_{nl}\left(\frac{\Delta\epsilon(\nu_{nl}^{\text{obs}}) - \mathcal{F}(\nu_{nl}^{\text{obs}})}{s_{nl}^{\text{obs}}}\right)^2, \tag{8.50}$$

where N is the total number of modes, $s_{nl}^{\text{obs}} = \sigma_{nl}^{\text{obs}}/\Delta\nu^{\text{obs}}$, and σ_{nl}^{obs} is the uncertainty on ν_{nl}^{obs}. A small value of χ^2 generally implies that the residuals are a function of frequency alone and hence the interior structure of the models matches that of the star.

We demonstrate the l independence of $\Delta\epsilon$ for two models that differ only in the outer layers, using two standard solar models constructed with different atmospheric models. The frequency differences between the two models are shown in top-left panel of Figure 8.10, and the relative sound-speed difference is shown in the top-right panel of Figure 8.10. The left panel in Figure 8.18 shows ϵ for the models, and the right panel shows the the difference in ϵ. As can be seen clearly in the right

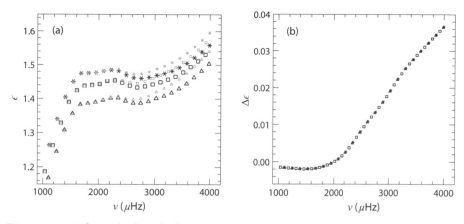

Figure 8.18. Left panel: Plot of ϵ for two solar models, one shown in black and the other in gray. The $l = 0$ modes are represented as asterisks, $l = 1$ as squares, and $l = 2$ as triangles. Right panel: Difference in ϵ calculated at a given frequency. The style of points is the same as in the left panel. Note that all points fall on the same curve.

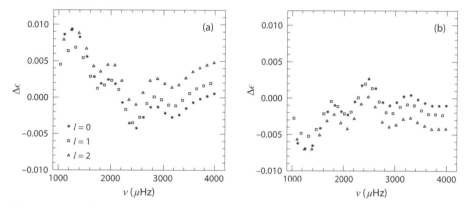

Figure 8.19. Differences in ϵ between an [Fe/H] $= 0$ model and an [Fe/H] $= -0.1$ model (left panel) and an [Fe/H] $= +0.1$ model (right panel). All models have the same mass, radius, T_{eff}, and mixing-length parameter. The frequency differences between the models are shown in the right panel of Figure 8.6.

panel, the points for all modes fall on a single curve. The l dependence of ϵ for models that differ in their interior structure is shown in Figure 8.19.

Neither phase matching nor ϵ matching guarantees that we will match the effective temperature and metallicity of a star, and hence a criterion to evaluate the goodness of fit to those quantities must be added. Since we still need to make models to apply the ϵ matching process, the different schemes used to make models that were discussed in Section 8.4 still need to be used. The only difference is the goodness-of-fit criterion and the fact that we do not need to assume a surface term.

8.8 ANALYSIS OF ROTATION

Stellar rotation is determined using rotational splittings, described in Section 3.5 and discussed in an observational context in Chapter 6. While it is unlikely that the data can tell us much about latitudinal differential rotation, at least for some stars one can estimate a measure of radial differential rotation. Such estimates rely on the fact that different splittings are most sensitive to the rotation rate at different radii.

The sensitivity of a given rotational splitting depends on the kernel that represents the splitting. One way of determining the radius at which the rotational kernel (and hence splitting) of a given mode is most sensitive is by defining a weighted radius,

$$r_{nl} \equiv \frac{\int_0^R r\, K_{nl}(r)\mathrm{d}r}{\int_0^R K_{nl}(r)\mathrm{d}r}, \tag{8.51}$$

where $K_{nl}(r)$ are the kernels defined by Eq. 3.153. Small values of r_{nl} indicate that the splitting carries information from deep inside the star, while large values show higher sensitivity in the outer layers. Another way of determining the sensitivity of the kernel to rotation at different radii is to look at the cumulative integral of the kernel as a function of radius:

$$\mathcal{I}(r) = \int_0^r \beta_{nl} K_{nl}(r)\mathrm{d}r, \tag{8.52}$$

where β_{nl} has been defined in Eq. 3.154. The radius at which the integral increases sharply is the radius at which the mode is most sensitive.

The usual method of analysis is a forward analysis: a given rotation rate is assumed, mode splittings are calculated for that rotation rate using the rotational kernels described earlier, and the observed and calculated splittings are compared. Very often, a two-zone model is used to reduce the number of free parameters, each zone rotating with a different rate. The rotation rates of the two zones and the radius that separates the two zones are free parameters. While the radius that demarcates the two zones can be kept as a free parameter, it is usual to assume that the convective envelope rotates with one rate and the interior with another. Although this is a very simple model, the uncertainties in the data have not yet warranted fits to a more complicated rotation profile. Inversion techniques discussed in Chapter 10 might prove more useful in determining whether such severe restrictions on the rotation profile are justified.

8.8.1 Main Sequence Stars

The rotation kernels of all dipole, quadrupole, and octopole modes for main sequence stars show essentially the same radial dependence. The radial sensitivity of the kernels as determined using the quantity $\mathcal{I}(r)$ looks very similar; a few examples are shown in Figure 8.20. The gentle rise of the curves implies that there is no radius at which the kernels show the most sensitivity, except perhaps the near-surface layers.

8.8.2 Subgiants

The presence of mixed modes in subgiants can help detect the presence of radial differential rotation in these stars. The kernels for mixed modes show high sensitivity in the core. The integrals $\mathcal{I}(r)$ for two models are shown in Figure 8.21. The jump in the value of the integral close to the core shows that the kernels have considerable

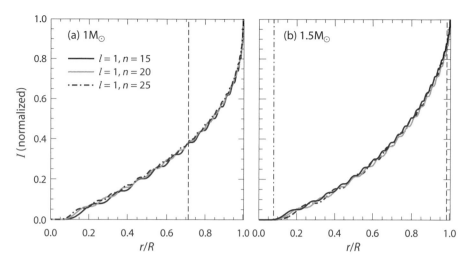

Figure 8.20. Normalized cumulative integral of a few rotation kernels of a $1M_\odot$ main sequence model (left panel), and a $1.5M_\odot$ main-sequence model (right panel). In both panels, the vertical dashed line marks the position of the base of the convective envelope. The dot-dashed line in the right panel marks the boundary of the convective core.

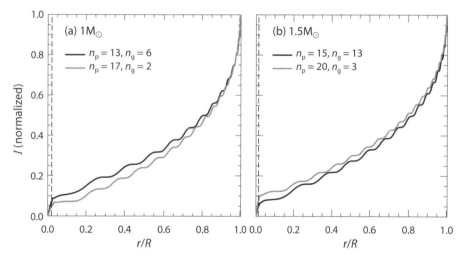

Figure 8.21. The same as Figure 8.20 but for two subgiants. The vertical dashed lines mark the edge of the helium core.

sensitivity there. On the early part of the subgiant branch, where only one or two mixed modes are present, deriving a precise measure of differential rotation may not be possible. However, as the star evolves and an increasing number of $l = 1$ mixed modes are seen, fitting a two-zone model presents little difficulty.

Given that the main motivation to determine stellar internal rotation is to test models of angular-momentum transport, the boundary of the two-zone model is usually kept at the edge of the core. Since the core continues to contract throughout the subgiant phase, conservation of angular momentum dictates that the core will rotate faster than the envelope, and that the rotation rate should speed up as a the star evolves. The discrepancy between the expected and inferred rotation rates is a test of angular-momentum transfer models.

8.8.3 Red Giants

As with subgiants, two-zone models for red giants encompass the shrinking core and the expanding envelope. The contrast between core and envelope rotation rates is expected to be much larger in this case. Although in principle, determining the contrast between core and envelope rotation should be simple—after all, red giants have many mixed modes—in practice, it can become difficult when core rotation becomes large. The problem lies mainly in determining the frequency splittings. Recall that red giants show a forest of dipole modes. Each mode in an $l = 1$ multiplet will undergo rotational splitting, and the splitting can become larger than the frequency separation between adjacent $l = 1$ modes in a multiplet. And as we saw in Section 6.5.2, this can make disentangling the underlying frequency separation (and period spacing) from the rotational splittings quite difficult.

We also saw in Section 6.5.2 that the mixed-mode rotational splittings can be expressed asymptotically as a combination of p and g mode splitting contributions (Eq. 6.158), both of which can be written using the appropriate kernels. Once the splittings are determined properly, the task of determining the rotation rate is then similar to the main sequence and subgiant cases.

8.9 FURTHER READING

The most common ways of fitting model stars when their mode frequencies are available have been described in:

- Silva Aguirre, V. et al., 2015, *Ages and Fundamental Properties of* Kepler *Exoplanet Host Stars from Asteroseismology, Mon. Not. R. Astron. Soc.*, **452**, 2127.
- Deheuvels, S., and Michel, E., 2011, *Constraints on the Structure of the Core of Sub-Giants via Mixed Modes: The Case of HD 49385, Astron. Astrophys.*, **535**, A91.

Corrections for the surface term are an important part of asteroseismic analyses that rely on frequency matching. We refer the reader to the following papers for discussions of the different forms of surface-term corrections:

- Kjeldsen, H., Bedding, T. R., and Christensen-Dalsgaard, J., 2008, *Correcting Stellar Oscillation Frequencies for Near-Surface Effects, Astrophys. J.*, **683**, L175.
- Ball, W. H., and Gizon, L., 2014, *A New Correction of Stellar Oscillation Frequencies for Near-Surface Effects, Astron. Astrophys.*, **568**, A123.
- Schmitt, J. R., and Basu, S., 2015, *Modeling the Asteroseismic Surface Term across the HR Diagram, Astrophys. J.*, **808**, 123.
- Roxburgh, I. W., 2015, *A Note on the Use of Surface Offset Corrections in Asteroseismic Model Fitting, Astron. Astrophys.*, **581**, A58.

More about phase and ϵ matching can be found in the following papers:

- Christensen-Dalsgaard, J., and Perez Hernandez, F., 1992, *The Phase Function for Stellar Acoustic Oscillations. I—Theory, Mon. Not. R. Astron. Soc.*, **257**, 62.
- Roxburgh, I. W, and Vorontsov, S. V., 1996, *An Asymptotic Description of Solar Acoustic Oscillation of Low and Intermediate Degree, Mon. Not. R. Astron. Soc.*, **278**, 940.
- Roxburgh, I. W, and Vorontsov, S. V., 2000, *Semiclassical Approximation for Low-Degree Stellar p Modes—I. The Classical Eigenfrequency Equation, Mon. Not. R. Astron. Soc.*, **317**, 141.
- Roxburgh, I. W, and Vorontsov, S. V., 2001, *Semiclassical Approximation for Low-Degree Stellar p Modes—III. Acoustic Resonances and Diagnostic Properties of the Oscillation Frequencies, Mon. Not. R. Astron. Soc.*, **322**, 85.
- Roxburgh, I. W, 2010, *Asteroseismology of Solar and Stellar Models, Atrophys. Space Sci.*, **328**, 3.
- Roxburgh, I. W., 2015, *Surface Layer Independent Model Fitting by Phase Matching: Theory and Application to HD 49933 and HD 177153 (aka Perky), Astron. Astrophys.*, **574**, A45.
- Roxburgh, I. W., 2016, *Asteroseismic Model Fitting by Comparing ϵ_{nl} Values, Astron. Astrophys.*, **585**, A63.

Asteroseismic research on stellar rotation has a relatively short history. The following papers discuss different aspects:

- Deheuvels, S., et al. 2012, *Seismic Evidence for a Rapidly Rotating Core in a Lower-Giant-Branch Star Observed with Kepler, Astrophys. J.*, **756**, 19.
- Lund, M. N., Miesch, M. A., and Christensen-Daalsgaard, J., 2014, *Differential Rotation in Main-Sequence Solar-Like Stars: Qualitative Inference from Astero-seismic Data, Astrophys. J.*, **790**, 121.
- Benomar, O., et al., 2015, *Nearly Uniform Internal Rotation of Solar-Like Main-Sequence Stars Revealed by Space-Based Asteroseismology and Spectroscopic Measurements, Mon. Not. R. Astron. Soc.*, **452**, 2654.
- Deheuvels, S., et al., 2014, *Seismic Constraints on the Radial Dependence of the Internal Rotation Profiles of Six Kepler Sub-Giants and Young Red Giants, Astron. Astrophys.*, **564**, A27.
- Marques, J. P., et al., 2013, *Seismic Diagnostics for Transport of Angular Momentum in Stars. I. Rotational Splittings from the Pre-Main Sequence to the Red-Giant branch, Astron. Astrophys.*, **549**, 74.
- Goupil, M. J., et al., 2013, *Seismic Diagnostics for Transport of Angular Momentum in Stars. II. Interpreting Observed Rotational Splittings of Slowly Rotating Red Giant Stars, Astron. Astrophys.*, **549**, A75.
- Mosser, M., et al., 2015, *Period Spacings in Red Giants. I. Disentangling Rotation and Revealing Core Structure Discontinuities, Astron. Astrophys.*, **584**, 50.

8.10 EXERCISES

1. The directory `Models` contains the "observed" frequencies and other properties of ten synthetic stars. Determine the properties of the stars from the frequency, metallicity, and temperature estimates. Explore how changing the model parameters can change the fits.

2. File `BiSON.txt` contains low-degree solar frequencies obtained by the Birmingham Solar-Oscillations Network (BiSON). File `ssm_freqs.txt` contains the frequencies of several standard solar models. Using the ϵ-matching method, determine whether it is possible to say which of these models is closest in structure to the Sun.

3. Files `rotker.m1.20_2.78Gyr.txt` and `rotker.m1.5_2.5Gyr.txt` contain the rotation kernels $\beta K(r)$ for a $1.20 M_\odot$ main sequence model and a $1.5 M_\odot$ evolved model, respectively. Using the data, calculate the weighted radius (Eq. 8.51) for each model. What do the results imply about determining the radial rotation profiles given in Exercise 4 in Chapter 3 using frequency splittings?

9 Interpreting Frequencies of Individual Modes: Other Diagnostics

Once we have the frequencies of individual modes of a star, comparing them with frequencies of models is only one way of determining stellar properties. There are other techniques, some of which do not require ad hoc surface-term corrections. In this chapter we discuss some of the other ways mode frequencies may be used, and what else they can tell us about the star.

We start by discussing how specific combinations of mode frequencies can be used to determine stellar properties. We also discuss how one can use signatures of acoustic glitches to infer properties of the star in question. Period spacings of red giants are potentially very useful, and we discuss how to make use of them.

9.1 LEARNING FROM COMBINATIONS OF FREQUENCIES

We introduced the small frequency separation $\delta \nu_{l,l+2}$ (Eq. 3.68) in Chapter 3. The small separation is an example of a frequency combination that is sensitive to a specific property of a star, in this case the sound-speed gradient in the core. One can understand the added sensitivity of the combination by looking at the resultant kernels; the kernels of p modes are most prominent near the surface, and their amplitudes (and hence sensitivities) decrease deeper inside the star. In contrast, the kernels for small separations have much lower amplitudes at the surface, and the amplitude near the core is actually slightly higher. This is illustrated in Figure 9.1.

Such frequency combinations as $\delta \nu_{02}$ and $\delta \nu_{13}$ are less sensitive to surface effects than the frequencies themselves, a result of subtracting two frequencies that are very similar (Figure 9.2). However, there is still a residual effect. A better way is to use ratios of frequency combinations. To make $\delta \nu_{02}$ and $\delta \nu_{13}$ less sensitive to the surface term than they already are, one can define

$$r_{02}(n) = \frac{\delta \nu_{02}(n)}{\Delta \nu_1(n)}, \quad \text{and} \quad r_{13}(n) = \frac{\delta \nu_{13}(n)}{\Delta \nu_0(n+1)}, \tag{9.1}$$

where $\Delta \nu_1(n)$ is the separation between $l = 1$ modes of order n and $n - 1$, and $\Delta \nu_0(n+1)$ is the separation between $l = 0$ modes of order $n+1$ and n. The modes

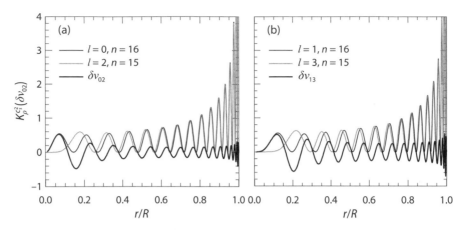

Figure 9.1. Sound-speed kernels for small separations $\delta\nu_{02}$ and $\delta\nu_{13}$ plotted as a function of radius (heavy black lines). The thin black and gray lines show the kernels for the modes that were used to calculate the small separations. Note that unlike the kernels of the modes themselves, kernels for the small separations have low amplitudes at the surface. This makes them good probes of the interior.

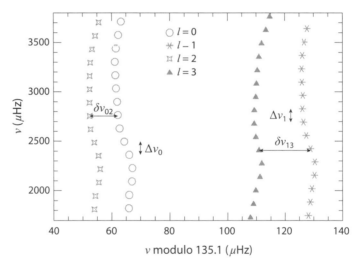

Figure 9.2. Échelle diagram of observed solar modes showing example of small separations $\delta\nu_{02}$ and $\delta\nu_{13}$ and the corresponding large separations Δ_1 and Δ_0 used to calculate the frequency ratios r_{02} and r_{13}.

used to calculate $\delta\nu_{02}$ have frequencies that lie between the frequencies of the modes used to calculate $\Delta\nu_1(n)$, and the modes used to calculate $\delta\nu_{13}$ have frequencies that lie between the frequencies of the modes used to calculate $\Delta\nu_0(n+1)$. This is illustrated in Figure 9.2. The quantities r_{02} and r_{13} are thus ratios of quantities with similar frequencies and hence similar surface terms, which divide out. Another advantage of using ratios is that the $\sqrt{GM/R^3}$ scaling in the frequencies is divided out (since both the numerator and the denominator have the same scaling); this makes it easier to compare the ratios of models with somewhat different masses and

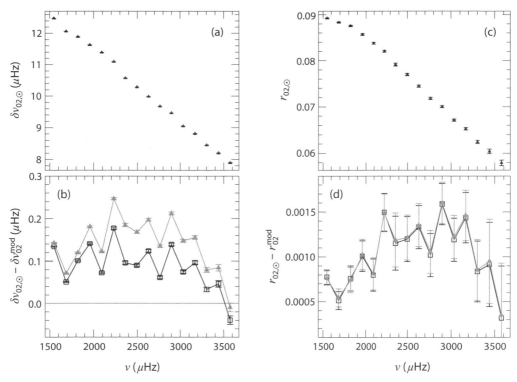

Figure 9.3. Top-left panel: Observed small frequency separation $\delta\nu_{02}$ of the Sun plotted as a function of frequency. The error bars are the size of the points and hence cannot be seen easily. Top-right panel: Observed frequency ratio r_{02} for the Sun plotted as a function of frequency. Bottom-left panel: The difference in $\delta\nu_{02}$ between the Sun and two standard solar models, constructed with different formulations of convection but otherwise identical physics. The frequency differences between the two models can be seen in the bottom-left panel of Figure 8.10. Although the frequency differences between the two models at $3{,}000\,\mu$Hz is of order $8\,\mu$Hz, the difference in $\delta\nu_{02}$ is less than $1\,\mu$Hz. Bottom-right panel: Difference in r_{02} between the Sun and the same two models as in the bottom-left panel. The r_{02} values for the two models are almost identical.

radii. It should be noted that separation ratios are intimately related to the interior phases described in Section 8.7.1. One can show that

$$r_{02}(\nu) = \frac{1}{\pi}[\delta_{l=2}(\nu) - \delta_{l=0}(\nu)]. \tag{9.2}$$

Figure 9.3 compares $\delta\nu_{02}$ and r_{02} of the two solar models and the Sun. The frequency differences between the two models are shown in the bottom-left panel of Figure 8.10. The small separations $\delta\nu_{02}$ for the models differ by much more than the observed errors, though the differences are not as large as the frequency differences between the models as seen in the bottom-left panel in Figure 8.10. The r_{02} differences between the models and the Sun are, however, almost identical. These two models only differ at the surface, and even though this difference leaks into $\delta\nu_{02}$, r_{02} is largely immune to the near-surface differences between the models. Thus a comparison of r_{02} yields more robust results.

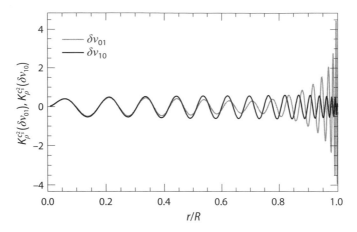

Figure 9.4. Kernels for δv_{01} and δv_{10} for $n = 15$.

While δv_{02} and δv_{13} are motivated directly by the asymptotic expression of p mode frequencies (Eq. 3.66), other not-so-obvious combinations are useful, too. In particular the $l = 1$ and $l = 0$ modes can be used to construct the small separations δv_{01} and δv_{10}, which we defined in Eq. 3.74. A sample of the kernels is shown in Figure 9.4, and as can be seen, the combination again reduces the amplitude at the surface.

When plotted against frequency, δv_{01} and δv_{10} are not usually smooth, thus when possible, even higher-order differences are used, for example,

$$dd_{01}(n) = \frac{1}{8}(v_{n-1,0} - 4v_{n-1,1} + 6v_{n,0} - 4v_{n,1} + v_{n+1,0}),$$

and

$$dd_{10}(n) = -\frac{1}{8}(v_{n-1,1} - 4v_{n,0} + 6v_{n,1} - 4v_{n+1,0} + v_{n+1,1}), \qquad (9.3)$$

with the corresponding frequency ratios being

$$rr_{01}(n) = \frac{dd_{01}(n)}{\Delta v_1(n)},$$

and

$$rr_{10}(n) = \frac{dd_{10}(n)}{\Delta v_0(n+1)}. \qquad (9.4)$$

Since $l = 1$ modes penetrate deeper inside a star than do $l = 2$ modes, r_{01} and r_{10} can often capture differences that r_{02} cannot. As an example, Figure 9.5 shows the frequency ratios for two models with exactly the same age, but one with a core convection zone and one without. The r_{02} for the models are indistinguishable, but r_{01} and r_{10} capture the differences in the core. These two ratios are useful when searching for evidence of a small convective core. Just like the average value of δv_{02},

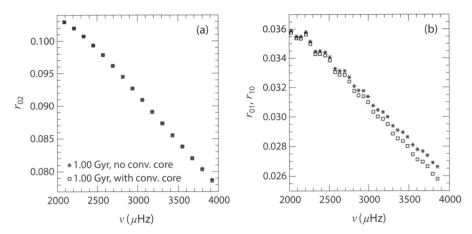

Figure 9.5. The frequency ratios r_{02} (left panel), and r_{01} and r_{10} (right panel) for two $1.16M_\odot$ models of the same age and metallicity. One model was constructed to have a convective core by assuming mixing in the core. Note that for the model with a convective core, the ratios r_{01} and r_{10} have a steeper slope with respect to frequency.

the average value of r_{02} is often used as an age diagnostic. Additional information about the conditions of the core can be obtained from the slopes of the r_{01}, r_{10}-ν relation. It can be shown that for models with the same value of $\Delta\nu$, the slope is larger for models with a convective core than for those without (see Figure 9.5). For models with core overshoot, one can show that the ratios r_{01} and r_{10} are actually sensitive to the sharp discontinuity in the chemical abundance (and sound speed) at the edge of the convective core. The absolute value of the slope of the r_{01}, r_{10}-ν relation increases as a function of the size of the chemical discontinuity (and therefore with decreasing central hydrogen content).

The importance of the presence (or otherwise) of a convective core and of core overshoot has led to the development of somewhat more complicated frequency combinations that can be used as diagnostics. The idea behind these is to use the fact that $l = 1, l = 2$, and $l = 3$ modes sample different parts of the core, and hence their frequency differences can probe the regions at the interface of convective cores and the regions affected by overshoot. The process starts with the small separations $\delta\nu_{02}$ and $\delta\nu_{13}$ that are scaled to remove their l dependence by means of

$$D_{l,l+2}(n) = \frac{\delta\nu_{l,l+2}(n)}{4l+6} = \frac{\nu_{n,l} - \nu_{n-1,l+2}}{4l+6}. \tag{9.5}$$

The quantity

$$\Delta D(n) = D_{02}(n) - D_{13}(n) \tag{9.6}$$

is sensitive to the discontinuity in the sound speed produced at the edge of a convective core. A more robust estimate is obtained, as usual, by dividing by the large separation:

$$\Delta R(n) = \frac{D_{02}(n)}{\Delta v_1(n)} - \frac{D_{13}(n)}{\Delta v_0(n+1)}. \tag{9.7}$$

Since $l = 3$ frequencies can be hard to detect (and when they are, they usually have large uncertainties), averaging the quantities can help reduce propagated errors. One quantity that works well is

$$\eta(n_1, n_2) = \left(\frac{\Delta v_\odot}{\Delta v}\right)^2 \langle[D_{13}(n) - D_{02}(n)]^2\rangle_{n=n_1 \text{ to } n_2}, \tag{9.8}$$

where n_1 and n_2 mark the limits of n over which the quantities are estimated. The multiplication by the ratio of the large separations is not strictly required; however, it helps put η for all models and stars on the same frequency scale. It is worth noting that $\Delta D(n)$, $\Delta R(n)$, $\eta(n_1, n_2)$, and $D_{13}(n)$ require $l = 3$ modes to be observed, something that is often not possible.

One factor to keep in mind while using frequency combinations and ratios is that the errors of adjacent data points will be correlated. Thus one needs to consider the full error-covariance matrix to determine the goodness of fit. Another important issue is that when comparing separation ratios of two models, or of a model with that of a star, the comparison should be done by interpolating to the same frequency, rather than the same value of n. This is more consistent with the theory governing small separations and separation ratios.

9.2 FITTING AND INTERPRETING THE SIGNATURE OF ACOUSTIC GLITCHES

The signatures of acoustic glitches, introduced in Section 3.4, can be used to derive some properties of a star. We saw that the usual glitches are those from the base of the convection zone and the helium ionization zones. With good data one can estimate the helium abundance in the region where ionization occurs (i.e., the convective envelope; stars without an outer convection zone do not show solar-like pulsations), and if the data allow, the amount of overshoot.

The change in frequencies caused by acoustic glitches is usually very small—fractions of a micro-Hertz or smaller. We have seen that one way to make the changes visible is to take the second difference of the frequencies:

$$\delta^2 v(n, l) = v_{n-1,l} - 2v_{n,l} + v_{n+1,l}. \tag{9.9}$$

This process removes the large, smooth, change of frequency with n and amplifies the oscillatory signatures. In Figure 9.6 we plot the second difference of the frequencies of some solar models, showing how they depend on changes in the extent of overshoot or the helium abundance. Fitting the signal of acoustic glitches does not require us to take second differences. However, since the process of taking the differences reduces the large, smooth background and increases the amplitude of the oscillatory term, it makes the process of fitting easier. The disadvantage of fitting the second differences

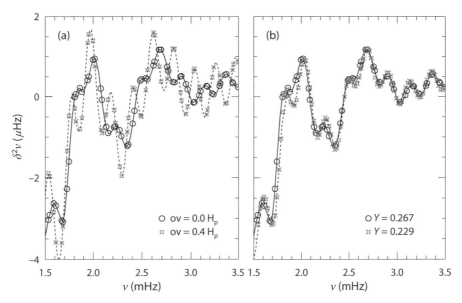

Figure 9.6. Second-differences of frequencies for two solar models with different amounts of overshoot (left panel), and different helium abundances (right panel). The solid and dotted lines are merely splines drawn through the points to guide the eye.

is that the errors of the points are not independent, and hence we have to take the full error-covariance matrix into account while fitting a model of the signatures to the data.

9.2.1 Simple Models of Glitch Signatures

Properties of a star, such as the acoustic depth of the convection-zone base, are determined by fitting models of the signatures left by acoustic glitches to the frequencies or the second differences. Of course, to do this we need to have a model of that signature, but there is no real consensus yet about the best model to fit. It is usual to use fairly simple models. The signatures of the glitches (specifically, their acoustic depths and amplitudes) can be obtained directly by fitting the frequencies or their second differences. Determining the amount of overshoot, or the helium abundance, requires calibration of the fitted amplitudes with those obtained for, respectively, models with known amounts of overshoot and known helium abundances. Figure 9.6 clearly shows that the amplitude of the oscillatory signal caused by acoustic glitches is frequency dependent and decreases with frequency. Thus whatever model we fit to the data needs to reproduce this feature. Intermediate- and high-degree modes are known to have l-dependent signatures, but that can be ignored when fitting low-degree modes that are available for stars other than the Sun.

One simple model that has been used to fit second differences is

$$\delta^2\nu = \left(a_1 + a_2\nu + \frac{a_3}{\nu}\right) + \left(b_1 + \frac{b_2}{\nu}\right)\sin(4\pi\nu\tau_{\text{He}} + \phi_{\text{He}})$$
$$+ \left(c_1 + \frac{c_2}{\nu}\right)\sin(4\pi\nu\tau_{\text{CZ}} + \phi_{\text{CZ}}). \tag{9.10}$$

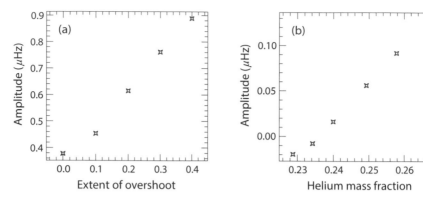

Figure 9.7. Left panel: Average amplitude of the convection-base signal in different solar models plotted as a function of the amount of overshoot. Right panel: Average amplitude of the helium signal plotted as a function of the helium abundance of different solar models of the same metallicity. The signals were fitted using Eq. 9.10, and the average amplitude was calculated between 1.5 and 3 mHz. Note that this is the amplitude in the second differences of the frequencies; the amplitude of the signal in the frequencies will be much smaller.

The terms in the first parentheses define the smooth trend in the second differences. The first sinusoidal term is the oscillatory signature of the helium ionization zone, and the second sinusoidal term is that of the convection-zone base. τ_{He} and τ_{CZ} are the two acoustic depths, ϕ_{He} and ϕ_{cz} are phases of the oscillatory signals. The amplitude of the helium signal is determined by b_1 and b_2, and that of the convection-zone base is determined by c_1 and c_2. The quantities a_1, a_2, a_3, b_1, b_2, c_1, c_2, τ_{He}, τ_{CZ}, ϕ_{He}, and ϕ_{cz} can be determined by a least-squares type fit.

We used Eq. 9.10 to fit the second differences of low-degree modes of solar models constructed with different amounts of overshoot and different helium abundances. The variation of the average amplitudes of the two oscillatory terms is shown in Figure 9.7. Results such as these, but using higher-l modes, have been used to determine the amount of overshoot below the solar convection zone.

Another model often used to fit the second differences is

$$\delta^2 \nu = a_0 + a_1 \nu + \frac{b}{\nu^2} \sin(4\pi \nu \tau_{CZ} + \phi_{CZ})$$
$$+ c_0 \nu \exp(-c_1 \nu^2) \sin(4\pi \nu \tau_{He} + \phi_{He}), \tag{9.11}$$

where we now have nine free parameters: a_0, a_1, b, τ_{CZ}, ϕ_{CZ}, c_0, c_1, τ_{He}, and ϕ_{He}. Unlike the model in Eq. 9.10, this model, particularly the term dealing with the signature of helium ionization, has a more physical basis.

The added complexity that arises when dealing with correlated errors can be avoided by fitting the glitch signatures in the frequencies directly. In this case Eq. 9.11 has to be modified to add a polynomial in frequency to account for the predominant smooth behavior of the frequencies. Thus we have

$$\nu_{nl} = P_{l,C}(n) + \frac{b}{\nu^2} \sin(4\pi \nu \tau_{CZ} + \phi_{CZ})$$
$$+ c_0 \nu \exp(-c_1 \nu^2) \sin(4\pi \nu \tau_{He} + \phi_{He}), \tag{9.12}$$

where $P_{l,C}(n)$ is a polynomial that represents the predominant smooth trend as a function of n in the frequencies of modes of a given degree and is represented as $P_{l,C}(n) = \sum_{i=0}^{4} A_{l,i} n^i$. Thus there are a total of $(l_{max} + 1) \times 5$ coefficients defining this term, where l_{max} is the highest degree available. There are seven other free parameters that need to be fitted. Equation 9.12 cannot be fitted using a simple least-squares fit. Instead one needs to smooth $P_{l,C}$, and this can be done using a regularized least-squares fit with a second-derivative smoothing.

Frequency ratios can also be used to determine the acoustic depth of glitches. In particular, $\delta\nu_{01}$ and $\delta\nu_{10}$ (defined in Eq. 3.74) and their corresponding ratios,

$$r_{01}(n) = \frac{\delta\nu_{01}(n)}{\Delta\nu_1(n)} \quad \text{and} \quad r_{10}(n) = \frac{\delta\nu_{10}(n)}{\Delta\nu_0(n+1)}, \tag{9.13}$$

can be used. Glitches, such as those discussed above, imprint their signatures on the ratios, and a Fourier transform of the residuals obtained after subtracting out a smooth global trend yields the position of the acoustic depth of the convection-zone base. The higher-order differences dd_{01} and dd_{10} (Eq. 9.3) and their corresponding ratios can be used to determine the acoustic depth of the base of the convection zone. However, this combination suppresses the signature of the HeII ionization zone.

All the expressions described above use p mode frequencies only. Mixed modes can confuse the signatures, and hence these methods work best for main sequence stars. For more evolved stars, the radial modes are the most promising, though sometimes $l = 2$ modes can be used as well.

9.2.2 Interpreting Results of the Fits

Since a real discontinuity exists at the base of the convection zone, the signature is easy to interpret. The acoustic depth τ_{CZ} can be determined unambiguously. In the absence of overshoot, it is simply the acoustic depth of the convection-zone base. When overshoot is present, the measured acoustic depth is that from the inner edge of the overshooting layer. There is often a systematic difference in the estimates of the acoustic depths of the glitches of models obtained by fitting the frequencies and those obtained from integrating the sound-travel time. This leads to an uncertainty in translating acoustic depths to actual depths. The difference between the two estimates is a result of the uncertainty in determining the acoustic surface of a star. An acoustic depth of zero is not the $r = R$ surface. In models, the surface defining an acoustic depth of zero is usually assumed to lie at the outermost grid point.

The offset between the two estimates of the acoustic depth of a glitch can be determined quite easily. If we consider a model without overshoot, the location of the glitch at the base of the convection zone is well defined—it is the point where $\nabla_{ad} = \nabla_{rad}$. The acoustic depth of this layer can be determined by fitting the signatures of the glitch and by integrating the sound-travel time. The difference between the two estimates can then be applied to the depths of all other glitches. Of course, this cannot be done for actual stars, though experiments with solar models indicate that correcting the offset in the models is enough to give a better comparison of the models with observations. One way to reduce the problems caused by the offset is to use the acoustic radius $T = \tau_0 - \tau$, where τ_0 is the total acoustic depth of the star. The acoustic radii of the base of the convection zone and the helium ionization zone can then be expressed respectively as

$$t_{CZ} = \int_0^{r_{cz}} \frac{dr}{c_s},$$

(9.14)

and

$$t_{HeII} = \int_0^{r_{HeII}} \frac{dr}{c_s}.$$

(9.15)

The value of the total acoustic depth is the same as the total acoustic radius of a star, and the two quantities are related to the large frequency separation by

$$\tau_0 = T_0 \simeq \frac{1}{2\Delta\nu}.$$

(9.16)

Since $\Delta\nu$ can be measured robustly, comparing the acoustic radius of the convection-zone depth in models with those in stars gives more robust results.

Issues in fitting the convection-zone signal arise mainly from its small amplitude. In the presence of data errors, the signal is difficult to retrieve properly. A different issue that arises if the convection zone is very deep is that of aliasing: instead of seeing a signature at τ_{CZ}, we see one at $\tau_0 - \tau_{CZ}$. This problem is exacerbated when the uncertainties in the frequencies are large.

The signature of the helium ionization zone is a bit more difficult to interpret. There is no true acoustic glitch in the case of helium ionization, but rather the signature arises because the change in Γ_1 is not a smooth one (see in Figure 9.8) but has dips. It was initially believed that the fitted value of τ_{HeII} corresponded to the dip caused by HeII ionization (D3 in Figure 9.8). Numerical experiments show that the τ_{HeII} obtained from the signatures of the glitch instead arises from the vicinity of point P2.

Another issue of practical concern is that the depression in Γ_1 in the HeII ionization zone in low-mass stars is small compared to that in higher-mass stars, and it becomes difficult to extract the signatures from noisy data. The depth of the depression in Γ_1 increases with mass, and thus we would expect that it would become easier to fit the signal in higher-mass stars. In reality, fits for higher-mass stellar models are difficult, because the HeII ionization zone and the base of the convection zone are relatively close to each other, which confuses the fits to the two signals. An added difficulty arises because these stars have a large composition gradient at the boundary of the convective core, which results in a strong peak in the buoyancy frequency. This introduces additional effects that are not modeled by the fitting functions that are usually used. A mass range of 0.8–$1.2 M_\odot$ seems to be ideal for carrying out the analysis of the helium ionization zone.

Interpreting the amplitudes of the glitches requires calibration using models with known properties. For a star of a given mass, the amplitude of the oscillatory signal caused by the helium glitch depends on the amount of helium—the higher the helium abundance, the higher will be the amplitude of the signal. This trend can be used to determine the convection-zone helium abundance of a star. The amount of helium present is determined by constructing models of a star with different helium abundances and comparing the amplitude of the helium signature found in

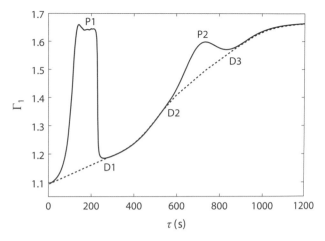

Figure 9.8. Adiabatic index Γ_1 as a function of acoustic depth of a solar model constructed with an extended atmosphere. The extended atmosphere ensures that the acoustic depth of the convection-zone base (as measured from the glitch signature) was the same as that calculated from the sound-speed profile. The dotted curve is a visual guide to the smooth change in Γ_1 that does not contribute to the oscillatory signal. D1 marks the dip due to the HI+HeI ionization zone, D3 is the dip due to the HeII ionization zone, and P1 and P2 are the peaks in between. Figure courtesy of K. Verma; see Verma et al., 2014, *A Theoretical Study of Acoustic Glitches in Low-Mass Main-Sequence Stars, Astrophys. J.,* **794**, 114, copyright American Astronomical Society.

the observed data with amplitudes from models. However, since we do not know the exact mass, radius, temperature, and metallicity of other stars, unlike the solar case shown in Figure 9.6, the calibration has to be done using models that differ in all these quantities. This is time-consuming but not difficult. Figure 9.9 shows the variation of the amplitude with helium for the the Sun-like stars 16 Cyg A and B. One point that should be noted is that the presence of noise in the real data usually causes an apparent increase in the amplitude of the signal; as a result, it is advisable to add random realizations of noise to the frequencies of the models before determining the amplitude of the helium signal and comparing with observations.

9.3 ANALYSIS OF RED GIANTS

The frequency spectra of red giants are sufficiently different from those of other stars that often very different diagnostics are used to determine their properties. Additionally, one major complication arises when studying these stars: there are two types of red giants. The first type are stars that are on the ascending part of the red giant branch (RGB stars). They have inert helium cores with hydrogen fusion occurring in a narrow shell around the core. The second type has cores with ongoing helium fusion as well as a shell of hydrogen fusion. High-metallicity stars of this type are usually referred to as red-clump (RC) stars, and in low-metallicity systems these stars form the horizontal branch in the HR diagram. Secondary clump stars are those that have high enough masses to begin helium fusion without going

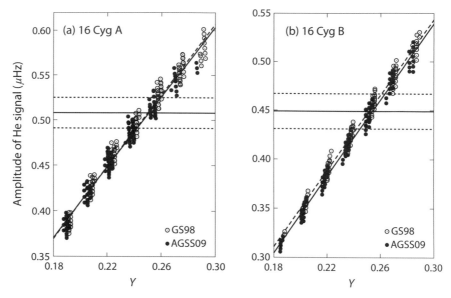

Figure 9.9. Amplitude of the helium signal (obtained using Eq. 9.11) as a function of Y for 16 Cyg A and B. The horizontal line is the observed amplitude, and the dashed lines flanking it represent 1σ uncertainty limits. The circles are the results for models. The solid line is a straight-line fit to the filled circles, and the dashed line is a fit to the open circles. The filled and open circles represent models constructed assuming two different solar metallicity estimates. Figure courtesy of K. Verma; see Verma et al., 2014, *Asteroseismic Estimate of Helium Abundance of a Solar Analog Binary System*, Astrophys. J., **790**, 138, copyright American Astronomical Society.

through a helium flash. Both types of stars share the region of the HR diagram with $\log(L/L_\odot) \simeq 1.5$–2, and surface gravities of about $\log g \simeq 2.5$. They also occupy the same range of effective temperatures. This makes it difficult to distinguish RGB stars from RC stars unless they are in clusters; yet it is important to do so, since the narrow range of luminosities for RC stars makes them good standard candles. Period spacings of $l = 1$ modes can help distinguish the two types of red giants.

Soon after the detection of mixed modes in red giants, it was found that the period spacing of the modes could be determined quite well. What the initial data showed is that these two types of stars have very different values of ΔP for $l = 1$ modes—the helium-burning stars having larger values than for the stars with inert cores. This is shown in Figure 9.10.

The reason for the difference in ΔP is quite easy to understand. ΔP is related to $\Delta \Pi_1$, which is given by Eqs. 3.79 and 3.80. The limits of the integral in Eq. 3.80 are different for inert-core red giants and red-clump stars. RGB stars have radiative cores, and thus for these stars, we have

$$\Delta \Pi_0(\text{RGB}) = 2\pi^2 \left[\int_0^{r_2} N \frac{dr}{r} \right]^{-1}. \tag{9.17}$$

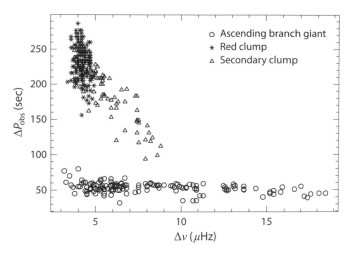

Figure 9.10. Observed average g mode-period spacing plotted as a function of $\Delta\nu$ for a sample of red giants observed by *Kepler*. The data show how red giants on the ascending branch have lower values of ΔP than do stars on the red clump. Secondary clump stars are those that have high enough masses to begin helium fusion without going through a helium flash. Data are from Bedding, T. R., et al., 2011, *Gravity Modes As a Way to Distinguish between Hydrogen- and Helium-Burning Red Giant Stars, Nature* **471**, 608.

However, stars undergoing helium fusion have convective cores. Thus the lower limit of the integral is the edge of the core r_c, so that

$$\Delta\Pi_0(\text{RC}) = 2\pi^2 \left[\int_{r_c}^{r_2} N \frac{dr}{r} \right]^{-1}. \tag{9.18}$$

Since the term in the denominator of Eq. 9.18 is smaller than that in Eq. 9.17, it follows that RC stars should have larger period spacings than do RGB stars.

9.3.1 Estimating Red-Giant Period Spacings

To determine properties of red giants using their p mode spacings, we need to estimate them either from their frequencies or directly from their power spectra.

The observed nonradial modes in red giants are mixed modes heavily dominated by g modes. However, we are not able to detect all the modes in red giants (i.e., some modes are either not excited or have very low amplitudes). The nonradial modes that are the easiest to detect are those that have the most p-like nature. This limits the sample of frequencies that is available to us and biases estimates of $\Delta\Pi$.

We can of course calculate all frequencies and period spacings for models. When we do so, the resulting calculated period spacings do indeed approach the asymptotic limit. The period spacings between observed modes are generally smaller than the asymptotic spacing, which is not surprising, since the observed frequencies are mixed modes that can be quite p mode-like in character. This is illustrated in Figure 9.11. Note how the modes with higher inertia (the g-dominated modes) have the largest period spacings, and the spacing is very close to that given by the asymptotic relation. The p-dominated modes, which have the lowest inertia and are more easily detected, have much lower period spacings.

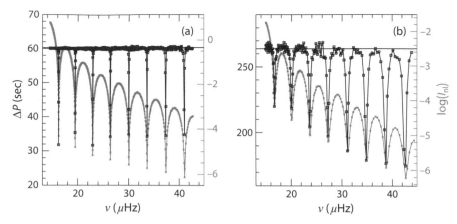

Figure 9.11. Period spacings between $l = 1$ modes for an RGB model (left panel), and an RC model (right panel) are plotted as black squares. The period spacing can be read using the axis labels on the left of each plot. Mode inertias are plotted in gray and can be read using the axes on the right of each plot. The black horizontal lines mark $\Delta\Pi_1$, the asymptotic period spacing for $l = 1$ modes.

However, the spacing between the visible modes can be related to the asymptotic spacing. For this we need to go back to the asymptotic expressions for g modes derived in Chapter 3. Recall that the frequency of a pure $l = 1$ g mode can be written as

$$\frac{1}{\nu_g} = (-n_g + \epsilon_g)\Delta\Pi_1, \qquad (9.19)$$

where the negative sign takes into account that n_g is by convention a negative number. The period of a mixed mode of order $n_m = n_p + n_g$ can be expressed in terms of what would be expected were it to be a pure g mode. Thus, the period P can be expressed as

$$P = n_m\Delta\Pi_1 + p, \qquad (9.20)$$

where the term p accounts for the fact that period spacing between modes is not strictly a constant. Our task is to determine the relationship between ΔP, the period spacing between two adjacent mixed modes, and the g mode period spacing $\Delta\Pi_1$. Using Eqs. 3.86, 3.88, and 3.89 and expressing frequencies in terms of periods, we get

$$\tan\left[\pi\left(\frac{1}{n_m}\frac{1}{\Delta\Pi_1 + p} - \frac{1}{2} - \epsilon_p\right)\right] = q\tan\left[\frac{n_m\Delta\Pi_1 + p}{\Delta\Pi_1} - \epsilon_g\right], \qquad (9.21)$$

where we have made the explicit assumption that we are working with $l = 1$ modes. Differentiating Eq. 9.21 with respect to n_m under the assumption that the p mode separation $\Delta\nu$ does not vary with frequency and noting that $\tan(n\pi) = 0$ for any n, we get

$$\frac{1}{\Delta\Pi_1}\frac{dP}{dn_m} = \left[1 + \frac{1}{q}\frac{\nu^2\Delta\Pi_1}{\Delta\nu}\frac{\cos^2\theta_g}{\cos^2\theta_p}\right]^{-1}. \qquad (9.22)$$

Note the implicit assumption that n_m is a continuous variable rather than an integer. But of course n_m is an integer, and if we assume $dn_m = \Delta n_m = 1$, then we can rewrite dP/dn_m as ΔP to obtain:

$$\frac{\Delta P}{\Delta \Pi_1} = \left[1 + \frac{1}{q} \frac{v^2 \Delta \Pi_1}{\Delta v} \frac{\cos^2 \theta_g}{\cos^2 \theta_p} \right]^{-1}. \tag{9.23}$$

Recalling Eq. 6.152, here we actually have

$$\frac{\Delta P}{\Delta \Pi_1} = \left[1 + \frac{f_1(v)}{q f_2(v)} \right]^{-1} \equiv \zeta. \tag{9.24}$$

Note that it is immediately obvious that $\Delta P < \Delta \Pi_1$. This explains the dip in the period spacings in Figure 9.11.

Recall from Section 6.5.2 that the quantity ζ measures the ratio of the inertia of the g mode part of the cavity to the inertia of the entire cavity, and it comes into play in determining rotational splittings for red-giant mixed modes (Eq. 6.158). As is clear from Eq. 9.24, this quantity also determines what the mixed-mode period spacing will be relative to the pure g mode spacing.

Since we cannot observe the pure g modes of a red giant, determining the asymptotic period spacing is nontrivial. Here we discuss two reasonably robust methods, beginning with one that makes use of individual frequencies that have already been extracted from the oscillation spectrum.

Determining $\Delta \Pi_1$ Using Observed Frequencies

This method is analogous to determining Δv for a main sequence star using the frequency échelle diagram. If the guessed value of Δv is correct, modes of a given degree align vertically in the $(v \bmod \Delta v)$–Δv plot; if the guessed value of Δv is incorrect, they will not. In the case of red giants, we use the fact that the g modes are equidistant in period rather than frequency and we make a guess for $\Delta \Pi_1$ and plot frequency as a function of $(P \bmod \Delta \Pi_1)$. If the modes line up, we have determined $\Delta \Pi_1$ correctly. This is shown in Figure 9.12, where we have plotted period échelle diagrams for $l = 1$ modes of one model, but for three different guessed values of $\Delta \Pi_1$. As can be seen, in only one case are the most g-like modes aligned vertically. If the modes do not line up, the period échelle diagram looks instead like the ones shown in the left and right panels of Figure 9.12. The problem of course is that we cannot generally detect the modes that line up to form the vertical structure. Thus the question becomes how we can objectively determine the value of $\Delta \Pi_1$ that gives the best vertical alignment. Here we introduce a method called the giant period spacing (GPS) finder.

The first step in the GPS finder is to establish a lower threshold of frequencies used in the process. This step is necessary because the asymptotic period spacing is strictly valid only for high-order (i.e., low-frequency) g modes. At low frequencies the period spacing is closer to the asymptotic value, though somewhat different from the period spacing of the higher-frequency modes. This manifests as curvature in the period échelle and can be seen clearly in Figure 9.12. This curvature can interfere with the process, and hence low-frequency modes should be avoided. One consequence is

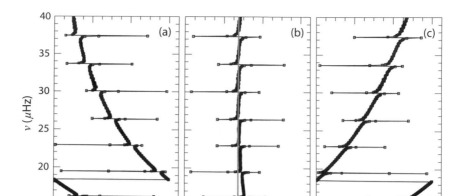

Figure 9.12. Period échelle diagrams for $l = 1$ modes of the model shown in the left panel Figure 9.11 for different values of $\Delta\Pi_1$. The gray vertical line in the middle panel (b) has been plotted to guide the eye. Note that only one of the three guessed values of $\Delta\Pi_1$ results in a clean vertical alignment.

that the value of $\Delta\Pi_1$ obtained by this method can be slightly lower than the true value. In the case of the model shown in Figures 9.12 and 9.13, the value of $\Delta\Pi_1$ calculated using the buoyancy frequency was 60.25 seconds, a value that does not give us a vertical alignment. The lower-frequency threshold depends on the star in question and is related to its ν_{max}. Numerical experiments with models show that the following set of thresholds ensures robust results:

$$\nu_{threshold} = \begin{cases} 7.5\,\mu\text{Hz}, & \text{for } \nu_{max} < 15\,\mu\text{Hz} \\ 0.5\nu_{max}, & \text{for } 15\,\mu\text{Hz} \leq \nu_{max} \leq 160\,\mu\text{Hz} \\ 80\,\mu\text{Hz}, & \text{for } \nu_{max} \geq 160\,\mu\text{Hz}. \end{cases} \quad (9.25)$$

For ease of further discussion we define for any mode

$$P_m = P \text{ modulo } \Delta\Pi_1, \quad (9.26)$$

where P is the period of that mode.

The second step in the GPS finder is to identify bands of $l = 1$ modes, where a band is defined as all $l = 1$ modes that lie between two adjacent $l = 0$ modes. These bands have been demarcated by the horizontal dashed lines in Figure 9.13. Not all bands are useful. For a band to be useful for determining $\Delta\Pi_1$, it must satisfy two conditions. First, P_m for the highest-frequency mode in the band should be lower than P_m for the lowest-frequency mode in the band; and second, there should be one, and only one, near-horizontal shift in P_m in any given band. This ensures that there are no missing modes that can affect results for ΔP and hence $\Delta\Pi_1$. Note that the bands shown in Figure 9.13 satisfy both conditions.

The third step in the method is to define a measure of the vertical alignment. For any given guess of $\Delta\Pi_1$, one first estimates P_m for the most g-like mode at the upper

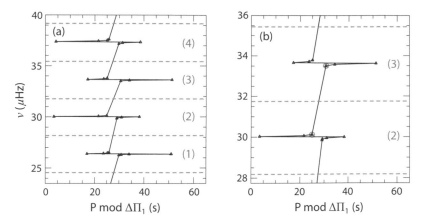

Figure 9.13. Left panel: Period échelle diagram shown in the center panel of Figure 9.12 but now showing only the modes we expect to observe. The horizontal dashed lines mark the positions of radial modes, and the numbers in gray labels the intervals (or bands) between two radial modes. Right panel: Close-up of two of the bands. The modes marked with circles are used to estimate the quantity P modulo $\Delta\Pi_1$ for the most g-like mode at the boundary between the bands.

boundary of any band i in the period échelle diagram; we call this $P^i_{m,g}$ for simplicity. Then one may calculate the dispersion between the values of $P_{m,g}$ for each band. The value of $\Delta\Pi_1$ that minimizes the dispersion is the best estimate of the asymptotic period spacing for the star. Thus to get the best vertical distribution possible, we minimize

$$\chi^2_{\text{vert}} = \sum_i^N \left(\frac{P^i_{m,g} - \langle P_{m,g} \rangle}{\sigma(P^i_{m,g})} \right)^2, \tag{9.27}$$

where $\langle P_{m,g} \rangle$ is the average $P_{m,g}$ over all bands, and $\sigma(P^i_{m,g})$ is the uncertainty in $P_{m,g}$.

Of course to carry out the minimization in Eq. 9.27, we need to know what $P^i_{m,g}$ is. This is defined as the arithmetic mean of P_m for the highest-frequency $l = 1$ mode in one band and that of the lowest-frequency mode in the band immediately above. These modes are marked with circles in the right panel of Figure 9.13 for the second band. These modes are the two immediate neighbors that demarcate the bands. There is one potential problem: the two $l = 1$ modes may not be equidistant in frequency from the $l = 0$ mode, which could skew the estimate of $P_{m,g}$. The solution to this problem is to find a weighted mean:

$$P^i_{m,g} = \frac{y^i P^{i+1,l}_m + y^{i+1} P^{i,h}_m}{y^i + y^{i+1}}, \tag{9.28}$$

where $P^{i+1,l}_m$ is P_m for the lowest-frequency mode in the $(i + 1)$th band, and $P^{i,h}_m$ is that for the highest-frequency mode in the ith band. The weights are then defined as

$$y^{i+1} = v^{i+1,l}_{l=1} - v^{i,i+1}_{l=0}, \tag{9.29}$$

and

$$y^i = v_{l=0}^{i,i+1} - v_{l=1}^{i,h}, \tag{9.30}$$

where $v_{l=0}^{i,i+1}$ is the frequency of the radial mode that demarcates the ith and $(i+1)$st bands, $v_{l=1}^{i+1,l}$ is the frequency of the lowest-frequency mode in the $(i+1)$th band and $v_{l=1}^{i,h}$ is the frequency of the highest-frequency mode in the ith band. When calculating the denominator of Eq. 9.27, matters are simplified if one assumes that the weights y in Eq. 9.28 do not have errors, and only errors in P and hence P_m are considered. Under that assumption, we have

$$\sigma(P_{m,g}^i) = \frac{1}{2}\sqrt{\sigma(P_m^{i+1,l})^2 + \sigma(P_m^{i,h})^2}, \tag{9.31}$$

and

$$\sigma(P_m^{i+1,l}) = \frac{\sigma(v_{l=1}^{(i+1,l)})}{(v_{l=1}^{(i+1,l)})^2}, \quad \text{and} \quad \sigma(P_m^{i,h}) = \frac{\sigma(v_{l=1}^{(i,h)})}{(v_{l=1}^{(i,h)})^2}. \tag{9.32}$$

One may also add another condition, that of symmetry of the observed modes around the central ridge. This is certainly true for high-Δv stars, but it becomes less and less accurate as Δv decreases. Although one may provide weightings for the condition of symmetry as a function of Δv, minimizing χ_{vert}^2 alone seems to work well.

Determining $\Delta\Pi_1$ Directly from the Frequency-Power Spectrum

Individual frequencies of red giants can only be determined precisely for stars that have data with high SNRs. The previous method of determining $\Delta\Pi_1$ depends on knowing the individual frequencies of many modes. Here we discuss a second method that is useful for lower-SNR data, which extracts an ensemble signature of the modes and does not rely on estimates of individual frequencies. This method makes use of the fact that the period spacing of observed $l = 1$ modes is related to the asymptotic period spacing, $\Delta\Pi_1$ (as per Eq. 9.24). For ease of discussion, this may be rewritten as

$$\Delta P = \zeta\Delta\Pi_1, \tag{9.33}$$

where ζ is defined Eq. 9.24. Thus the function ζ depends on the coupling constant q as well as the asymptotic spacing. Equation 9.33 shows that the spacing ΔP between the mixed modes is not constant. As ζ approaches unity, ΔP approaches $\Delta\Pi$. The function ζ for the model in the left panel of Figure 9.11 is shown in the left panel of Figure 9.14. To get a constant spacing between the mixed modes, we need to correct or *stretch* the frequency axis of the frequency-power spectrum until the observed mixed modes have equal-period spacings.

Each bin in the spectrum may be stretched according to ζ. We introduce τ as the independent variable for the modified spectrum, which we define as

$$d\tau = \frac{1}{\zeta}\frac{dv}{v^2}. \tag{9.34}$$

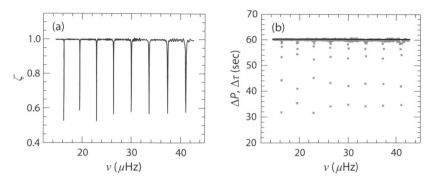

Figure 9.14. Left panel: Values of ζ as a function of frequency for $l = 1$ modes of the RGB model shown in Figure 9.11. Right panel: Period spacings (in gray) of the $l = 1$ modes, and the spacings in the stretched coordinate τ (black).

Clearly τ has dimensions of time. The factor $1/\zeta$ accounts for the stretch, and ν^{-2} is needed to change from the frequency to the time (or period) domain. Mathematically, the change of variables corresponds to a transformation between ν_{n_m} and $(n_m - n_0)\Delta\Pi_1$, where n_m is the order of the mixed mode, and n_0 is an arbitrary constant. This transformation allows us to ignore the actual order of the modes, which we could not do if we were to calculate mixed-mode frequencies using Eq. 9.21. Using the correct transformation, the mixed modes will be equidistant in $\Delta\tau$ even if ζ is only approximately known; this is shown in the right panel of Figure 9.14.

The first practical step toward estimating $\Delta\Pi_1$ is to determine $\Delta\nu$ and ν_{max}. Knowledge of ν_{max} allows us to focus on the most reliable part of the power spectrum, and knowing $\Delta\nu$ allows us to divide the spectrum into regions that contain the $l = 1$ modes and regions that contain the $l = 0$ and $l = 2$ modes. Since there are no $l = 0$ mixed modes, and since the $l = 2$ mixed-mode pattern is different from that of $l = 1$ modes, those regions of the frequency-power spectrum that contain the radial and quadrupole modes are removed. The signal from granulation can cause difficulties, and hence, the process is best applied to power spectra that have already been corrected for the background (i.e., by dividing out a best-fitting background model).

The next step is to choose an initial guess for $\Delta\Pi_1$. To do this, one can rely on the $\Delta\nu$–$\Delta\Pi_1$ results for stars for which $\Delta\Pi_1$ is already known. There is no ambiguity in the initial guess for $\Delta\Pi_1$ provided stars have $\Delta\nu > 9.5\,\mu$Hz (i.e., those stars that can only be on the ascending part of the RGB). Stars with lower values of $\Delta\nu$ may also be on the ascending branch, and hence have an inert helium core; or they could be RC or secondary-clump stars with helium-burning cores, which have a different $\Delta\nu$–$\Delta\Pi_1$ relation than those on the RGB. A guessed value of the coupling parameter is also needed. For RGB stars, q is about 0.17. It can be higher for RC stars. The initial guesses allow one to calculate ζ (see also Section 6.5.2), and the frequency spectrum may then be stretched accordingly.

To retrieve the underlying $\Delta\Pi_1$, the Fourier spectrum of the stretched spectrum is computed. This is of course analogous to the PSPS method (Section 6.2.1) for estimating $\Delta\nu$. Here we look for a strong peak at $\Delta\Pi_1$. The retrieved $\Delta\Pi_1$ is reasonably insensitive to the initial guesses for $\Delta\Pi_1$ and q; however, an iterative

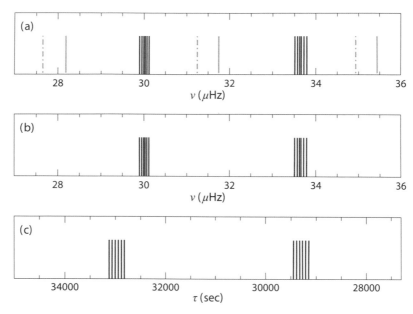

Figure 9.15. Top panel: Part of a pseudo-oscillation spectrum of a red giant. The gray solid and dot-dashed lines mark the radial and quadrupole modes respectively, and the black lines are the dipole modes. Middle panel: The spectrum after the radial and quadrupole modes have been removed. Bottom panel: The $l = 1$ modes plotted as a function of the stretched coordinate τ calculated with a guessed value of ζ. Note that the modes appear equidistant, and hence the Fourier transform should reveal the underlying spacing.

procedure is recommended to minimize any dependences. A schematic diagram of the process is shown in Figure 9.15.

Stretching the power spectrum makes it easier to determine other characteristics of a red giant, in particular rotational splittings. An échelle diagram against the stretched period τ will show multiple ridges corresponding to the different rotationally split components of the modes.

9.3.2 An Alternative Method for Distinguishing Between Red Giant and Red Clump Stars

The period spacing is not the only distinguishing quantity between helium-burning RC stars and RGB stars. The phase function ϵ determined around ν_{max} can also be used for this purpose.

The phase shift ϵ changes with the evolutionary state of a star, and that can be exploited to distinguish RC stars from stars on the RGB. The frequency-dependence of ϵ makes it somewhat difficult to use. However, the *central* value of ϵ—given by the three radial modes around ν_{max}—contains the necessary information.

When we consider the central, radial modes alone, the asymptotic expression for p modes can be written as

$$\nu_{c0} = \Delta\nu_c(n + \epsilon'_c), \tag{9.35}$$

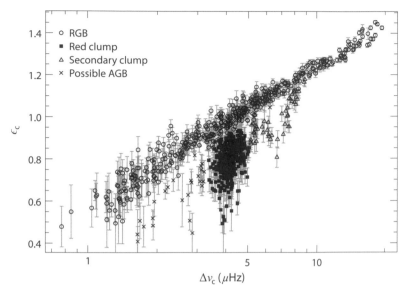

Figure 9.16. Central value of ϵ plotted against the large separation for a sample of giants observed by *Kepler*. Data from Kallinger et al., 2012, *Astron. Astrophys.* **541**, A51, courtesy of T. Kallinger.

where $\Delta \nu_c$ is the separation between the central modes. Knowing $\Delta \nu_c$ and ν_{c0}, one can determine ϵ_c using the procedure outlined in Eqs. 6.89 and 6.90, that is, here we will have:

$$\epsilon_c = \begin{cases} \epsilon'_c + 1, & \text{if } \epsilon'_c < 0.5 \text{ and } \Delta\nu > 3\mu\text{Hz} \\ \epsilon'_c & \text{otherwise,} \end{cases}$$

with

$$\epsilon'_c = \left(\frac{\nu_{c0}}{\Delta \nu_c}\right) \bmod 1. \tag{9.36}$$

The phase shifts of the central radial modes of a sample of *Kepler* stars are shown in Figure 9.16. When the evolutionary status of the stars is determined from their period spacings, RC stars clearly stand out (see Figure 9.10). However, the separation between RGB and RC stars is not as clear-cut when ϵ_c is used. But, it is much easier to determine ϵ_c than ΔP, since the former quantity is determined from $l = 0$ modes. Thus, in principle, the ϵ method can be used to determine the evolutionary state of lower-SNR stars where the closely spaced $l = 1$ modes are difficult to resolve. The large frequency separations of RC stars occupy a very narrow range of frequencies, which makes their identification in an ϵ_c–$\Delta\nu$ plot relatively easy.

9.3.3 Additional Diagnostic Potential of Period Spacings

Period spacings of red giants depend on their mass and evolutionary phase; thus the spacings have additional diagnostic potential. The variation of $\Delta\Pi_1$ with mass and

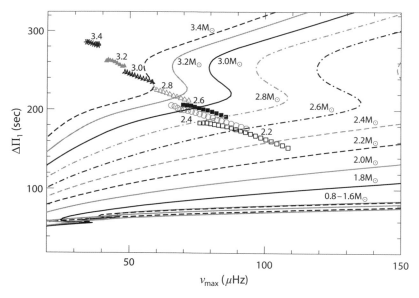

Figure 9.17. Change in $l = 1$ period spacings as a red giant evolves. The lines represent ascending-branch giants. The symbols are helium-burning stars. Note that the change in $\Delta\Pi_1$ is dramatic for higher-mass stars but much more subtle for the low-mass ones.

evolution can be seen in Figure 9.17. In this figure, stars evolve from the right to the left as they climb the RGB. They disappear along the left edge and then reappear as helium-burning stars. The figure does not show the position of true RC stars; recall these are stars that undergo a helium flash at the tip of the giant branch. The secondary-clump stars (i.e., stars with high enough masses to ignite helium gently) are also marked.

For stars of all masses, $\Delta\Pi_1$ slowly decreases as they ascend the RGB, and higher-mass stars have a higher value of $\Delta\Pi_1$ at a given value of ν_{max}. At the low-mass end, the curves are bunched together, and may occasionally intersect near the red-giant bump. The low-mass end can be seen better in the top panel of Figure 9.18. There are measurable differences, at least in the lower part of the RGB. How well period spacings can be used to determine masses for low-mass stars will thus depend on the uncertainties in the measurements of $\Delta\Pi_1$.

The period spacing has a small dependence on metallicity. We can see this in the top panel of Figure 9.18. However, since red giants of different metallicities occupy different temperature ranges, the change in $\Delta\Pi_1$ because of metallicity at a given temperature can be quite large, as can be seen in the bottom panel of Figure 9.18.

The period spacing contains information that is very different from that in $\Delta\nu$ and ν_{max}. The period spacing thus provides us with more information that can be used to infer the properties of red giants. In the simplest case, $\Delta\Pi_1$, along with the usual inputs of $\Delta\nu$, ν_{max}, T_{eff}, and metallicity, can be used in a grid-based search (Chapter 7) to infer stellar properties. This should be particularly useful, since the uncertainties in temperature, combined with the narrow range of temperatures that these stars have, means that temperature is not always a very useful quantity for determining properties of red giants.

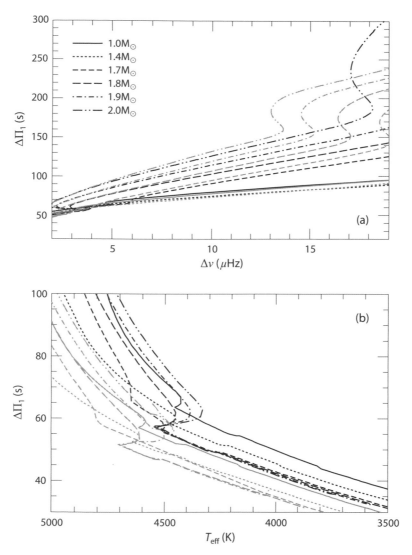

Figure 9.18. The value of $\Delta\Pi_1$ with $\Delta\nu$ (top panel), and T_{eff} (bottom panel) for models constructed with two different values of metallicity. The black lines are for [Fe/H] = 0 and the gray lines are for [Fe/H] = -0.5. The different line types denote different masses as indicated in the top panel. A different range of $\Delta\Pi_1$ is shown in the bottom panel to highlight what happens at low values of $\Delta\Pi_1$.

9.4 FURTHER READING

The interpretation of frequency combinations is based on the details of the properties of stellar oscillations that we have not gone through in detail. We direct all those interested in learning more about the theory behind the combinations to the following papers:

- Roxburgh, I. W, and Vorontsov, S. V., 1994, *The Seismology of Stellar Cores: A Simple Theoretical Description of the "Small Frequency Separations"*, Mon. Not. R. Astron. Soc., **267**, 297.

- Roxburgh, I. W, and Vorontsov, S. V., 2003, *The Ratio of Small to Large Separations of Acoustic Ocillations as a Diagnostic of the Interior of Solar-Like Stars*, Mon. Not. R. Astron. Soc., **411**, 215.
- Roxburgh, I. W, 2005, *The Ratio of Small to Large Separations of Stellar p-Modes*, Mon. Not. R. Astron. Soc., **434**, 665.
- Cunha, M. S., and Metcalfe, T. S., 2007, *Asteroseismic Signatures of Small Convective Cores*, Astrophys. J., **666**, 413.
- Roxburgh, I. W, and Vorontsov, S. V., 2013, *On the Use of The Ratio of Small to Large Separations in Asteroseismic Model Fitting*, Astron. Astrophys., **560**, A2

For detailed discussions of the practical use of frequency combinations to determine stellar properties, we refer readers to:

- Mazumdar, A. et al., 2006, *Asteroseismic Diagnostics of Stellar Convective Cores*, Mon. Not. R. Astron. Soc., **372**, 949.
- Cunha, M., and Metcalfe, T. S., 2007, *Asteroseismic Signatures of Small Convective Cores*, Astrophys. J., **666**, 413.
- Cunha, M., and Brandão, I. M., 2011, *Probing Tiny Convective Cores with the Acoustic Modes of Lowest Degree*, **529**, A10.
- Silva Aguirre, V. et al., 2011, *Constraining Mixing Processes in Stellar Cores Using Asteroseismology. Impact of Semiconvection in Low-Mass Stars*, Astron. Astrophys., **529**, A63.

Details of how glitches can be—and have been—used to determine stellar properties are found in the following papers:

- Basu, S., 1997, *Seismology of the Base of the Solar Convection Zone*, Mon. Not. R. Astron. Soc., **288**, 572.
- Monteiro, M. J. P. F. G., and Thompson, M. J., 1998, *On the Seismic Signature of the HeII Ionization Zone in Stellar Envelopes*, Proc. IAU Symp. **185**, 317.
- Basu, S., et al., 2004, *Asteroseismic Determination of Helium Abundance in Stellar Envelopes*, Mon. Not. R. Astron. Soc., **350**, 277.
- Houdek, G., and Gough, D. O., 2007, *An Asteroseismic Signature of Helium Ionization*, Mon. Not. R. Astron. Soc., **375**, 861.
- Mazumdar, A., et al., 2014, *Measurement of Acoustic Glitches in Solar-type Stars from Oscillation Frequencies Observed by* Kepler, Astrophys. J., **782**, 18.
- Broomhall, A.-M., et al., 2014, *Prospects for Asteroseismic Inference on the Envelope Helium Abundance in Red Giant Stars*, Mon. Not. R. Astron. Soc., **440**, 1828.
- Verma, K., et al., 2014, *A Theoretical Study of Acoustic Glitches in Low-Mass Main-Sequence Stars*, Astrophys. J., **794**, 114.
- Verma, K., et al., 2014, *Asteroseismic Estimate of Helium Abundance of a Solar Analog Binary System*, Astrophys. J., **790**, 138.

Our discussion of the estimation of the asymptotic g mode period spacing presents the methods in the following three papers:

- Datta, A., et al., 2015, *Automated Determination of g-Mode Period Spacing of Red Giant Stars*, Mon. Not. R. Astron. Soc., **447**, 1935.
- Mosser, B., et al., 2015, *Period Spacings in Red Giants. I. Disentangling Rotation and Revealing Core Structure Discontinuities*, Astron. Astrophys., **584**, 50.

- Vrard, M., Mosser, B., and Samadi, R., 2016, *Period Spacings in Red Giants. II. Automated Measurement, Astron. Astrophys.*, **588**, A87.

The use of ϵ to distinguish between RGB and RC models is based on the following paper:

- Kallinger, T. et al., 2012, *Evolutionary Influences on the Structure of Red-Giant Acoustic Oscillation Spectra from 600d of Kepler Observations, Astron. Astrophys.*, **541**, A51.

Finally, for an example of how period spacings can be used to provide additional information for analysis, see:

- Martig, M. et al., 2015, *Young α-Enriched Giant Stars in the Solar Neighborhood, Mon. Not. R. Astron. Soc.*, **451**, 2230.

9.5 EXERCISES

1. The archives `frequencies_ov0.tar.gz`, `frequencies_ov1.tar.gz`, and `frequencies_ov2.tar.gz` contain files with frequencies of $1.4M_\odot$ and $1.5M_\odot$ models with core overshoots of 0.0, 0.1, and $0.1H_p$, respectively. The global properties of the models are in files `properties_ov[0-2].txt`. Use the files to calculate $\delta\nu_{01}$, $\delta\nu_{10}$, r_{01}, r_{10}, $\delta\nu_{02}$ and r_{02} and explore the following.

(a) How do the average values of $\delta\nu_{02}$ and r_{02} change with age? Does overshoot have any effect on the results?

(b) How does the slope of the r_{01}, $r_{10}-\nu$ relation change with age for a given mass?

(c) For a similar stage of evolution (which on the main sequence can be assumed to be a similar value of the central hydrogen content X_c), how do $\delta\nu_{01}$, $\delta\nu_{10}$ and the corresponding ratios depend on mass? Does overshoot change this relationship?

2. The archive `frequencies_helium.tar` contains frequencies of a set of $1M_\odot$, $1R_\odot$ models constructed with different convection-zone helium abundances. There are two sets of models, one constructed with the OPAL equation of state and the other with the MHD equation of state. The information about the internal structure of the models is in the archive `structure_helium.tar.gz`.

(a) Fit the signatures of the helium ionization zone and the base of the convection zone for each model using the frequencies provided. Use only the observed mode set (file `BiSON.txt`) and give weights to the modes according to the observed uncertainties. For each case compare the acoustic depth determined from the frequencies with that obtained by integrating the sound-speed profile. What is the difference?

(b) Use four of the OPAL models as test models to determine how the amplitude of the helium signature changes as a function of the helium abundance. Use the result to determine the helium abundance of the fifth OPAL model.

(c) Repeat part (b) using the MHD models.

(d) Determine what happens when you use the MHD models to calibrate the helium signature in order to determine the helium-abundance of any OPAL model and vice versa.

(e) Use the solar frequencies (BiSON.txt) and the two sets of models to determine the helium abundance of the Sun. How well can you reproduce the solar helium abundance, which has been determined using high-degree solar modes?[1]

3. Files RGB_ModelA.txt, RGB_ModelB.txt, and RGB_ModelC.txt contain frequencies of three red giant models.

(a) Find $\Delta\Pi_1$ for the three models using all $l = 1$ frequencies.

(b) Find $\Delta\Pi_1$ using only those $l = 1$ frequencies that we expect to observe, and determine the influence of uncertainties in the frequencies in determining $\Delta\Pi_1$. Note that for the value of n_p, the order of the p mode, we only expect to observe between three and five of the modes with the lowest inertias

(c) All models have the same mass. Using Figure 9.10 to determine which of the three stars are ascending-branch red giants and which are RC stars.

(d) Estimate ν_{max} for the two ascending-branch stars using Figure 9.17.

---

[1] For solar helium abundance estimates, see Basu, 1998, *Mon. Not. R. Astron. Soc.*, **298**, 719.

10 Inverting Mode Frequencies

Most asteroseismic analyses are carried out in a manner that is familiar to all astrophysicists: one obtains data on an object, tries to make models that "fit" the data and then assumes that the properties of the best-fitting model represent the properties of the observed object. When studying the Sun using solar oscillation frequencies, it was found that the surface term makes model fitting difficult. Additionally, different physics inputs resulted in very different internal structure, and judging which one was better was often difficult. As discussed in Chapter 8, the surface term makes fitting stellar data difficult, too. And yet, the solar case is easier than that of other stars—it is a well constrained system. We know the Sun's mass, radius, luminosity, and age independently, and as a result we do not need to use pulsation data to determine its global properties. While this means that we should be able to use solar frequencies to better distinguish between models, this is often not the case, especially with modern solar models. Since the surface term changes with the physics of the models, comparing frequencies becomes meaningless. Recall that in the case of other stars, we often use the solar surface term to correct for this offset; this assumes we have a good solar model, but in the presence of a surface term, it is often difficult to distinguish a good solar model from one that is not so good.

The problem with comparing frequencies led to the development of inversion techniques that allow us to determine the structure of the Sun directly using solar frequencies almost independently of models. Inversions have been the mainstay of helioseismic studies now for several decades. It is not yet a part of the normal repertoire of asteroseismic analysis techniques, but that is likely to change as we extract more precise and larger numbers of mode frequencies from asteroseismic data.

10.1 THE INVERSION EQUATION

Inversions for solar and stellar structure start with Eq. 3.130:

$$\frac{\delta\omega_i}{\omega_i} = \int K^i_{c_s^2,\rho}(r)\frac{\delta c_s^2}{c_s^2}(r)\,dr + \int K^i_{\rho,c_s^2}(r)\frac{\delta\rho}{\rho}(r)\,dr. \qquad (10.1)$$

Recall that this equation was derived using the variational principle to represent the relative differences in frequencies, $\delta\omega/\omega$, that result from making changes to the sound speed and density. In the context of inversions, $\delta\omega/\omega$ is the relative frequency difference between the star and a model of the star; the model is usually called the *reference model*. The functions $\delta c_s^2/c_s^2$ and $\delta\rho/\rho$ are the relative sound-speed and density differences between the star and the reference model. The kernels $K_{c_s^2,\rho}$ and K_{ρ,c_s^2} are functions of the reference model and hence are known. The aim of the inversion is to use the known frequency differences to determine the relative sound-speed and density differences between the star and the model. Since the sound-speed and density profiles of the reference model are known, we can now determine the sound-speed and density profiles of the star. In Section 3.3 we discussed how kernels for sound speed and density may be converted to kernels for other variable pairs, such as ρ and Γ_1. We can use these kernels in the inversion equation, too.

Equation 10.1 does not take the surface term into account. The equation implies that we can invert the frequency differences between a star and its model provided we know how to make models of the star, and provided that the frequencies can be described using the equations for adiabatic oscillations. We know from Section 8.3 that this not the case, and part of the frequency arises from the near-surface layers that we cannot model. We have seen in Section 3.4 and particularly in Figure 3.17 that a near-surface perturbation gives rise to a frequency contribution that is quite smooth. It also can be shown that for modes that are not of very high degree (l less than about 200), the contribution is a function of frequency alone, once the effect of mode inertia has been taken into account. Since the near-surface contribution is a smooth term, it can be represented as a low-degree polynomial in frequency, and thus Eq. 10.1 can be modified to

$$\frac{\delta\omega_i}{\omega_i} = \int K_{c_s^2,\rho}^i(r)\frac{\delta c_s^2}{c_s^2}(r)\,dr + \int K_{\rho,c_s^2}^i(r)\frac{\delta\rho}{\rho}(r)\,dr + \frac{F(\omega_i)_{\text{surf}}}{I_i}, \qquad (10.2)$$

where $F(\omega_i)_{\text{surf}}$ is a slowly varying function of frequency that represents the surface term, and I_i is the mode inertia of the ith mode. The efficacy of using this form of the surface term can be judged by inverting frequency differences between pairs of models with differing surface terms.

If we have N observed modes, Eq. 10.2 represents N equations that need to be solved, or *inverted*, to determine the sound-speed and density differences. The data are $\delta\omega_i/\omega_i$ or equivalently, $\delta\nu_i/\nu_i$, and these are known. The kernels $K_{c_s^2,\rho}^i$ and $K_{\rho,c_s^2}^i$ are also known, and the task of the inversion is to estimate the two functions $\delta c_s^2/c_s^2$ and $\delta\rho/\rho$ after somehow accounting for the surface term.

There is an inherent problem with the inversion problem: no matter how many oscillation modes we observe, we only have a finite amount of data and hence a finite amount of information. The two unknowns, $\delta c_s^2/c_s^2$ and $\delta\rho/\rho$, are functions, and hence technically have an infinite amount of information, and there is no way to recover infinite information from a finite dataset. The best we can hope for is to find a representative average, hopefully a localized one, of the underlying function at a finite number of points. How localized the averages are depends on the data we have and the technique used for the inversion.

Because of the finite amount of data available we often have to add physically motivated constraints to the inversion process. One of the usual constraints is that the sound-speed and density profiles must be positive everywhere—this seems to be an obvious assumption, but the presence of data errors often means that without this explicit assumption, we can obtain negative (and hence unphysical) sound speeds and densities in certain parts of the star. Other constraints are also used; a common assumption when inverting solar oscillation frequencies is that mass is conserved, implying that the reference model and the Sun have the same mass, that is,

$$\int \frac{\delta \rho}{\rho} \rho(r) r^2 \mathrm{d}r = 0. \tag{10.3}$$

This assumption is usually applied by defining an additional mode with

$$\omega = 0, \quad K_{c_s^2, \rho} = 0, \quad \text{and} \quad K_{\rho, c_s^2} = \rho(r) r^2. \tag{10.4}$$

There are two popular methods of inverting Eq. 10.2: (1) regularized least squares (RLS), and (2) the method of optimally localized averages (OLA). The two techniques have different aims. The aim of RLS is to find $\delta c_s^2/c_s^2$ and $\delta \rho/\rho$ profiles that give best fits to the data (as measured by how small the residuals are) while keeping the uncertainties in the solution small. The aim of OLA is to find linear combinations of the frequency differences in such a way that the corresponding combination of kernels provides a localized average of the unknown function, again keeping the uncertainties small. OLA techniques do not try to fit the data at all and unlike RLS, do not provide an explicit approximation of the underlying functions. When OLA and RLS results agree, we can be reasonably confident that we have a good result.

10.2 THE REGULARIZED LEAST SQUARES TECHNIQUE

RLS inversions start by expressing the three unknown functions in Eq. 10.2 in terms of basis functions:

$$F(\omega) = \sum_{i=1}^{m} a_i \psi_i(\omega),$$

$$\frac{\delta c_s^2}{c_s^2} = \sum_{i=1}^{n} b_i \phi_i(r),$$

and

$$\frac{\delta \rho}{\rho} = \sum_{i=1}^{n} c_i \phi_i(r), \tag{10.5}$$

where $\psi_i(\omega)$ are suitable basis functions in frequency ω, and $\phi_i(r)$ are suitable basis functions in radius r. Then for N observed modes, Eq. 10.2 represents N equations of the form

$$b_1 \int \phi_1(r) K_{c^2}^i dr + b_2 \int \phi_2(r) K_{c^2}^i dr + \cdots + b_n \int \phi_n(r) K_{c^2}^i dr$$

$$+ c_1 \int \phi_1(r) K_\rho^i dr + c_2 \int \phi_2(r) K_\rho^i dr + \cdots + c_n \int \phi_n(r) K_\rho^i dr$$

$$+ a_1 \psi_1(\omega_i)/I_i + a_2 \psi_2(\omega_i)/I_i + \cdots + a_m \psi_m(\omega_i)/I_i = \delta\omega_i/\omega_i. \quad (10.6)$$

For ease of writing, $K_{c_s^2,\rho}^i$ has been contracted to $K_{c_s^2}^i$ and $K_{\rho,c_s^2}^i$ to K_ρ^i. To perform the inversion, we need to find coefficients a_i, b_i, and c_i such that we get the best fit to the data $\delta\omega_i/\omega_i$. This is done by minimizing

$$\chi^2 = \sum_{i=1}^{N} \left(\frac{\frac{\delta\omega_i}{\omega_i} - \Delta\omega_i}{\sigma_i} \right)^2, \quad (10.7)$$

where $\Delta\omega_i$ represents the left-hand side of Eq. 10.6.

The minimization does give us a solution for the three unknown functions, but the solutions are often highly oscillatory and unphysical. This is a result of the amplification of the data errors and, more fundamentally, a result of trying to recover more information than we are putting in. To ensure a more physical profile, regularization or smoothing techniques need to be applied. Smoothing a solution is tantamount to adding the constraint (and hence information) that a physically valid profile does not oscillate wildly. The usual technique is to assume that the second derivative of the function is smooth; this automatically reduces the large swings in the solution. Thus instead of minimizing Eq. 10.7, we minimize

$$\chi_{\text{reg}}^2 = \chi^2 + \|L\|^2$$

$$= \sum_{i=1}^{N} \left(\frac{\frac{\delta\omega_i}{\omega_i} - \Delta\omega_i}{\sigma_i} \right)^2 + \alpha^2 \int_0^R \left[\left(\frac{d^2}{dr^2} \frac{\delta\rho}{\rho} \right)^2 + \left(\frac{d^2}{dr^2} \frac{\delta c_s^2}{c_s^2} \right)^2 \right] dr, \quad (10.8)$$

where α is the regularization or smoothing parameter. Setting $\alpha = 0$ recovers Eq. 10.7. If we so wish, we could keep separate smoothing parameters for sound speed and density, but that merely adds to the number of parameters we need to determine. Often it is desirable to smooth $\delta c_s^2/c_s^2$ and $\delta\rho/\rho$ preferentially in one part of the star. To do this, a function of r is included in the smoothing term so that smoothing is higher wherever the function is larger:

$$\chi_{\text{reg}}^2 = \sum_{i=1}^{N} \left(\frac{\frac{\delta\omega_i}{\omega_i} - \Delta\omega_i}{\sigma_i} \right)^2 + \alpha^2 \int_0^R q(r)^2 \left[\left(\frac{d^2}{dr^2} \frac{\delta\rho}{\rho} \right)^2 + \left(\frac{d^2}{dr^2} \frac{\delta c_s^2}{c_s^2} \right)^2 \right] dr, \quad (10.9)$$

where $q(r)$ is the smoothing profile. We show the influence of the smoothing parameter in Figure 10.1. The surface term is not usually smoothed explicitly; instead we restrict the number and type of basis functions $\phi(\omega)$ so that the function is forced to be smooth.

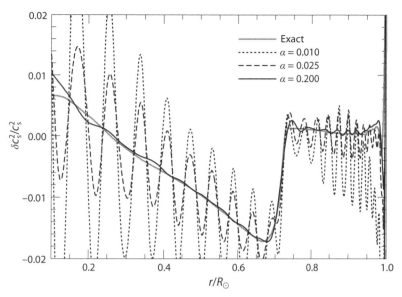

Figure 10.1. Sound-speed differences between two solar models obtained by inverting their frequency differences using the RLS technique. The heavy gray line is the true difference between the models; other lines are inversion results with different values of the smoothing parameter. Note how the solution becomes less oscillatory as the value of α is increased. To be realistic, we used only those modes for which solar frequencies are available, and in particular, we use the modes in the BiSON-13 mode set. The BiSON-13 mode set is a combination of low-degree modes obtained by the Birmingham Solar Oscillation Network and intermediate-degree modes determined by data from the Michelson Doppler Imager. For details of the set, see Basu *et al.*, 2009, *Astrophys. J.*, **699**, 1403. The frequencies of the test model were perturbed to mimic the uncertainties in the observed modes.

10.2.1 Implementing RLS

Equation 10.6 clearly gives us N equations in $k = 2n + m$ unknowns. These equations can thus be represented as

$$\mathbf{Ax} = \mathbf{d}, \tag{10.10}$$

where $\mathbf{A}$ is a $N \times (2n + m)$ matrix; $\mathbf{x}$ is a vector of length $(2n + m)$ consisting of the unknown coefficients a_i, b_i, and c_i; and $\mathbf{d}$ is the vector with the data (i.e., $\delta\omega/\omega$). To take data errors into account, each row i of each matrix is divided by σ_i, where σ_i is the error for the ith data point. We have to determine the elements of vector $\mathbf{x}$. Naïvely, one would think that all we need to do is to find the inverse of $\mathbf{A}$, and then the solution is simply

$$\mathbf{x} = \mathbf{A}^{-1}\mathbf{d}. \tag{10.11}$$

The problem is that $\mathbf{A}$ is not a square matrix, and besides, we are looking for a least-squares type of solution. One way of inverting $\mathbf{A}$ is to use a singular value decomposition (SVD).

An SVD of matrix $\mathbf{A}$ results in the decomposition of $\mathbf{A}$ into three matrices such that

$$\mathbf{A} = \mathbf{U}\Sigma\mathbf{V}^{\mathrm{T}}, \tag{10.12}$$

where $\mathbf{U}$ and $\mathbf{V}$ have the property that $\mathbf{U}^{\mathrm{T}}\mathbf{U} = \mathbf{I}$, $\mathbf{V}^{\mathrm{T}}\mathbf{V} = \mathbf{V}\mathbf{V}^{\mathrm{T}} = \mathbf{I}$, $\mathbf{I}$ being the identity matrix and $\mathbf{U}^{\mathrm{T}}$ the transpose of $\mathbf{U}$. The matrix $\Sigma = \mathrm{diag}(s_1, s_2, \cdots, s_{2n+m})$, where $s_1 \geq s_2 \geq \cdots s_{2n+m}$ are the singular values of $\mathbf{A}$. Thus the given set of equations can be reduced to

$$\mathbf{U}\Sigma\mathbf{V}^{\mathrm{T}}\mathbf{x} = \mathbf{d}, \tag{10.13}$$

and hence,

$$\mathbf{x} = \mathbf{V}\Sigma^{-1}\mathbf{U}^{\mathrm{T}}\mathbf{d}. \tag{10.14}$$

The solution for $\mathbf{x}$ obtained in this manner is a least-squares solution. Since the equations have already been normalized by the uncertainties, the uncertainty on any component x_k of vector $\mathbf{x}$ is given by

$$e_k^2 = \sum_{i=1}^{2n+m} \frac{v_{ki}^2}{s_i^2}. \tag{10.15}$$

We have not yet talked about implementing the smoothing condition: this can be done by replacing the integrals in Eq. 10.9 by a sum over M uniformly spaced points and adding the corresponding equations to the system of equations to be solved. Thus we add

$$\frac{\alpha}{\sqrt{M}} q(r_j) \left(\frac{d^2}{dr^2} \frac{\delta c_s^2}{c_s^2} \right)_{r=r_j} = \frac{\alpha}{\sqrt{M}} q(r_j) \sum_i^n b_i \frac{d^2}{dr^2} \phi_i(r_j) = 0, \tag{10.16}$$

for $i = 1, \ldots, n$ and $j = 1, \ldots, M$; and

$$\frac{\alpha}{\sqrt{M}} q(r_j) \left(\frac{d^2}{dr^2} \frac{\delta\rho}{\rho} \right)_{r=r_j} = \frac{\alpha}{\sqrt{M}} q(r_j) \sum_i^n c_i \frac{d^2}{dr^2} \phi_i(r_j) = 0, \tag{10.17}$$

for $i = n+1, \ldots, 2n$ and $j = M+1, \ldots, 2M$ to the matrix $\mathbf{A}$. The corresponding elements of $\mathbf{x}$ are set to zero. We now have $N' = N + 2M$ equations in $k = 2n + m$ unknowns, and we can again use the SVD decomposition technique to solve the equations.

10.2.2 Choosing Inversion Parameters

The art of doing inversions lies in being able to determine the inversion parameters properly. The inversion parameters depend on the function being inverted for, as well as the size of the mode set and the precision of the mode data. To avoid complications caused by the small datasets available for asteroseismic studies, we assume in this section that we have the usual datasets that are available for solar inversions (i.e, modes from $l = 0$ to $l = 200$ and frequencies from 1 to 4 mHz). The case of asteroseismic inversions is discussed in more detail in Section 10.4. Additionally, we demonstrate the process for sound-speed inversions. The same steps can be followed to determine the parameters for density inversions. Although Eq. 10.9 suggests that we can invert for sound speed and density simultaneously, better results are obtained by tuning the parameters for each function separately.

To do an RLS inversion, we need to decide on

1. the type of basis function;
2. the number of basis functions in radius to describe the two functions of radius;
3. the number of basis functions in frequency to describe the surface term; and
4. the smoothing parameter α and optionally, the smoothing function $q(r)$.

It is advantageous to have well-localized basis functions, so that the solution at one radius is sufficiently independent of the solution farther away. Early helioseismic inversions used a set of top-hat functions as the basis function:

$$f(x) = \begin{cases} 1, & \text{if } x_1 \leq x < x_2 \\ 0, & \text{otherwise.} \end{cases} \tag{10.18}$$

A set of points $x_1, x_2, \ldots, x_n$ is defined, and the top-hat is defined for each pair of points. While this is easy to implement, it requires a very large number of top-hats to represent the full function to the level of smoothness needed. The resulting function is piecewise linear, which makes defining derivatives—particularly the second derivative—difficult.

An example of a well-localized basis function with continuous first and second derivatives is the cubic B-spline. Each B-spline is defined over five points known as knots, and for m basis functions, one has to define $m + 2$ knots. Each B-spline is defined with four cubic polynomials. Over knots x_{-2}, x_{-1}, x_0, x_1, and x_2, we define the B-spline $B_0(x)$ as

$$B_0(x) = \begin{cases} 0, & x \leq x_{-2} \\ P_1(x), & x_{-2} \leq x \leq x_{-1} \\ P_2(x), & x_{-1} \leq x \leq x_0 \\ P_3(x), & x_0 \leq x \leq x_1 \\ P_4(x), & x_1 \leq x \leq x_2 \\ 0, & x \geq x_2 \end{cases} \tag{10.19}$$

where

$$P_1(x) \equiv a_1(x - x_{-2})^3,$$
$$P_2(x) \equiv a_1(x_{-1} - x_{-2})^3 + 3a_1(x_{-1} - x_{-2})^2(x - x_{-1})$$
$$\quad + 6a_1(x_{-1} - x_{-2})(x - x_{-1})^2 + a_3(x - x_{-1})^3,$$
$$P_3(x) \equiv a_2(x_1 - x_2)^3 + 3a_2(x_1 - x_2)^2(x - x_1)$$
$$\quad + 6a_2(x_1 - x_2)(x - x_1)^2 + a_4(x - x_1)^3,$$

and

$$P_4(x) \equiv a_2(x - x_2)^3, \tag{10.20}$$

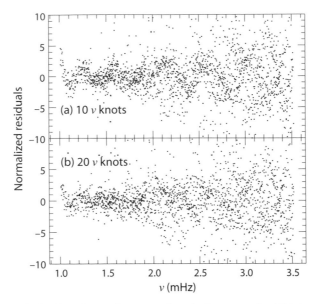

Figure 10.2. Residuals plotted as a function of mode frequency for RLS inversions with a different number of ν knots. The mode set is the same as in Figure 10.1. The residuals have been normalized by the uncertainties in the mode frequencies. Note that the residuals obtained when only 10 knots are used has a definite structure as a function of frequency. The structure is less clear in the 20-knot case.

and a_1, a_2, a_3, and a_4 are determined such that $P_2(x_0) = P_3(x_0) = 1$, $P_2'(x_0) = P_3'(x_0)$, and $P_2''(x_0) = P_3''(x_0)$. B_0 is nonzero over four intervals, and B_0, B_0', and B_0'' are continuous at each of the knots. If we define $B_k(x)$ to be a B-spline centered at x_k and $B_{-1}(x)$ as the one centered at x_{-1} and so on, then any function $F(x)$ in the interval $[x_0, x_N]$ can be written as

$$F(x) = \sum_{k=-1}^{N+1} a_k B_k(x). \tag{10.21}$$

The sum goes from -1 to $N + 1$ since B_{-1} is nonzero between x_0 and x_1, and B_{N+1} is nonzero between x_{N-1} and x_N. At most four B-splines contribute to $F(x)$ at any arbitrary point x in the interval over which $F(x)$ is defined.

The B-spline formulation becomes much simpler if the knots are equidistant; however, for helio- and asteroseismic inversions, knots in r are usually defined to be equidistant in the acoustic depth τ. This takes into account the fact that modes spend more time in regions of lower sound speed, making those regions easier to resolve. Without a large number of knots in these regions of low sound speed, we will not achieve the resolution that the data allow us; knots that are equidistant in τ achieve this quite naturally. If B-splines are used as the basis function to define the surface term, then the knots in frequency are usually kept equidistant in ν or ω.

It is usual to determine the number of knots in frequency first. This is done by fixing the number of r knots to a reasonably high value and then increasing the number of ν knots gradually until the residuals plotted as a function of frequency show no organized structure and are essentially random. The process is shown in Figure 10.2. A statistical test for randomness, such as the well-known *run test*, is often helpful.

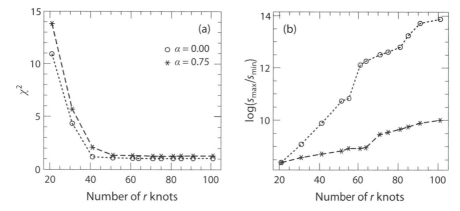

Figure 10.3. Determining the number of knots in r. Left panel: How the χ^2 per degree of freedom changes as a function of the number of knots. Right panel: The corresponding change in the condition number of the matrix **A**. The number of knots in frequency was held constant at 30. Results are for the unsmoothed case and one case with moderate smoothing. The results in the left panel show that the number of r-knots should be at least 40, while the right panel suggests that the number should be larger than 65, the point where the condition number increases suddenly.

To determine how many knots in r are required, we have to examine how the reduced χ^2 behaves as a function of the number of r knots for a reasonable regularization parameter. We also have to examine how the *condition number* of the matrix **A** in Eq. 10.10 changes as the number of r knots is changed. The condition number is the ratio of the largest singular value to the smallest singular value. A large condition number implies an ill-conditioned matrix; a singular matrix has a condition number of infinity. The value of χ^2 decreases sharply as the number of knots is increased, but it becomes almost flat beyond a certian number of knots. The number of knots chosen should be from the flat part of the curve after the sharp decrease. The condition number has the opposite behavior: it rises slowly and then jumps. Inversion results are better when the number of knots is chosen so that the condition number has gone though the jump. The process of selecting r-knots is illustrated in Figure 10.3.

To determine the smoothing parameter α, we use the so-called L-curve. This is a plot of the smoothing constraint $||L||^2$, defined in Eq. 10.8 as a function of the χ^2 per degree of freedom, obtained by changing the smoothing parameter. The smoothing parameter increases the mismatch between the data and the solution, and thus $\alpha = 0$ gives the highest $||L||^2$ (since the oscillations in the solution are not constrained at all) but the lowest χ^2. A large value of α will iron out all oscillations and leave a linear solution (since the second derivative of the function is being minimized), but the fit to the data will be bad, and hence the χ^2 will be large. The optimum value of smoothing should lie somewhere in between these two extremes. There is usually an "elbow" in the L-curve, and the value of smoothing is chosen to be somewhere in the elbow, usually on the low-χ^2 side. The L-curve for the inversion results in Figure 10.1 are shown in Figure 10.4.

The smoothing parameter also determines the uncertainties in the solutions that arise from the propagation of data errors. The higher the smoothing, the lower the uncertainties will be and vice versa. Thus in RLS solutions, the goal of suppressing

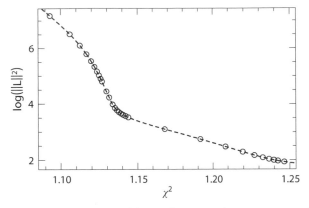

Figure 10.4. The change in amplitude of the oscillations in the inversion results change as a function of the reduced χ^2 when α is increased. The best values of α lie in the "elbow" or "L" of this L-curve. Experience suggests that a slightly lower value of α than that at the elbow works best.

errors is equivalent to suppressing oscillations in the solution. Smoothing does not necessarily remove systematic errors; in fact in some cases, it can add to them.

We reiterate that the number of knots and the smoothing parameter selected depend on the type of basis functions used. They also depend on the number of modes available for inversion, as well as on the uncertainties in the data. Hence the parameters obtained for one set of modes cannot automatically be used for another set. Changing the reference model can also change the inversion parameters. One should always determine the inversion parameters first for a few pairs of models, using only the observed mode set and data uncertainties before inverting the observed frequencies.

10.3 THE METHOD OF FORMING OPTIMALLY LOCALIZED AVERAGES

A very different method of inverting oscillation frequencies is the one that aims to determine localized averages of the underlying function. The method is usually referred to as the optimally localized average (OLA) method, after its goal. The method was originally developed by geophysicists and later co-opted by helioseismologists. The OLA inversion technique forces us to invert for each function separately; as before, we assume that we are inverting for the sound-speed profile. The process is the same for other variables.

OLA inversions proceed by determining coefficients c_i, called *inversion coefficients*, such that the sum

$$\mathcal{K}(r_0, r) = \sum_i c_i(r_0) K^i_{c_s^2, \rho}(r) \tag{10.22}$$

is well-localized around $r = r_0$. If $\int \mathcal{K}(r_0, r) dr = 1$, then

$$\left\langle \frac{\delta c_s^2}{c_s^2} \right\rangle (r_0) = \sum_i c_i(r_0) \frac{\delta \omega_i}{\omega_i} \tag{10.23}$$

is a localized average of $\delta c_s^2 / c_s^2$, and the inversion result at radius r_0, as long as

$$\mathcal{C}(r_0, r) = \sum_i c_i(r_0) K_{\rho, c_s^2}^i, \tag{10.24}$$

is small, and the surface-term contribution

$$\mathcal{F} = \sum_i c_i(r_0) \frac{F_{\text{surf}}(\omega_i)}{I_i} \tag{10.25}$$

is small, too. The error in the solution will be $e^2(r_0) = \sum_i c_i^2(r_0) \sigma_i^2$, where σ_i is the relative uncertainty associated with the frequency of mode i. The radius r_0 is usually referred to as the *target radius*. One may or may not be able to form an averaging kernel localized at a specified r_0. This depends on whether we have a sufficient number of modes whose lower-turning point lies in the vicinity of r_0. For instance, if we only have modes of $l \leq 3$, we should not expect to get a localized averaging kernel at $r = 0.8R$.

The function $\mathcal{K}(r_0, r)$ is the *averaging kernel* or the *resolution kernel* at $r = r_0$. The term $\mathcal{C}(r_0, r)$ is the *cross-term kernel* and measures the contribution of the second variable (in this case density) on the sound-speed inversions. The averaging kernel represents the region over which the underlying function is averaged, and its width is a measure of the resolution of the inversions. The aim of OLA is to obtain the narrowest possible averaging kernels allowed by the data, while keeping the uncertainty in the results to an acceptable level.

There are two widely used variants of OLA. The first, the original one, is usually known as *multiplicative* optimally localized averages (MOLA). The reason for the name will become clear later. This is a fairly computation-intensive method that requires a matrix to be set up and inverted at each target radius r_0. An alternative, faster technique is the variant commonly referred to as *subtractive* optimally localized averages, (SOLA). SOLA requires just one inversion. The idea behind this innovation is that one can decide what the averaging kernels should look like; that is, we define the *target kernel* and then determine the inversion coefficients such that the sum in Eq. 10.22 results in averaging kernels that are similar to the target.

10.3.1 Implementing OLA

The inversion coefficients for MOLA are determined by minimizing

$$\int \left(\sum_i c_i K_{c_s^2, \rho}^i \right)^2 J(r_0, r) \mathrm{d}r + \beta \int \left(\sum_i K_{\rho, c_s^2}^i \right)^2 \mathrm{d}r + \mu \sum_{i,j} c_i c_j E_{ij}, \tag{10.26}$$

where $J = (r - r_0)^2$ [often J is defined as $12(r - r_0)^2$], and E_{ij} are elements of the error-covariance matrix. This implementation is called the multiplicative OLA because the function J multiplies the averaging kernel in the first term of Eq. 10.26. The parameter β ensures that the cross-term kernel is small, and μ ensures that the error in the solution is small. The constraint of unimodularity of the averaging kernel $\mathcal{K}$ is applied through a Lagrange multiplier. The surface term is usually expressed as a polynomial,

$$F_{\text{surf}}(\omega) = \sum_{j=1}^{\Lambda} a_j \Psi_j(\omega), \tag{10.27}$$

the $\Psi_j(\omega)$ being suitable basis functions, and we apply a constraint of the form

$$\sum_i c_i \frac{\Psi_j(\omega_i)}{I_i} = 0, \quad j = 1, \ldots, \Lambda, \tag{10.28}$$

to ensure that the surface term contribution is small (Λ is the number of basis functions used). Note that unlike RLS inversions, we do not recover what the surface term looks like.

SOLA proceeds by minimizing

$$\int \left(\sum_i c_i K^i_{c_s^2, \rho} - \mathcal{T} \right)^2 dr + \beta \int \left(\sum_i K^i_{\rho, c_s^2} \right)^2 dr + \mu \sum_{i,j} c_i c_j E_{ij}, \tag{10.29}$$

where $\mathcal{T}$ is the target-averaging kernel. The difference between the averaging and target kernels in the first term of Eq. 10.29 is why we call this method subtractive OLA. Again we have to apply the condition of unimodularity for the averaging kernels and the surface constraints.

The choice of the target kernel is often dictated by what the purpose of the inversion is. It is quite common to use a Gaussian and define

$$\mathcal{T}(r_0, r) = A \exp\left(-\left[\frac{r - r_0}{\Delta(r_0)}\right]^2\right), \tag{10.30}$$

where A is a normalization factor that ensures that $\int \mathcal{T} dr = 1$. However, this form has the disadvantage that for small r_0, $\mathcal{T}$ may not be equal to 0 at $r = 0$, where the kernels $K_{c_s^2, \rho}$ and K_{ρ, c_s^2} are zero. This leads to a forced mismatch between the target and the resultant averaging kernel. One solution to this problem is to force the target kernels to go to zero at $r = 0$ using a modified Gaussian, such as

$$\mathcal{T}(r_0, r) = A r \exp\left(-\left[\frac{r - r_0}{\Delta(r_0)} + \frac{\Delta(r_0)}{2r_0}\right]^2\right), \tag{10.31}$$

where A is the normalization factor that ensures that the target is unimodular.

The averaging kernel obtained for MOLA inversions at a given target radius depends on the amount of information available for that radius. With the usual solar mode sets, one usually finds that the widths are smaller toward the surface than toward the core, which is a result of the fact that p mode kernels have larger amplitudes in the shallower layers. In SOLA inversions, the widths of the target kernels are made a function of the target radius by defining the width, Δ_f, of the target kernel at a fiducial target radius $r_0 = r_f$ and then specifying the widths at other locations as $\Delta(r_0) = \Delta_f c_s(r_0)/c_s(r_f)$. This ensures that we get narrow kernels toward the surface and progressively wider kernels as we go deeper.

The equations that result from the minimization of Eqs. 10.26 and 10.29, once the different constraints are applied, can be written as a set of linear equations of the form

$$\mathbf{Ac} = \mathbf{v}. \tag{10.32}$$

For M observed modes the matrix elements A_{ij} for SOLA inversions are given by

$$A_{ij} = \begin{cases} \int K^i_{c_s^2} K^j_{c_s^2} dx + \beta \int K^i_{\rho} K^j_{\rho} dx + \mu E_{ij}, & (i, j \leq M) \\ \int K^i_{c_s^2} dx, & (i \leq M, j = M+1) \\ \int K^j_{c_s^2} dx, & (j \leq M, i = M+1) \\ 0, & (i = j = M+1) \\ \psi_j(\omega_i)/I_i, & (i \leq M, M+1 < j \leq M+1+\Lambda) \\ \psi_i(\omega_j)/I_j, & (M+1 < i \leq M+1+\Lambda, j \leq M) \\ 0, & (\text{otherwise}), \end{cases} \tag{10.33}$$

where x is the fractional radius.

The vectors $\mathbf{c}$ and $\mathbf{v}$ have the form

$$\mathbf{c} = \begin{pmatrix} c_1 \\ c_2 \\ \cdot \\ \cdot \\ \cdot \\ c_M \\ \lambda \\ 0 \\ \cdot \\ \cdot \\ 0 \end{pmatrix}, \quad \text{and} \quad \mathbf{v} = \begin{pmatrix} \int K^1_{c_s^2} \mathcal{T} dx \\ \int K^2_{c_s^2} \mathcal{T} dx \\ \cdot \\ \cdot \\ \cdot \\ \int K^M_{c_s^2} \mathcal{T} dx \\ 1 \\ 0 \\ \cdot \\ \cdot \\ 0 \end{pmatrix}, \tag{10.34}$$

where λ is the Lagrange multiplier.

For MOLA, the elements of A are the same as those in Eq. 10.33 except for elements for which $i, j \leq M$. These elements are given by

$$A_{ij} = \int (x - x_0)^2 K^i_{c_s^2} K^j_{c_s^2} dx + \beta \int K^i_{\rho} K^j_{\rho} dx + \mu E_{ij}. \tag{10.35}$$

And as far as the vector $\mathbf{v}$ is concerned, the $(M+1)$th element for MOLA is the same as that for SOLA, but all other elements are equal to zero. The column vector $\mathbf{c}$ is identical in both cases.

One can see that the main difference in the implementation of MOLA and SOLA lies in the matrix $\mathbf{A}$. It depends on r_0 in MOLA, and thus needs to be set up and inverted at each r_0. SOLA is faster, since $\mathbf{A}$ is independent of r_0. In addition, r_0 appears implicitly in $\mathbf{v}$ via the target radius, and $\mathbf{v}$ does not have to be inverted.

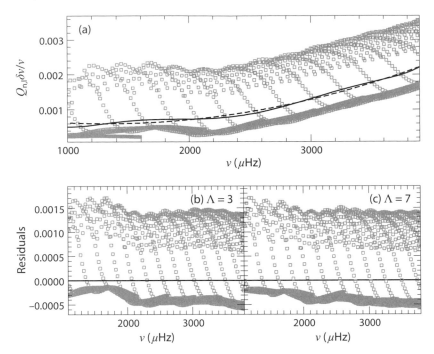

Figure 10.5. Top panel: Scaled frequency differences between two solar models (symbols) fitted with the first three Legendre polynomials (dashes) and the first seven Legendre polynomials (solid line). At any given frequency we can see that $\delta v / v$ has a range of values; the range depends on the differences in structure between the two models. The surface term is the predominant frequency-dependent difference that forms the upper and lower envelopes. Bottom panels: Residuals for the two cases. We can see that subtracting out the polynomial reduces most of the frequency dependence of the residuals, though there is some remaining structure when only three polynomials are used.

This makes SOLA much faster than MOLA. One should be aware that the matrix **A** can be quite ill conditioned, and as a result, one has to be careful when selecting an algorithm to solve Eq. 10.32.

10.3.2 Determining Inversion Parameters

Three parameters—the number of surface terms Λ, the error suppression parameter μ, and the cross-term suppression parameter β—need to be determined when doing a MOLA inversion. A fourth parameter, Δ_f, the width of the fiducial target kernel, is needed for SOLA. Unlike RLS, there are no simple graphs and curves that can help determine the parameters.

As with RLS, the number of basis functions needed for the surface term is the first parameter to be determined. This is done by fitting a polynomial with different numbers of basis functions Ψ. The number of basis functions is increased until there is no large-scale structure left in the residuals. This is shown in Figure 10.5.

The width of the target kernels in SOLA is chosen so as to minimize the mismatch between averaging kernels obtained and the target. The mismatch is

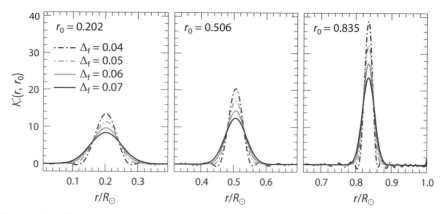

Figure 10.6. Behavior of the averaging kernels at three target radii as the width of the target kernel is decreased. Note how small negative side-lobes appear as Δ_f is decreased. Be aware that the results will change if the mode set were to change, or even if the distribution of uncertainties of modes in a given mode set change.

usually defined as

$$\chi(r_0) = \int [\mathcal{K}(r_0, r) - \mathcal{T}(r_0, r)]^2 \mathrm{d}r. \tag{10.36}$$

The target kernels as defined by either Eq. 10.30 or Eq. 10.31 are all positive. However, if the selected target width is too small, the resultant averaging kernels have negative side-lobes. This is shown in Figure 10.6. As we shall show soon, narrow averaging kernels also increase the uncertainty in a solution for a given error-suppression parameter μ. One generally chooses the smallest width that avoids negative side-lobes. Choosing a larger width than is necessary would result in an unnecessary degradation of the resolution of the inversions.

In the case of MOLA, the error-suppression parameter μ and the cross-term parameter β determine the averaging kernel. Although no mismatch can be defined in this case, one can try to minimize the negative side-lobes by ensuring that the quantity

$$\chi'(r_0) = \int_0^{r_A} \mathcal{K}^2(r_0, r)\mathrm{d}r + \int_{r_B}^R \mathcal{K}^2(r_0, r)\mathrm{d}r \tag{10.37}$$

is small. Here, r_A and r_B are defined in such a way that the averaging kernel $\mathcal{K}$ has its maximum at $(r_A + r_B)/2$, and its full width at half maximum is $(r_B - r_A)/2$.

The errors between the solutions obtained at different radii are correlated, and the error correlation between solutions at two radii r_1 and r_2 is

$$E(r_1, r_2) = \frac{\sum c_i(r_1)c_i(r_2)\sigma_i^2}{\left[\sum c_i^2(r_1)\sigma_i^2\right]^{1/2}\left[\sum c_i^2(r_2)\sigma_i^2\right]^{1/2}}. \tag{10.38}$$

$E(r_1, r_2)$ has values between ± 1. A value of $+1$ implies complete correlation, and a value of -1 implies complete anticorrelation. Correlated errors can introduce

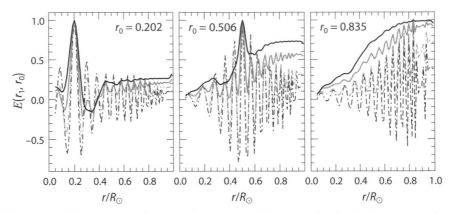

Figure 10.7. How propagated errors at three target radii are correlated with the errors at other radii for sound-speed inversions using the same solar mode set as before. In particular, note how the correlations change with the target width. The legend for the line styles is the same as in Figure 10.6. Note that the correlation increases substantially when the averaging kernels are forced to be wide.

features into the solution on the scale of the order of the correlation function's width. The primary cause of the correlated error is simply the fact that solutions at different radii are obtained using the same dataset. The correlations are enhanced because of the surface term. The surface term is a function of frequency, and modes with the same frequency propagate to different radii. As a result, removing the surface term correlates errors at all radii that form the propagation cavity for modes that have similar frequencies. Although wide averaging kernels reduce $\chi(r_0)$ and $\chi'(r_0)$, and also reduce uncertainties in the results, they can increase error correlations. Small negative side-lobes can be advantageous as far as error correlations are concerned— they suppress a part of the correlation. Consequently it is usual to make the kernels slightly narrower than what the data will allow. The influence of negative side-lobes on the error correlations in SOLA inversions can be seen in Figure 10.7.

The conservation of mass applied to density inversions forces the solution for any one part of the star to be affected by the solution at others. This also causes the errors to be correlated. There is an anticorrelation of errors between the core (where density is the highest) and the outer layers (where density is the lowest), and the crossover occurs around the radius at which $r^3\rho$ has the largest value. The error correlations for density inversions are shown in Figure 10.8.

For a given error-suppression parameter β, changing the width of the averaging kernel also changes the cross-term kernels, and hence the contribution of the second function to the inversion results of the first function. The contribution of the cross-term kernels can be gauged using the quantity

$$C(r_0) = \sqrt{\int \mathcal{C}^2(r_0, r)dr}. \tag{10.39}$$

We need to aim for small $C(r_0)$ to get correct results. Figure 10.9 shows how the error in the solution, the mismatch of the averaging kernels, and the measure of the

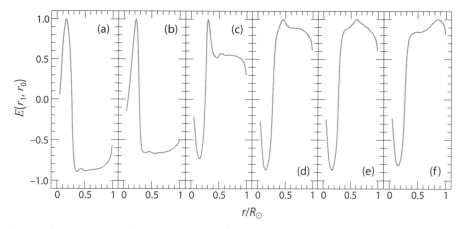

Figure 10.8. Error-correlation function for density inversions at a few target radii. Note the difference in shape of these functions compared with those shown in Figure 10.7.

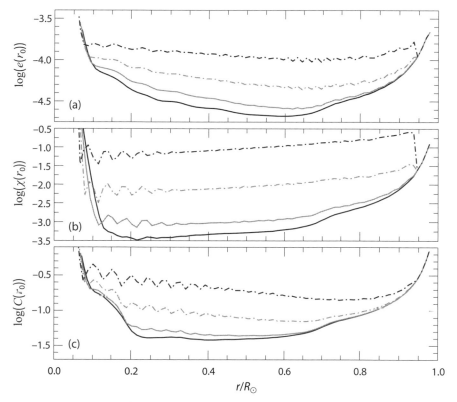

Figure 10.9. Error $e(r_0)$ in the solution, the mismatch between target and obtained averaging kernels $\chi(r_0)$, and the influence of the cross-term kernels $\mathcal{C}(r_0)$ plotted as a function of target radius for different values of the target width. The parameters μ and β have been kept fixed. The legend for the styles is the same as in Figure 10.6.

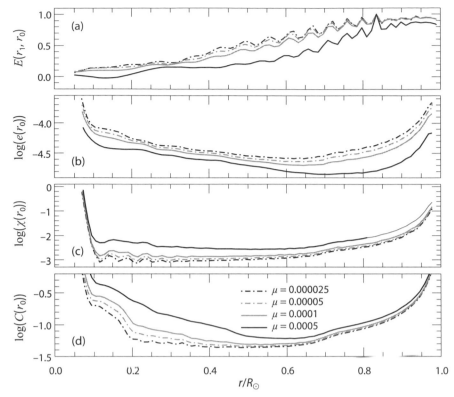

Figure 10.10. Error correlation $E(r_1, r_2)$ between the solution at $r_0 = 0.835 R_\odot$ and that at other radii, error $e(r_0)$ in the solution, the mismatch between target and obtained averaging kernels $\chi(r_0)$, and the influence of the cross-term kernels $\mathcal{C}(r_0)$ plotted as a function of target radius for different values of the error suppression parameter μ. The parameters Δ_f and β have been kept fixed.

cross-term change when the width of the target kernel is changed. As can be seen from this figure and others, the requirement that the error correlation be small acts against the other requirements. Thus the choice of Δ_f needs to be a balance of all these quantities.

Two other parameters need to be determined: the error-suppression parameter μ and the cross-term suppression parameter β. Increasing μ decreases the error correlation $E(r_1, r_0)$ and decreases the uncertainty in the solution $e(r_0)$ as well. However, increasing μ increases $\chi(r_0)$, the mismatch between the target kernels and the averaging kernel. The influence of the cross-term as measured by $C(r_0)$ also increases. Increasing β increases $E(r_1, r_0)$, $e(r_0)$, and $\chi(r_0)$, but decreases $C(r_0)$. Although $\chi(r_0)$ is undefined for MOLA, we can use $\chi'(r_0)$ instead, and this quantity has the same variation with μ and β as $\chi(r_0)$ does for SOLA inversions. The influence of μ and β are shown in Figs. 10.10 and 10.11, respectively.

As we can see, determining inversion parameters for MOLA and SOLA is not straightforward. Given that increasing β decreases systematic errors, at the cost of increasing random error, it is usual to try to reduce the cross-term contribution $C(r_0)$, even if that results in a somewhat larger uncertainty in the solution. Again,

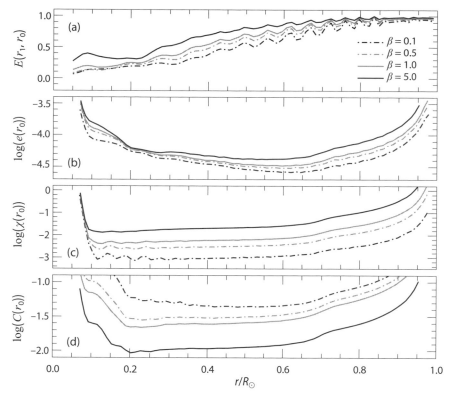

Figure 10.11. The same as Fig. 10.10, but for different values of the cross-term suppression parameter β. Again, Δ_f and μ have been kept fixed.

before inverting observations, it is best if one inverts frequency differences between pairs of models to evaluate the effects of the parameters.

10.4 APPLYING INVERSIONS TO STARS

We need to confront two major issues when trying to invert oscillation frequencies of stars other than the Sun:

1. The mode set is smaller. Typically we only have frequencies of modes of degrees $l = 0, 1$, and 2 (and if we are lucky $l = 3$). This automatically restricts the radius over which we can do the inversion. It also means that we do not generally have enough information to handle the surface term, as we do in the case of the Sun.
2. We do not have independent measurements of the mass and radius of the star being studied. This means that the reference model used may not have the same mass and radius as that of star, which can introduce systematic errors.

The effect of the mode set. The small mode sets that are generally available for stars other than the Sun limit what variables we can use. To demonstrate this, we use the solar low-degree mode set observed by BiSON. Using the BiSON mode set [i.e., the observed (n, l) values] and associated frequency uncertainties, we invert for

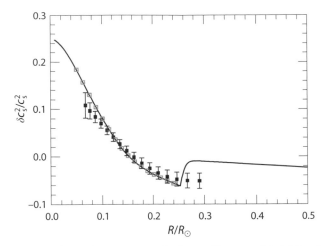

Figure 10.12. The sound-speed inversion result obtained between Model S and a solar model with an artificially mixed core using the solar mode set and errors obtained by BiSON. The continuous line is the true difference, the filled boxes are the inverted results, and the gray open boxes show the radii at which we were attempting to obtain inversion results. Note that we cannot invert properly very close to the core or at radii larger than $0.25R_\odot$.

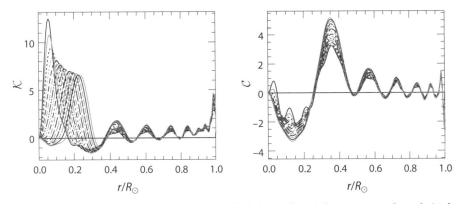

Figure 10.13. A sample of the averaging kernels (left panel) and the cross-term kernels (right panel) for the inversions shown in Figure 10.12.

the differences between Model S[1] and a solar model constructed with an artificially mixed core. These models have large differences in the core, which is what we expect to see between stars and their models. The results of SOLA inversions are shown in Figure 10.12. These are inversions for sound-speed differences with density as the second variable. As can be seen, the positions of the target kernels and the actual averaging kernels are quite different. The averaging kernels and the cross-term kernels are not very good either. The averaging kernels are clearly not well localized, and the cross-term kernels are large (Figure 10.13).

---

[1] Christensen-Dalsgaard et al, 1996, *Science*, **272**, 1286.

One way out is to add more information, and one promising avenue is to assume that we know the equation of state of stellar matter perfectly and use the (u, Y) kernel pair instead. Here u is as usual the squared isothermal sound speed, and Y is the helium abundance. The advantage of this choice is that Y kernels are nonzero only in the helium ionization zones, and thus the cross-term kernel can be suppressed easily.

To calculate kernels for the (u, Y) variable pair, we need the kernels for the (ρ, Γ_1) pair of variables, along with derivatives of Γ_1 calculated using the equation of state. In particular we need

$$
\Gamma_{1,p} \equiv \left(\frac{\partial \ln \Gamma_1}{\partial \ln P} \right)_{\rho, Y} ; \quad \Gamma_{1,\rho} \equiv \left(\frac{\partial \ln \Gamma_1}{\partial \ln \rho} \right)_{P, Y} ; \quad \Gamma_{1,Y} \equiv \left(\frac{\partial \ln \Gamma_1}{\partial Y} \right)_{P, \rho} .
$$

The kernels are given by

$$
K_{Y,u} = \Gamma_{1,Y} K_{\Gamma_1,\rho}, \tag{10.40}
$$

and

$$
K_{u,Y} = P \frac{d(P^{-1}\psi)}{dr} + \Gamma_{1,p} K_{\Gamma_1,\rho}, \tag{10.41}
$$

where ψ is obtained by solving the differential equation

$$
\frac{d\psi(r)}{dr} - 4\pi G \rho r^2 \int_r^R \frac{\rho}{r^2 P} \psi \, dr = - \left[K_{\rho,\Gamma_1} + \left(\Gamma_{1,p} + \Gamma_{1,\rho} \right) K_{\Gamma_1,\rho} \right] . \tag{10.42}
$$

Inversions with the (u, Y) kernel pair are indeed much better, as we can see in Figure 10.14. We can actually capture a part of the large jump in the sound speed at the edge of the core. The averaging kernels and cross-term kernels are much better too (Figures 10.15 and 10.16).

The results obtained with solar data are, of course, the best we can expect to get. The mode sets of other stars will be far more limited, and the use of u rather than c_s^2 is more likely to be necessary. The best nonsolar dataset available at the time of writing is that of 16 Cyg A. Figure 10.17 shows how the mode-set of 16 Cyg A obtained with 30 months of *Kepler* data compares with the solar one.

The u inversion results using the 16 Cyg A mode set are shown in Figure 10.18. The results are worse than in the solar case, as are the averaging kernels (Figure 10.19). But the results are acceptable, implying that we should be able to perform such inversions.

The effect of differences in M and R. Using the fitting techniques described in Chapter 8, we can make models of a given star. However, because of uncertainties in the data, it is very likely that the total mass and radius of the model is not the same as that of the star. This introduces systematic errors in the inversion results.

As mentioned in Chapter 3, the dimensionless squared sound speed varies with total mass and radius as R/M; dimensionless u has the same dependence. Inversion kernels are almost invariably calculated in terms of dimensionless quantities.

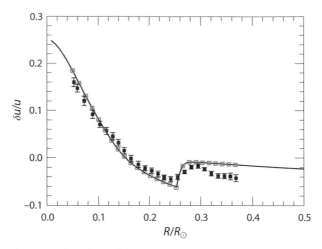

Figure 10.14. The squared isothermal sound-speed inversion result obtained between Model S and a solar model with an artificially mixed core using the solar mode set and errors obtained by BiSON. The continuous line is the true difference, the filled boxes are the inverted results, and the gray open boxes show the radii at which we were attempting to obtain inversion results.

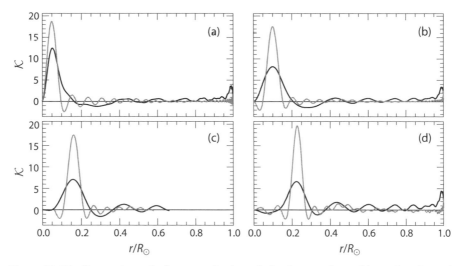

Figure 10.15. Comparison of a few averaging kernels for the sound-speed inversion shown in Figure 10.12 and the u inversion in Figure 10.14. The black lines are the averaging kernels for the c_s^2 inversion, while the gray lines are those for the u inversion.

A difference in M and R between the reference model and the star results in a systematic error in the results. This is illustrated in Figure 10.20, Which shows the results of inverting the frequency differences between a solar model and two $1.1 M_\odot$ models, one with a radius of $1.1 R_\odot$ and the other with a radius of $1.16 R_\odot$. A large set of modes was used—all $l = 0, 1, 2$, and 3 modes from $n = 2$ to $n = 31$, assuming uniform errors. This was done to eliminate uncertainties due to limitations of the mode set. Note than when M/R is the same, we recover the difference

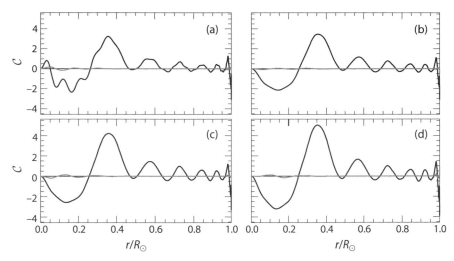

Figure 10.16. The same as Figure 10.15, but showing cross-term kernels. The cross-term kernels for the u inversions are shown in gray and they have amplitudes that are close to zero. The cross-term kernels for the c_s^2 inversion are shown in black.

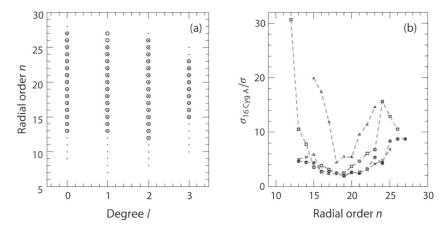

Figure 10.17. Left panel: The points mark l and n of the BiSON solar mode set, while circles mark the mode set of 16 Cyg A. Note that there are fewer low-n (i.e., low-frequency) modes for 16 Cyg A. Right panel: Comparison of the uncertainties in the frequencies of 16 Cyg A and the Sun. The circles, crosses, squares, and triangles are for $l = 0, 1, 2,$ and 3, respectively. The dashed lines joining the points are to guide the eye.

in u (the difference being calculated at constant fractional radius), but there is a clear offset when M/R is different. The fact that it is indeed the scaling that is causing the problems in the right panel of Figure 10.20 can be confirmed by plotting the results against $\delta u'/u' = \delta u/u + \delta R/R - \delta M/M$, where δR and δM are respectively the difference in radius and mass between the reference and test models (see Figure 10.21). We shall show in the next section that for certain inversions it is easy to put in a correction for differences in M and R, but that is not the case for all variables.

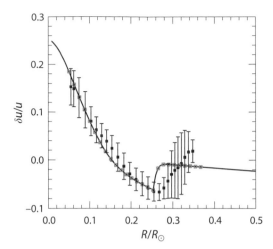

Figure 10.18. The same as Figure 10.14 but for results obtained with the mode set (but not the frequencies) of 16 Cyg A.

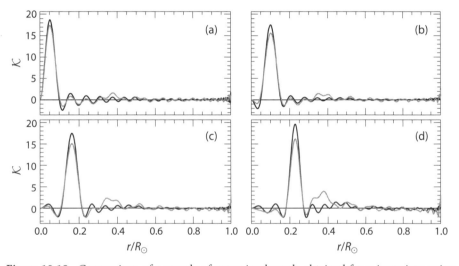

Figure 10.19. Comparison of a sample of averaging kernels obtained for u inversions using the solar mode set (black) with those obtained using the mode set of 16 Cyg A (gray).

10.5 INVERSIONS FOR GLOBAL QUANTITIES

The difficulties in inverting for structure differences as a function of radius between stars and their models have led to the development of inversion techniques for some global parameters, such as mean density and acoustic radius. These inversions generally rely on tuning the target averaging kernels $\mathcal{T}$ in SOLA inversions to get the desired result. Inversions for global quantities require less information than inversions for radial profiles, and hence in principle we should be able to get more precise results. Additionally, for some of these quantities, one can apply constraints

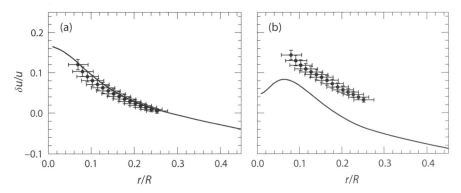

Figure 10.20. Left panel: Inversion results for u differences when the frequency differences between a $1.1\text{M}_\odot$, $1.1\text{R}_\odot$ model and a solar model were inverted. The line shows the true difference. Right panel: Inversion results when the frequency differences between a $1.1\text{M}_\odot$, $1.6\text{R}_\odot$ model and a solar model were inverted. The constant difference in the relative frequency differences $\delta\omega/\omega$ of the models caused by differences in $\sqrt{M/R^3}$ was removed in the process of inverting.

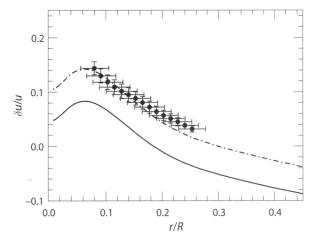

Figure 10.21. Inversion results shown in Figure 10.20 compared with the relative difference of the actual u (continuous line) as well as the relative difference between the dimensionless u of the two models (dot-dashed line).

during inversions to avoid systematic errors caused by differences in total mass and radius.

10.5.1 Mean Density

The average density of a star is simply $\bar\rho = 3M/(4\pi R^3)$, M and R being the total mass and radius of a star. The aim is to determine the relative difference in average density $\delta\bar\rho/\bar\rho$ between the star and its reference model. While the relative differences between large frequency separations in principle gives us the relative differences in mean density, deviations in the $\Delta\nu$-$\bar\rho$ scaling relations mean that other methods must be examined. If the mean density differs between two stars, there is usually a difference between the total mass of the stars. For the moment let us assume that we have a star and its model, and that they have the same radius but different masses. The difference in mass between the star and its model can be written as

$$\delta M = \int_0^R 4\pi r^2 \delta\rho \, dr. \tag{10.43}$$

It follows that the relative difference in mean density is

$$\frac{\delta\bar\rho}{\bar\rho} = \int_0^1 4\pi x^2 \frac{\rho}{\rho_R} \frac{\delta\rho}{\rho} dx, \tag{10.44}$$

where x is the fractional radius. The quantity ρ_R is M/R^3 of the reference model and is used to make the equation dimensionless. One fact that is immediately obvious from Eq. 10.44 is that if $\delta\rho/\rho = 0$ at all radii, then $\delta\bar\rho/\bar\rho$ will be zero. Similarly if $\delta\rho/\rho = 1$ at all radii, $\delta\bar\rho/\bar\rho$ will be unity. Thus we will get

$$\int_0^1 4\pi x^2 \frac{\rho}{\rho_R} \frac{\delta\rho}{\rho} dx = 1. \tag{10.45}$$

We want to determine $\delta\bar\rho/\bar\rho$ as a linear combination of the frequency differences,

$$\frac{\delta\bar\rho}{\bar\rho} = \sum_i c_i \frac{\delta\omega_i}{\omega_i}, \tag{10.46}$$

and for this we start by rewriting Eq. 10.1 in terms of (ρ, Γ_1) kernels to get

$$\frac{\delta\omega_i}{\omega_i} = \int K^i_{\Gamma_1,\rho}(x) \frac{\delta\Gamma_1}{\Gamma_1}(x) dx + \int K^i_{\rho,\Gamma_1}(x) \frac{\delta\rho}{\rho}(x) dx. \tag{10.47}$$

Thus we have

$$\sum_i c_i \frac{\delta\omega_i}{\omega_i} = \int \left(\sum_i c_i K^i_{\Gamma_1,\rho}(x) \right) \frac{\delta\Gamma_1}{\Gamma_1}(x) dx, + \int \left(\sum_i c_i K^i_{\rho,\Gamma_1}(x) \right) \frac{\delta\rho}{\rho}(x) \, dx, \tag{10.48}$$

or

$$\sum_i c_i \frac{\delta\omega_i}{\omega_i} = \int \mathcal{K}(x) \frac{\delta\rho}{\rho}(x) dx + \int \mathcal{C}(x) \frac{\delta\Gamma_1}{\Gamma_1}(x) dx, \tag{10.49}$$

where as usual $\mathcal{K}$ is the averaging kernel, and $\mathcal{C}$ is the cross-term kernel. Comparing Eq. 10.49 with Eq. 10.44, we can immediately see that for $\sum_i c_i \delta\omega_i/\omega_i$ to be equal to $\delta\bar\rho/\bar\rho$, the averaging kernel $\mathcal{K}$ should be as close as possible to $4\pi x^2(\rho/\rho_R)$, and $\mathcal{C}$ should be as close to zero as possible. In other words, instead of being the usual

Gaussian or modified Gaussian, the target kernel $\mathcal{T}$ should be

$$\mathcal{T}(x) = 4\pi x^2 \frac{\rho(x)}{\rho_R}, \tag{10.50}$$

and we can thus proceed with the inversion by minimizing

$$\int \left(\sum_i c_i K^i_{\rho,\Gamma_1} - \mathcal{T} \right)^2 dx + \beta \int \left(\sum_i K^i_{\Gamma_1,\rho} \right)^2 dx + \mu \sum_{i,j} c_i c_j E_{ij}, \tag{10.51}$$

after applying the usual surface-term corrections as discussed in Section 10.3.1.

We have not discussed what happens if there is a difference in total radius. Density scales as M/R^3, and frequencies scale as $\sqrt{M/R^3}$. Thus if the mean density difference is $\delta\bar{\rho}/\bar{\rho}$, then the frequency differences will be $0.5\delta\bar{\rho}/\bar{\rho}$, and hence Eq. 10.46 implies $\sum_i c_i = 2$. This condition can be explicitly added to the inversion process through a Lagrange multiplier to account for the differences in radius.

10.5.2 Acoustic Radius

Inversions for the acoustic radius follow the same philosophy as for the mean density: we determine the form that the averaging kernel will take, except in this case the cross-term kernel also needs to have a specific form (as we shall see below).

The acoustic radius of a star is defined as $\tau = \int dx/c_s$, in other words, the relative difference in τ is

$$\frac{\delta\tau}{\tau} = \frac{1}{\tau} \int \frac{-1}{c_s} \frac{\delta c_s}{c_s} dx = \int \frac{-1}{2c_s\tau} \frac{\delta c_s^2}{c_s^2} dx. \tag{10.52}$$

As usual, we want to find coefficients c_i such that $\delta\tau/\tau = \sum_i c_i \delta\omega_i/\omega_i$. Thus comparing with Eq. 10.1, is it immediately clear that the averaging kernel in this case should be $-1/(2c_s\tau)$:

$$\mathcal{T} = \frac{-1}{2c_s\tau}. \tag{10.53}$$

Although this is correct, there is a practical problem to overcome: we need to minimize the contribution of density in the inversions, and that is difficult. To avoid this problem, we resort to using the (ρ, Γ_1) pair instead of the (c_s^2, ρ) pair of kernels. Since

$$\frac{\delta\rho}{\rho} = \frac{\delta c_s^2}{c_s^2} - \frac{\delta\Gamma_1}{\Gamma_1} - \frac{\delta P}{P}, \tag{10.54}$$

and

$$P(x) = \int_x^1 \frac{m(y)\rho}{y^2} dy, \tag{10.55}$$

where y is a dummy variable of integration denoting fractional radius, and

$$m(x) = \int_0^x 4\pi x^2 \rho \, dx, \tag{10.56}$$

the target averaging kernel will now have the form

$$\mathcal{T}_{av} = \frac{1}{2c_s\tau} - \frac{m(x)\rho}{x^2} \left[\int_0^x \frac{1}{2c_s\tau P} dy \right] - 4\pi x^2 \rho \left[\int_x^1 \left(\frac{\rho}{y^2} \int_0^y \frac{1}{2c_s\tau P} dz \right) \right] dy, \tag{10.57}$$

where y and z are dummy variables representing fractional radius. The fact that $K_{\Gamma_1,\rho} = K_{c_s^2,\rho}$ means that we also have a nonzero target for the cross-term kernel, and

$$\mathcal{T}_{cr} = \frac{-1}{2c_s\tau}. \tag{10.58}$$

Thus the inversion coefficients c_i can be determined by minimizing

$$\int \left(\sum_i c_i K_{\rho,\Gamma_1}^i - \mathcal{T}_{av} \right)^2 dx + \beta \int \left(\sum_i K_{\Gamma_1,\rho}^i - \mathcal{T}_{cr} \right)^2 dx + \mu \sum_{i,j} c_i c_j F_{ij}, \tag{10.59}$$

after applying the usual surface-term corrections.

To account for differences in M and R between a star and its models, we note that the acoustic radius is basically the inverse of a frequency. The relative differences in τ are therefore just minus those in the frequencies, and hence we need to apply the condition $\sum_i c_i = -1$. As usual, this condition can be applied through a Lagrange multiplier when minimizing Eq. 10.59.

10.5.3 Age Indicators

The acoustic radius is very sensitive to surface uncertainties because of the $1/c_s$ dependence. A better variable to work with is the gradient of the sound speed. Recall that the small separations δ_{02} depend on the gradient of the sound-speed near the core, and since the gradient changes with evolution, it is a good indicator of age on the main sequence. This motivates the definition of an age indicator t as

$$t = \int_0^1 \frac{1}{x} \frac{dc_s}{dx} dx, \tag{10.60}$$

which can be determined by inverting mode frequencies.

Equation 10.60 leads to

$$\frac{\delta t}{t} = \frac{1}{t} \int_0^1 \frac{1}{x} \frac{d\delta c_s}{dx} dx = \frac{1}{t} \int_0^1 \frac{1}{x} \frac{dc_s}{dx} \frac{d\delta c_s/dx}{dc_s/dx}. \tag{10.61}$$

The multiplication and division by dc_s/dx is required to offset the effect of the $1/x$ term at the center. This means that if we can derive kernels for dc_s/dx, we can define a target averaging kernel for t inversions as

$$\mathcal{T}_t = \frac{\frac{1}{x}\frac{dc_s}{dx}}{\int_0^1 \frac{1}{x}\frac{dc_s}{dx}dx}. \tag{10.62}$$

We now need to determine kernels for dc_s/dx. The first term in Eq. 10.1 can be rewritten as

$$\int_0^1 K_{c_s^2,\rho}\frac{\delta c_s^2}{c_s^2}dx = -\int_0^1 \left(\int_0^x \frac{2K_{c_s^2,\rho}}{c_s}dy\right)\frac{dc_s}{dy}\frac{d\delta c_s/dx}{dc_s/dx}dx$$
$$+ \left[\left(\int_0^x 2\frac{K_{c_s^2,\rho}}{c_s}dy\right)\delta c_s\right]_0^1. \tag{10.63}$$

The second term of the above expression is exactly zero at the center because of the behavior of the kernels $K_{c_s^2,\rho}$. At the surface the term is much smaller (by factors of between 60 and 150) than the first term, and hence can be ignored to first order. Thus one can define the kernels for the sound-speed gradient as

$$K_{dc_s/dx,\rho} = -\frac{dc_s}{dx}\int_0^x \frac{2K_{c_s^2,\rho}}{c_s}dy, \tag{10.64}$$

and

$$K_{\rho,dc_s/dx} = K_{\rho,c_s^2}. \tag{10.65}$$

This allows us in principle to do the usual SOLA inversion. But it turns out that keeping the first term in the minimization in the usual form (i.e., $\int(\mathcal{K}-\mathcal{T})^2dx$) is not adequate because of the structure of the kernels—the inversion outputs are not robust to small changes in the inputs. Better and more stable results are obtained by using the integrals of the averaging and target kernels and minimizing the following expressions with respect to the inversion coefficients:

$$\int_0^1 \left[\int_0^x \mathcal{T}(y)dy - \int_0^x \mathcal{K}(y)dy\right]^2 dx + \beta\int_0^1 \mathcal{C}^2(x)dx + \beta\sum_{ij} c_i c_j E_{ij}. \tag{10.66}$$

Of course we have to apply the usual surface-term conditions.

To account for M and R changes, note that the derivative of the sound speed scales exactly the same way as the square root of density. Thus the extra constraint that we should apply is $\sum_i c_i = 1$.

10.6 ALTERNATIVE METHODS

The inversion techniques described in Sections 10.1, 10.4, and 10.5 rely on explicitly linearizing the equation relating stellar oscillation frequencies to stellar structure. An alternative means of inversion is a nonlinear iterative process that has as its basis the phase-matching technique described in Section 8.7.1. Recall that the phase-shift function $\mathcal{G}(l, v)_{nl}$ could be used to examine whether the internal structure of a stellar model matched that of a star whose frequencies are known (see Eq. 8.31). The alternative "inversion" technique discussed here is simply a variant of that in Section 8.7.1.

The idea is to start with a model with a structure that is reasonably close to that of the star being studied, and to iteratively modify the structure until a much better fit is obtained. This method uses the fact that we can determine all the seismically relevant quantities of a model if we know the density ρ as a function of radius, and we either assume that $\Gamma_1 = 5/3$ (as is the case for an ideal gas) or use the Γ_1 profile of a suitable model. The density profile can be used to determine pressure assuming hydrostatic equilibrium; density, pressure, and Γ_1 can then be used to determine c_s^2. This allows us to solve for the phase function $\mathcal{G}(l, v)_{nl}$ using the equations in Section 8.7.1.

The inversion starts as before with a reference model. Its structure is parameterized. This can be done in different ways. One could parameterize the structure using N_i parameters D_i defined by $D_i \equiv \Gamma_{1,i} d \log \rho_i / d \log P_i$ over a set of mesh points in fractional mass $q_i = m_i / M$, where m_i is the mass enclosed by mesh-point i. Alternatively one can simply parameterize the density profile by N_k parameters $C_k \equiv d \log \rho_k / dx_k$, where x_k represents the fractional radius at each mesh point k. Also needed is a suitable interpolation routine. The parameterized model is used to calculate the phase shift $\mathcal{G}^{\text{mod}}(l, v_{\text{obs}})$ at the observed frequencies v_{obs}. As discussed in Section 8.7.1, if the internal structure of the reference model matches that of the star, $\mathcal{G}^{\text{mod}}(l, v_{nl}^{\text{obs}})$ should be independent of degree l. This is tested as before by determining the goodness of fit to a function of frequency $\mathcal{A}(v_{nl})$ that can be expressed in terms of different basis functions in frequency. We therefore calculate

$$\chi^2(C_k) = \frac{1}{N - N_a} \sum_k \left(\frac{\mathcal{G}^{\text{mod}}(l, v_{nl}^{\text{obs}}) - \mathcal{A}(v_{nl}^{\text{obs}})}{e_{nl}} \right)^2, \qquad (10.67)$$

where N_a polynomials define $\mathcal{A}$, e_{nl} is the error on $\mathcal{G}$ as described in Section 8.7.1, and $\chi^2(C_k)$ denotes the χ^2 per degree of freedom of the model parameterized by parameters C_k. The next step is to correct the parameters C_k, get a new model, and repeat the process until a minimum in χ^2 is reached. Obviously, the implementation requires an optimization routine to actually calculate the corrections and perform the iterations. The results of such an exercise are shown in Figure 10.22.

A complication can arise if the star has a convective core, as in the case shown in Figure 10.22. The sound-speed profile of such a star will have a discontinuity at the edge of the convective core. Smoothing over the discontinuity will produce a bad match. Thus in addition to the mesh points and the already defined parameters to represent structure, we need another free parameter, the fractional mass or radius at the edge of the convection zone; the position of this extra mesh point needs to be varied in the minimization process.

It should be noted that the solution obtained in this case may not be unique. For this method, all we can say is that all models that have χ^2 per degree of freedom around unity are a solution. As usual, the number of data points and the data errors play a role. An independent constraint, Such as stellar luminosity, can help select among the allowed solutions.

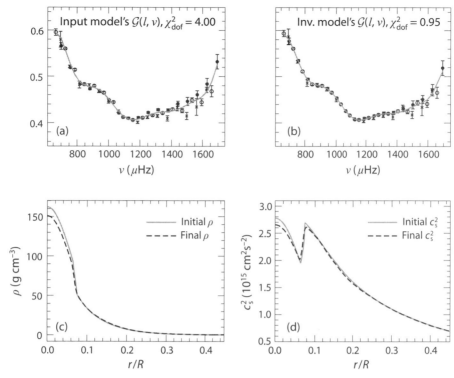

Figure 10.22. Top-left panel: $\mathcal{A}(\nu_{nl}^{\mathrm{obs}})$ for the initial reference model plotted as a function of frequency. The symbols show $\mathcal{G}^{\mathrm{mod}}(l, \nu_{nl}^{\mathrm{obs}})$ at the observed frequencies. The filled circles are $l = 0$, open circles $l = 1$, and crosses $l = 2$. Top-right panel: The same as in the top-left panel but for the final model. Note the considerable reduction in the χ^2. Bottom-left panel: Density profiles of the initial and final models. Bottom-right panel: Profile of the squared sound speed for the initial and final models. Data and results are courtesy of Ian Roxburgh.

10.7 INVERSIONS FOR ROTATION

Inversions for rotation are in principle much simpler than inversions for structure. In the case of slow rotation, which we assume, the inversion problem is linear to begin with. Unlike the structure inversion case, we do not have to rely on the linearization of the equation around a reference model. A model is required only to calculate the kernels. Another advantage of rotation inversions is that there is no cross-term influence to worry about. And yet another advantage is the absence of a surface term. The complexity of rotation inversions arises from the two-dimensional nature of rotation. Rotation makes frequencies m dependent as explained in Chapter 3, and it is this m dependence that we exploit to perform rotation inversions.

In the case of the Sun we know that the rotation rate depends both on radius and on latitude. Consequently, two-dimensional inversions are the norm. For other stars, however, it is unlikely that two-dimensional inversions will be feasible. The limitation lies in the data. For stars other than the Sun, we can only observe modes of $l = 0$, 1, and 2, and if we are lucky, $l = 3$. Modes of $l = 0$, often the most precisely measured modes, are not affected by rotation. Modes of $l = 1$ split into three, but all components may not be visible because of the star's inclination (e.g., see Sections 4.1 and 6.4.5). Similarly, modes with $l = 2$ split into at most five components, and again, not all may be visible. Thus for stars, concentrating on determining the radial variation of rotation is perhaps the more promising avenue than trying to find latitudinal variations.

We have seen in Chapter 3 that the difference between the frequency of a mode with a given (n, l, m) from that of the mode with the same n and l, but $m = 0$ can be expressed as

$$\delta\omega_{nlm} = m\beta_{nl} \int_0^1 \Omega(x)K_{nl}(x)dx, \tag{10.68}$$

where as usual x is the fractional radius, Ω the rotation rate in the same units as $\delta\omega_{nlm}$, and the K_{nl} are the rotation kernels. The kernels are given by Eq. 3.153. As can be seen clearly, Eq. 10.68 is much simpler compared to the equation for structure inversions (Eq. 10.1)—there is only one term, and we do not have to bother with surface-term, corrections. Standard inversion techniques can be applied very easily.

To do an RLS inversion, the rotation rate is expressed in terms of suitable basis functions in r or x,

$$\Omega(x) = \sum_{i=1}^M c_i \phi_i(x), \tag{10.69}$$

and the coefficients c_i are determined by minimizing

$$
\begin{aligned}
\chi_{\text{reg}}^2 &= \chi^2 + ||L||^2 \\
&= \sum_{i=1}^N \frac{1}{\sigma^2}\left(\delta\omega_{nlm} - \sum_{j=1}^M c_j \int \phi_i(x)dx\right)^2 + \alpha^2 \int \left(\sum_{j=1}^M c_i \frac{d^2\phi_i}{dx^2}\right)^2 dx,
\end{aligned} \tag{10.70}
$$

where N is the total number of data points, M is the number of basis functions, α is the regularization parameter, and σ the uncertainty on $\delta\omega_{nml}$. This is completely analogous to Eq. 10.8.

For a SOLA inversion, we try as usual to determine inversion coefficients c_i such that the averaging kernel at any point x_0, $\mathcal{K}(x_0) \equiv \sum_i c_i K_{nl}$, is a well-localized function, so that the inverted rotation rate $\Omega_{\text{inv}} = \sum_i c_i \delta\omega_i$ is a localized average of the underlying rotation rate. Thus if $\mathcal{T}$ is the target radius, whether Gaussian or modified Gaussian, then we can determine coefficients c_i by minimizing

$$\int_0^1 (\mathcal{K} - \mathcal{T})^2 dx + \mu \sum_{ij} c_i c_j E_{ij}, \tag{10.71}$$

where μ is again the error-suppression parameter, and E_{ij} is the error covariance matrix. As mentioned earlier, unlike the structure case, we do not need a term to suppress the influence of a second variable; nor do we need to worry about the surface term. It is worth noting that the SOLA technique was first developed for solar rotation inversions, and only later was it applied to the problem of structure inversions.

Given the dearth of data, instead of trying to determine the stellar rotation rate as a function of radius, one can assume, for example, a two-zone model and try to determine the rotation rate in each zone, that is, one assumes that

$$\Omega(r) = \begin{cases} \Omega_c, & \text{for } 0 \le r/R \le r_1 \\ \Omega_c + \Delta\Omega_c, & \text{for } r_1 < r/R \le 1. \end{cases}$$

This allows the study of the contrast in rotation rate, $\Delta\Omega_c$, between the core and the envelope. This too can be difficult when only p modes are observed. Frequency splittings on mixed modes can help immensely.

It has been suggested recently that to determine whether radial differential rotation is present in solar-like main sequence stars, an ensemble inversion method may be more promising. This would involve inverting the splittings of all stars together while assuming an average two-zone model.

10.8 FURTHER READING

The very early history of helioseismic inversions can be found in the following papers:

- Christensen-Dalsgaard, J., and Gough, D. O., 1976, *Towards a Heliological Inverse Problem*, Nature, **259**, 89.
- Gough, D., 1985, *Inverting Helioseismic Data*, Solar Phys, **100**, 65.

The following review gives a good description of different types of inversions, including some that have not been discussed in this chapter, because they cannot be applied to stars. The article also discusses how to change from one set of kernels to another:

- Gough, D. O., 1993, *Linear Adiabatic Stellar Pulsation, Astrophysical Fluid Dynamics—Les Houches 1987*, eds. Zahn, J.-P., and Zinn-Justin, J., Elsevier Science.

Worked-out examples of constructing kernels can be found in:

- Elliott, J. R., 1996, *Equation of State in the Solar Convection Zone and the Implications of Helioseismology*, Mon. Not. R. Astron. Soc., **280**, 1244.
- Kosovichev, A. G., 1999, *Inversion Methods in Helioseismology amd Solar Tomography*, J. Comp. Appl. Math., **109**, 1.

Most inversions require a set of linear equations to be solved by inverting matrices. Using an SVD is a common way. To learn more about the properties of SVD and the type of solution obtained with SVD, as well as algorithms to find the SVD of a matrix, we refer the reader to:

- Golub, G. H., and van Loan, C. F. *Matrix Computations*, Johns Hopkins University Press.

Details of the statistical run test, which can be used to determine whether the residuals from an RLS inversion are randomly distributed, can be found in the following or other similar texts:

- Miller, I., and Freund, J., 1977, *Probability and Statistics for Engineers*, Prentice-Hall.

For some details of the L-curve, see:

- Hansen, P. C., 1992, *Numerical Tools for Analysis and Solution of Fredholm Integral Equations of the First Kind, Inverse Probs,* **8**, 849.

For the original papers describing OLA, we point the reader to the following:

- Backus, G., and F. Gilbert, 1968, *The Resolving Power of Gross Earth Data, Geophys. J. Int.,* **16**, 169.
- Backus, G., and F. Gilbert, 1970, *Uniqueness in the Inversion of Inaccurate Gross Earth Data, R. Soc. London Phil. Trans.,* **266**, 123.

The development of the SOLA method is a result of the following papers:

- Pijpers, F. P., and Thompson, M. J., 1992, *Faster Formulations of the Optimally Localized Averages Method for Helioseismic Inversions, Astron. Astrophys.,* **262**, 33.
- Pijpers, F. P., and Thompson, M. J., 1994, *The SOLA Method for Helioseismic Inversion, Astron. Astrophys.,* **281**, 231

Readers who want to know more about parameter selection and associated pitfalls are referred to the following papers. The first one deals with RLS inversions, the second with OLA-type inversions.

- Basu, S., and Thompson, M. J., 1996, *Constructing Seismic Models of the Sun, Astron. Astrophys.,* **305**, 631.
- Rabello-Soares, M. C., Basu, S., and Christensen-Dalsgaard, J., 1999, *On the Choice of Parameters in Solar-Structure Inversion, Mon. Not. R. Astron. Soc.,* **309**, 35.

For details of global inversions and tests to show how well or badly they work, we refer readers to the following papers:

- Reese, D. R., et al. 2012, *Estimating Stellar Mean Density Through Seismic Inversions, Astron. Astrophys.,* **539**, A63.
- Buldgen, G., et al. 2015, *Stellar Acoustic Radii, Mean Densities, and Ages from Seismic Inversion Techniques, Astron. Astrophys.,* **574**, A42.
- Buldgen, G., Reese, D. R., and Dupret, M. A., 2015, *Using Seismic Inversions to Obtain an Internal Mixing Processes Indicator for Main-Sequence Solar-Like Stars, Astron. Astrophys.,* **583** , A62.

An example of nonlinear inversions can be found in:

- Roxburgh, I. W., and S. V. Vorontsov, 2002, *Inversion for the Structure of a star of 1.45 M_{solar} Using the Internal Phase Shift,* Proc. Stellar Structure and Habitable Planet Finding, eds. B. Battrick et al., ESA SP **485**, 341, European Space Agency.

The sensitivity of rotation inversion to data uncertainties can be found in:

- Schunker, H., Schou, J., and Ball, W. H., 2016, *Asteroseismic Inversions for Radial Differential Rotation of Sun-Like Stars: Sensitivity to Uncertainties, Astron. Astrophys.*, **586** A24.

The utility of mixed modes in rotation inversions was first demonstrated in the following paper, which also discusses different schemes of inverting data on mode splittings:

- Deheuvels, S., et al., 2012, *Seismic Evidence for a Rapidly Rotating Core in a Lower-Giant-Branch Star Observed with* Kepler, *Astrophys. J.*, **756**, 19.

The concept of ensemble inversions is discussed in the following:

- Schunker, H., et al., 2016, *Asteroseismic Inversions for Radial Differential Rotation of Sun-Like Stars: Ensemble Fits, Astron. Astrophys.*, **586**, A79.

For those interested in techniques used to perform two-dimensional inversions to determine solar rotation, we recommend:

- Howe, R., 2009, *Solar Interior Rotation and Its Variation, Living Rev. Solar Phys.*, **6**, 1.

10.9 EXERCISES

1. Folder `Inversions_data` contains files with kernels for Model S.[2] Kernels have been calculated for different pairs of variables. Also in the directory is a file containing some of the most relevant properties of the model and a file with the frequencies of the model. Using the kernels, do the following.

  (a) Assuming you have the solar mode set and uncertainties, attempt to construct localized averaging kernels using the c_s^2, ρ and u, Y pairs of kernels. How does the assumption of an added surface term change the results? Note that the frequencies themselves do not come into play, except to determine the surface term.
  (b) How do the results change if you assume that the uncertainties for the solar frequencies are a factor of three larger?
  (c) What happens if you consider the mode set of 16 Cyg A as given in file `16Cyg_freqs.txt`?
  (d) Invert the frequency differences between model AGS05 and Model S. The frequencies of the AGS model are in file `freqs_ags05_diffusion.txt`, and the structural properties are in file `solarmodel_ags05_diffusion.txt.gz`. Use the different mode sets used in Exercise 1(a)—(c) for this. How well do the inversion results match the actual c_s^2 and u differences between the models?

---

[2]Christensen-Dalsgaard et al., 1996, *Science*, **272**, 1286.

2. Attempt to invert for the mean acoustic radius of the Sun using the kernels of Model S and the BiSON mode set.

3. Use the kernels in files `rotker.m1.20_2.78Gyr.txt`, and `rotker.m1.5_2.5Gyr.txt` to examine whether one can obtained localized averaging kernels and invert the splittings calculated for Exercise (5) in Chapter 3. Be aware of the fact that not all modes listed in the files will be observed.

Index